Remaining Issues in the Decommissioning of Nuclear Powered Vessels

NATO Science Series

A Series presenting the results of scientific meetings supported under the NATO Science Programme.

The Series is published by IOS Press, Amsterdam, and Kluwer Academic Publishers in conjunction with the NATO Scientific Affairs Division

Sub-Series

I. Life and Behavioural Sciences	IOS Press
II. Mathematics, Physics and Chemistry	Kluwer Academic Publishers
III. Computer and Systems Science	IOS Press
IV. Earth and Environmental Sciences	Kluwer Academic Publishers
V. Science and Technology Policy	IOS Press

The NATO Science Series continues the series of books published formerly as the NATO ASI Series.

The NATO Science Programme offers support for collaboration in civil science between scientists of countries of the Euro-Atlantic Partnership Council. The types of scientific meeting generally supported are "Advanced Study Institutes" and "Advanced Research Workshops", although other types of meeting are supported from time to time. The NATO Science Series collects together the results of these meetings. The meetings are co-organized bij scientists from NATO countries and scientists from NATO's Partner countries – countries of the CIS and Central and Eastern Europe.

Advanced Study Institutes are high-level tutorial courses offering in-depth study of latest advances in a field.
Advanced Research Workshops are expert meetings aimed at critical assessment of a field, and identification of directions for future action.

As a consequence of the restructuring of the NATO Science Programme in 1999, the NATO Science Series has been re-organised and there are currently five sub-series as noted above. Please consult the following web sites for information on previous volumes published in the Series, as well as details of earlier sub-series.

http://www.nato.int/science
http://www.wkap.nl
http://www.iospress.nl
http://www.wtv-books.de/nato-pco.htm

Series IV: Earth and Environmental Sciences – Vol. 22

Remaining Issues in the Decommissioning of Nuclear Powered Vessels

Including Issues Related to the Environmental Remediation of the Supporting Infrastructure

edited by

A.A. Sarkisov

Nuclear Safety Institute,
Russian Academy of Sciences,
Moscow, Russia

and

L.G. LeSage

Engineering Research,
Argonne National Laboratory,
Argonne, IL, U.S.A.

Kluwer Academic Publishers

Dordrecht / Boston / London

Published in cooperation with NATO Scientific Affairs Division

Proceedings of the NATO Advanced Research Workshop on
Scientific Problems and Unresolved Issues Remaining in the Decommissioning of Nuclear Powered Vessels and in the Environmental Remediation of their Supporting Infrastructure,
Moscow, Russia
April 22–24, 2002

A C.I.P. Catalogue record for this book is available from the Library of Congress.

ISBN 1-4020-1353-1

Published by Kluwer Academic Publishers,
P.O. Box 17, 3300 AA Dordrecht, The Netherlands.

Sold and distributed in North, Central and South America
by Kluwer Academic Publishers,
101 Philip Drive, Norwell, MA 02061, U.S.A.

In all other countries, sold and distributed
by Kluwer Academic Publishers,
P.O. Box 322, 3300 AH Dordrecht, The Netherlands.

Printed on acid-free paper

Printed in the Netherlands.

TABLE OF CONTENTS

INTERNATIONAL COOPERATION IN NUCLEAR VESSEL DECOMMISSIONING AND UTILIZATION

RADIOECOLOGY, RADIATION, AND SAFETY ISSUES

SPENT NUCLEAR FUEL ISSUES

RADIOACTIVE WASTE MANAGEMENT

SUBMARINES WITH LIQUID METAL COOLANT

SPECIAL TOPICS IN NUCLEAR SUBMARINE DECOMMISSIONING AND UTILIZATION

PREFACE

The Russian NATO Advanced Research Workshop on "Scientific Problems and Unresolved Issues Remaining in the Decommissioning of Nuclear Powered Vessels and in the Environmental Remediation of Their Supporting Infrastructure," was held in Moscow, Russia at the Presidium of the Russian Academy of Sciences on April 22-24, 2002. This was the third in this series of North Atlantic Treaty Organization (NATO) sponsored workshops in Moscow on nuclear vessel decommissioning. The first one was in June 1995 and served to focus international attention on the problems of nuclear vessel decommissioning in Russia and elsewhere. The second one was in November 1997 and it focused on the risks associated with nuclear vessel decommissioning. Attendance at the current workshop was approximately 100 with participants form Russia, United States, Norway, France, Denmark, Germany, Japan, Korea, NATO, and the European Union.

The Workshop was sponsored and funded by the Security-Related Civil Science and Technology Program of the Scientific and Environmental Affairs Division of NATO. Within Russia, the Workshop was sponsored and supported by the Russian Academy of Sciences, Minatom of Russia, Rossudostroenie, Ministry of Industry and Science of Russia, and the Russian Navy. Within the U.S., the Workshop was supported by the U.S. Department of Energy. The sponsorship and support of all of the above organizations are gratefully acknowledged.

The Workshop was organized in Russia by the Nuclear Safety Institute of the Russian Academy of Sciences (IBRAE). In addition to flawless arrangements, which we have come to expect from IBRAE, they developed a technical program that was broad in scope, technically deep, challenging, relevant and interesting. The efforts of many individuals from IBRAE in producing the Workshop are especially recognized.

The given collection of articles of the 2002 Russian NATO Advanced Research Workshop, taken together with earlier published papers of the two previous Russian NATO ARWs of 1995 and 1997, represents the most complete set of materials on the issues related to nuclear submarine utilization and safety when managing spent nuclear fuel and radioactive wastes. We do hope that this book will be of use to scientists, engineers, shipyard technicians and decision-makers in the fields of both radioecology and the management of different-purpose decommissioned radiation-dangerous facilities.

OPENING ADDRESS

OPENING ADDRESS

M. BARSKOV
Vice-Admiral, Chief of Shipbuilding, Armaments and Armament Maintenance, Second-in-Command in the Russian Navy
Moscow, Russia

The Russian Navy began Nuclear Submarine (NS) decommissioning and multi-purpose utilization in 1986. In keeping with the initial plans, all NSs, for which the service life was up, were to be decommissioned during the following 5 to 6 years, and the needed standard-legal, organizational, technical and industrial basis to ensure their multi-purpose utilization was to be organized and constructed in the course of the same period.

However, due to different circumstances, at the beginning of the 1990s mass decommissioning of Russian NSs started. Consequently, neither industrial enterprises nor Navy shipyards were prepared to perform such enormous NS utilization-ensuring work. The Russian Federation (RF) Ministry of Defense (Navy) was forced to spend more and more financial resources to solve different NS utilization-related issues. As a result, the problem of both NS multi-purpose utilization and Spent Nuclear Fuel (SNF) & Radioactive Wastes (RW) management overstepped the intra-branch level of the RF Ministry of Defense and developed into a national-scale task. As of April 2002, 191 Russian NSs were decommissioned (Arctic Navy – 115 units, Pacific Navy - 76 units) including:

- 96 de-fueled NSs;
- 95 NSs stored with fuel;
- 67 block-modules prepared for long-term waterborne storage (made of 96 de-fueled NSs).

In keeping with the RF Decree n° 518 of May 28, 1998 "On Measures to Accelerate Utilization of the Decommissioned Nuclear Submarines & Nuclear Surface Vessels and Perform Environmental Rehabilitation of Radiation-Dangerous Units & Facilities Owned by the Navy," the Russian Navy transferred:

- 4 Coastal Servicing Enterprises (CSEs) to RF Minatom to perform their environmental rehabilitation; and
- 95 NSs to enterprises to carry out utilization operations.

According to an agreed schedule, 96 NSs, Floating Servicing Vessels (FSVs) performing NS reactor-reloading operations and Technical Pouring Tankers (TPTs) decommissioned from the Russian Navy are to be transferred to utilization by 2007. Until 2007 the Russian Navy should ensure servicing operations for the above nuclear vessels.

From the environmental safety viewpoint waterborne storage of NSs of both the first- & the second- generation (whose service life exceeds 30 years) present the greatest hazard. Such NSs are based in: Gremikha-settlement & Vidiaevo-settlement (the Northwest Region) and in Petropavlovsk-Kamchatski & Soviet Gavan' (the Far East Region). These NSs need to be transferred to enterprises performing utilization within the next two years.

A.A. Sarkisov and L.G. LeSage (eds.),
Remaining Issues in the Decommissioning of Nuclear Powered Vessels, 3–5.

An increase of RW volumes results from:

- mass decommissioning of NSs & Nuclear Maintenance Support Vessels (NMSVs);
- toughening of requirements to ensure radiation safety in keeping with the present-day international agreements; and
- further increase of RW amounts in the near future due to planned NS utilization operations;

Predetermined Russian participation in a number of international programs, such as:

1. Updating of a LRW processing facility at Service & Repair Enterprise (SRE) "Atomflot" invested by the USA and Norway (the international part of the Project is known under the name of "The Murmansk Initiative – the Russian Federation").
2. Constructing a floating Liquid RW (LRW) processing complex "Landysh" in Primorskiy kray funded by the USA and Japan.
3. Realizing the project "Development, Construction and Putting into Operation of a Mobile Module Facility to Process LRW of Complex Physical & Chemical Composition" within the framework of the Arctic Military Environmental Cooperation Program (AMEC Program).

However, only via commissioning of the above facilities, the problem of LRW management in Russia cannot be resolved in full because of the following reasons:

- due to some constructional peculiarities new LRW processing installations allow for handling only low-level LRW (up to 10^{-6} Ci/l activity);
- the cost of LRW-processing operations is rather high; and
- LRW processing results in producing considerable amounts of Solid RWs (SRWs).

Actually, for lack of appropriate capacities, SRWs are not processed in Russia and are stored within special storage sites. The construction of SRW processing facilities is hindered by underinvestment; consequently, at present only project documentation is available.

International cooperation is in progress between Ministries of Defense of Russia, Norway and the USA on different environment-protection tasks in the Arctic Region resulting from the activities of the above Ministries (AMEC Program). At present 6 radioactive contamination-related projects and 2 "non-radiation" projects are under realization. Further development of military environmental projects depends directly on increasing the AMEC Program funding: the needed sum is estimated at 2 million USD.

Regarding SRW management, the AMEC Program provides for:

- constructing sites & prefabricated storage facilities at the Federal State Unitary Enterprise (FSUE) "The 10th Shipyard of the RF Ministry of Defense" in towns of Poliarnyi and Murmansk (are to be commissioned in 2003);
- ordering and making 400 containers ('UZK1A-6'-type) at FSUE "Zvezdochka" to collect and temporarily store SRWs (the work has been already completed);
- constructing mobile module facilities on primary processing of LRW & SRW to be installed at FSUE "The 10th Shipyard" (the work is actually in progress). The first facility is to be put into operation by the end of 2002.

After the facility commissioning the problem of SRW primary processing in Murmansk region will be partly resolved. In keeping with a Joint Memorandum, the facility is to ensure the needs of the whole Murmansk region.

When the above problem issues are solved:

- The Russian Navy is to be ultimately released form such alien functions as nuclear vessel utilization operations and work on processing, long-term storage and disposal of RW & SNF produced during nuclear vessel operation & utilization;
- Russia will be able to: -stabilize funding sources and diminish expenses for RW processing; -reduce RW amounts via implementing technological and other improvements; -bring Russian rules & standards regarding RW & SNF handling to conformity with international regulations;
- the RF will be ready to develop the needed specialized industrial environmentally appropriate capacities to solve issues of nuclear vessel utilization as well as of SNF & RW processing, long-term storage and disposal;
- the RF will accede to Resolution LC.51(16) of the London Convention, 1972, prohibiting marine pollution by dumping of radioactive wastes and other materials.

OVERVIEW AND STATUS OF ISSUES RELATED TO NUCLEAR VESSEL DECOMMISSIONING AND UTILIZATION

UTILIZATION OF RUSSIAN NUCLEAR SUBMARINES: CONTENTS OF THE PROBLEM, REVIEW OF THE ACTUAL STATUS, ANALYSIS OF THE RELATED RISKS AND INTERNATIONAL COOPERATION

ASHOT A. SARKISOV
Academician, Advisor of the Russian Academy of Sciences
Moscow, Russia

Nuclear Submarines (NSs) constitute an important component of the Russian Navy's military potential. But at the same time, their consideration is absolutely necessary when resolving different military-origin tasks regarding environment protection.

In the late 1980s and during the 1990s the Russian Federation (RF) faced a new problem related to mass decommissioning of Russian NSs as a direct consequence of their entering operation in the 1960s and 1970s.

Altogether in the former USSR and in the RF over 240 different-design NSs were built equipped with more than 400 reactors the cores of which comprised 200 to 300 fuel assemblies.

In the course of the same period a developed RF Navy servicing infrastructure was created. It comprised: -Coastal Servicing Enterprises (CSEs) and Floating Servicing Vessels (FSVs) to perform reactor reloading operations; -storage facilities for Liquid Radioactive Wastes (LRWs) & Solid Radioactive Wastes (SRWs) and -other installations.

Among radiation-dangerous facilities of the Russian Navy infrastructure, the decommissioned NSs constitute the main source of potential nuclear & radiation risk. In Figure 1 the dynamics of the Russian NS decommissioning process is depicted. So far, about 200 NSs are decommissioned including 120 NSs in the Northwest region and 80 NSs in the Pacific region.

A.A. Sarkisov and L.G. LeSage (eds.),
Remaining Issues in the Decommissioning of Nuclear Powered Vessels, 9–26.

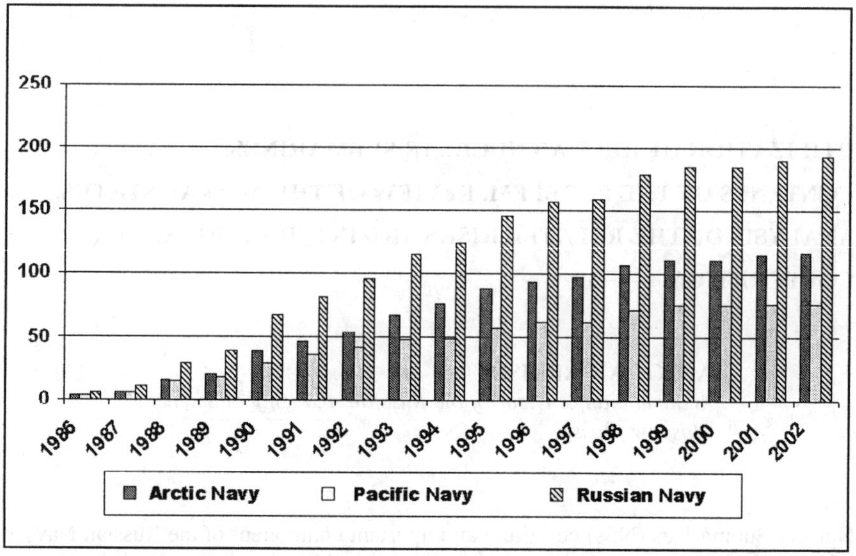

Figure 1. Dynamics of the Russian Navy Nuclear Submarine Decommissioning

This diagram virtually retraces a plot of NS construction in the former USSR with a right time shift of about 20-30 years.

The following data give an idea of the scale of NS utilization process in Russia:

- Integral activity of NS Spent Nuclear Fuel (SNF) exceeds 600 million Ci;
- Total weight of NS contaminated structures to be utilized is over 150 000 t;
- Total weight of NS metal to be cut out equals about 1.5 million t.

Non-preparedness of the infrastructure to the tasks of large-scale NS utilization considerably complicates the problem: in the period of rapid NS construction only a few funds were assigned to create an appropriate industrial base of utilization operations.

In addition, the onset of NS decommissioning concurred with the period of economic depression in Russia resulting from unskillful attempts to reform the Russian economy.

Because of all the above reasons NS utilization in Russia is being performed using a temporary flowsheet the distinctive feature of which can be generalized as follows: after NS Reactor Compartments (RCs) are cut out they are not immediately transported to coastal long-term storage centers.

At the same time the initial stages after NS decommissioning are the traditional ones. They comprise, first, NS waterborne 'waiting-for-de-fueling' stage, next, NS hauling to the enterprise utilization-executor, SNF unloading & management, waterborne storage of de-fueled NSs and, finally, NS cutting operations.

However the following NS utilization stages have their special features. During NS cutting at slip conditions three-compartment units are formed comprising the reactor compartment itself and two adjacent compartments. Such three-compartment (in some cases multi-compartment) units are stored within temporary waterborne storage areas.

The dynamics of NS cutting out and three-compartment unit forming is depicted in Figure 2.

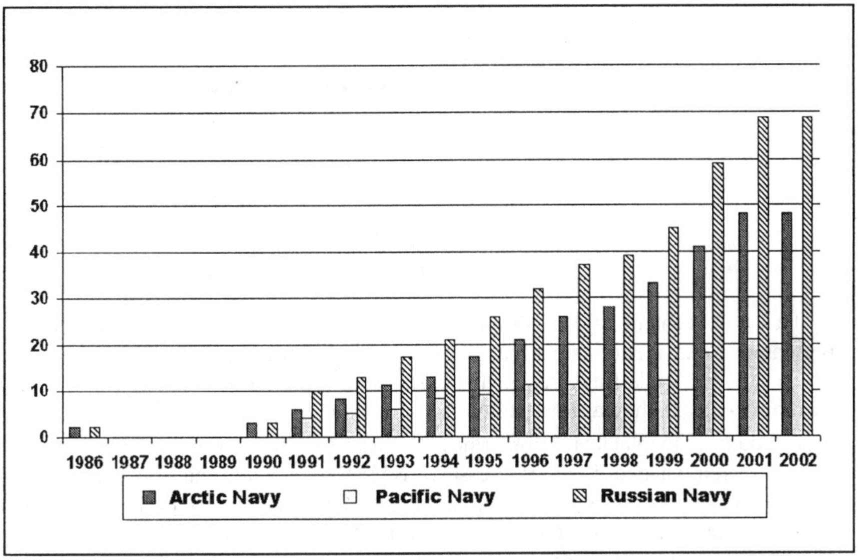

Figure 2. Dynamics of NS Cutting out and Three-Compartment Unit Forming

As the needed infrastructure is ready, RCs will be cut out of the above three-compartment units and transported to coastal long-term storage centers.

In Figure 3 a curve of the integral activity decay in RCs of the second-generation NSs is depicted. From these data it follows that the period of 80 to 100-year RC storage is enough to restart the subsequent cutting operations at shipyards aimed at reaching a stage of NS ultimate disposal.

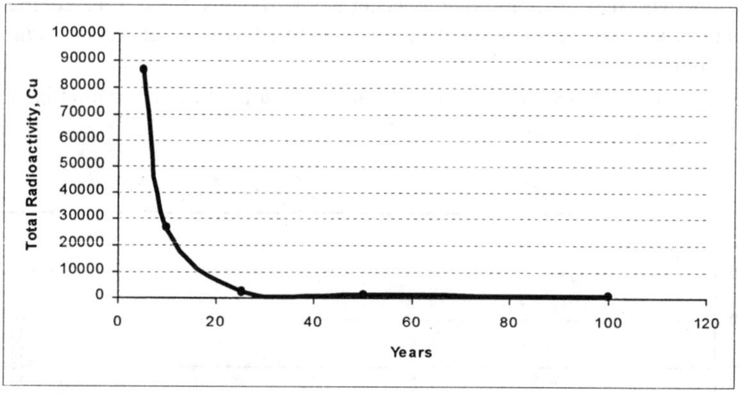

Figure 3. Integral Radioactivity (Ci) in RCs of Different-type NSs and the Dynamics of Its Decay during 100-year Storage

To create centers of RC long-term storage in the Northwest and the Far East regions of Russia, some sites have been chosen and technical proposals developed. These are the Saida-Bay area in the Northwest region and a territory adjacent to Razboinik-Bay in the Far East region.

In addition to Nuclear Submarines, Nuclear Coastal Servicing Enterprises and Nuclear Maintenance Support Vessels (NMSVs) carrying out nuclear fuel and Radioactive Waste (RW) handling operations (such as: FSVs performing reactor reloading and special Technical Pouring Tankers (TPTs)) should be also considered as radiation-dangerous facilities and potential sources of environment contamination.

The lifetime (30 years) of all FSVs is up. Due to high levels of both irradiation and contamination, their repair is extremely difficult. The actual technical condition of FSVs does not meet the requirements of present-day nuclear & radiation safety standards and, thus, these FSVs cannot guarantee the needed safety of work.

The actual technical status of TPTs is even worse. To keep them operable for performing the relevant operations, important funds are necessary.

Thus, the problem of NMSV utilization represents a separate and a rather specific task.

NS utilization operations are accompanied with different risks, which can be classified as follows:

1. Radiation impact on the environment.
2. Nuclear safety-related risks.
3. Risks due to potential terrorist attacks regarding radiation-dangerous facilities.
4. Risks resulting from fissile material proliferation.
5. Risks related to chemical pollution when performing NS dismantling operations.

Below every type of the above-listed NS utilization-related risks is considered (except for chemical pollution risks because of their specific nature).

1. Radiation Impact on the Environment

Despite the huge territory of the Russian Federation, the Russian atomic fleet is based within rather restricted areas of the Northwest region and the Far East region. This circumstance results in the concentration of the potential impact on the environment of both the decommissioned NSs and their supporting infrastructure.

In Figure 4 all utilization stages & infrastructure facilities presenting real and potential hazard to the environment are demonstrated. Units & installations containing SNF (being the most hazardous element from the viewpoint of the environment contamination) are shaded.

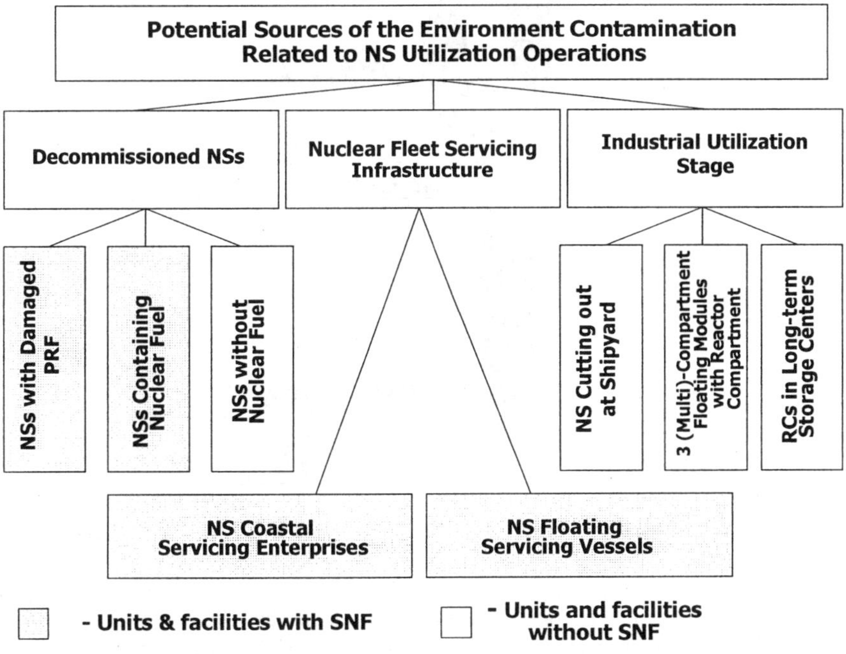

Figure 4. Potential Sources of the Environment Contamination Due to NS Utilization Operations

When analyzing the radioecological risks, real contamination of the environment by technogenic radionuclides and the risks resulting from the whole potential of accumulated radioactive materials should be assessed separately.

The real contribution of different technogenic sources to radioactive contamination of the concerned regions is demonstrated in Figure 5.

From the presented diagrams it follows that radionuclide fallout resulting from nuclear weapons tests (performed in the atmosphere before 1963) constitutes the principal source of environment contamination in both regions. Releases of the radiochemical plant in Sellafield, the UK, are the second most important contamination source in the Arctic. Radionuclides transferred via waters of Siberian rivers are ranked third.

In both regions the facilities involved into the NS utilization process make the least contribution to the environmental contamination among other sources; if considering absolute values, their contribution is equally low.

The largest contribution of NS utilization-related units & installations to the environmental contamination is due to some infrastructure facilities among which CSEs should be mentioned first because of their poor technical status.

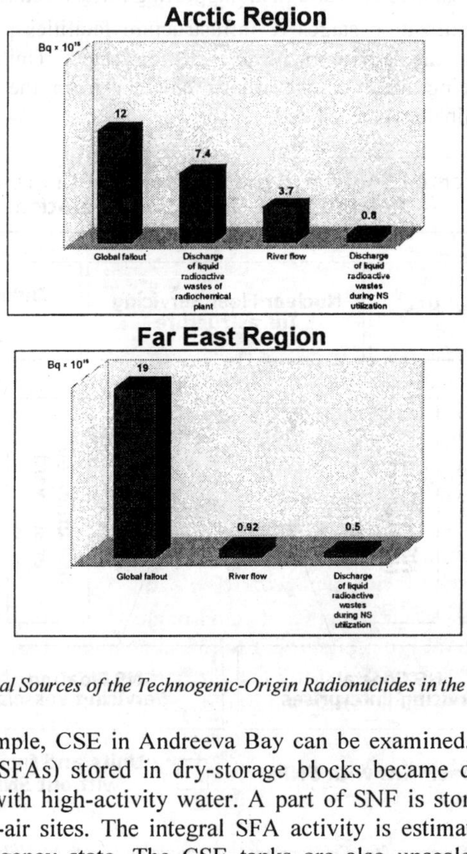

Figure 5. Real Sources of the Technogenic-Origin Radionuclides in the Environment

By way of example, CSE in Andreeva Bay can be examined. Here many Spent Fuel Assemblies (SFAs) stored in dry-storage blocks became depressurized, their cells being filled with high-activity water. A part of SNF is stored within old-type containers at open-air sites. The integral SFA activity is estimated at 10^7 Ci. One facility is in emergency state. The CSE tanks are also unsealed and contain the remaining parts of fuel structures with the integral activity of $\sim 4{\cdot}10^3$ Ci.

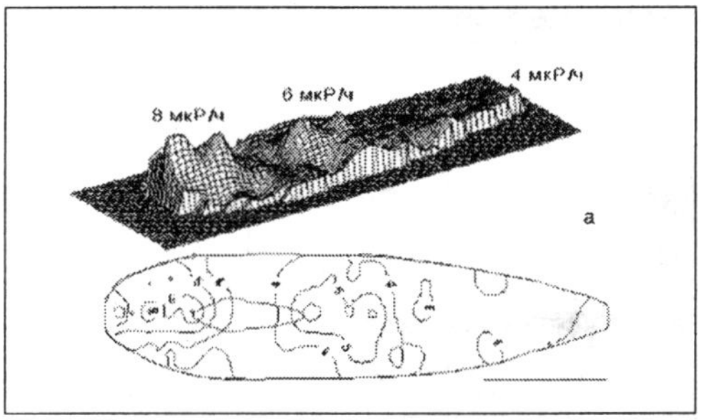

Typical nuclear submarine containing fuel

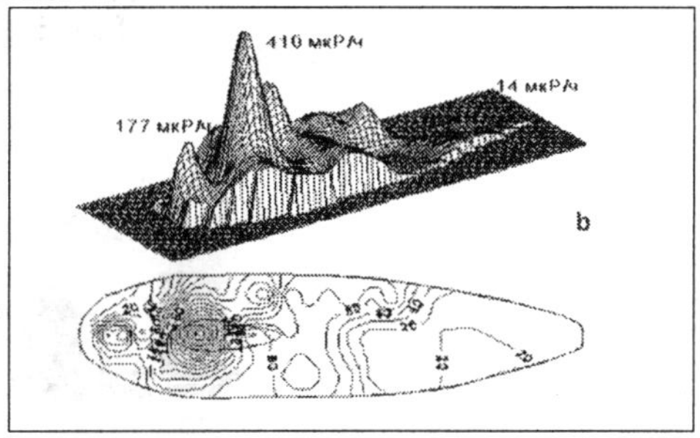

Nuclear submarine with damaged power reactor facility

Figure 6. Gamma-Fields Regarding: -Typical Decommissioned NSs under Waterborne Storage with fuel and – NS with Damaged Power Reactor Facility

A part of the SRW storage facility at the Andreeva Bay CSE is also unsealed and contains water. Some LRW tanks are equally unsealed. As a result, the environment around damaged facilities has a high contamination level.

Figure 6 gives an idea of the environmental effects issuing from NS themselves (here are depicted gamma-fields close to: -a typical decommissioned NS stored afloat with fuel and -a NS stored with damaged PRF).

However, when assessing the potential environment contamination risk, one faces quite a different pattern. From Figure 7 it follows that most of the radiation potential (in both the Northwest region and the Far East region) is concentrated within SNF housed by the decommissioned NSs and stored at CSEs.

The integral fuel amount of the decommissioned Russian NSs makes up 75 000 kg, (about 15 000 kg of ^{235}U, and about 60 000 kg of ^{238}U) with the radiation potential above 300 million Ci.

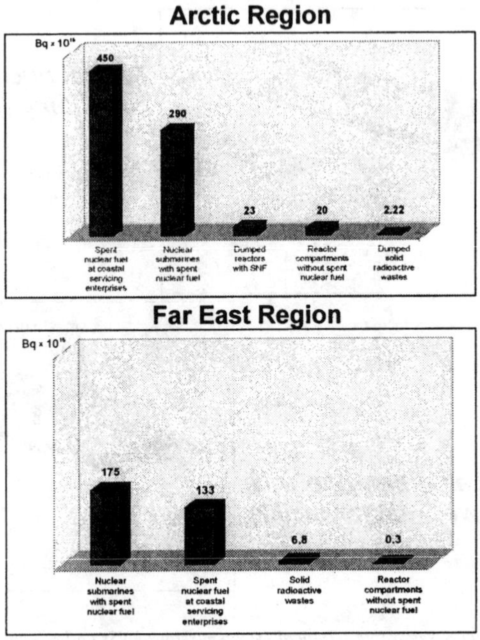

Figure 7. Potential Sources of Technogenic-Origin Radionuclides in the Environment of the Russian Northwest Region & the Far East Region

When considering issues of nuclear & environmental safety, one should be aware that the most critical circumstance is due to the fact that over 60% of the decommissioned NSs still store SNF. The dynamics of NS de-fueling is depicted in Figure 8.

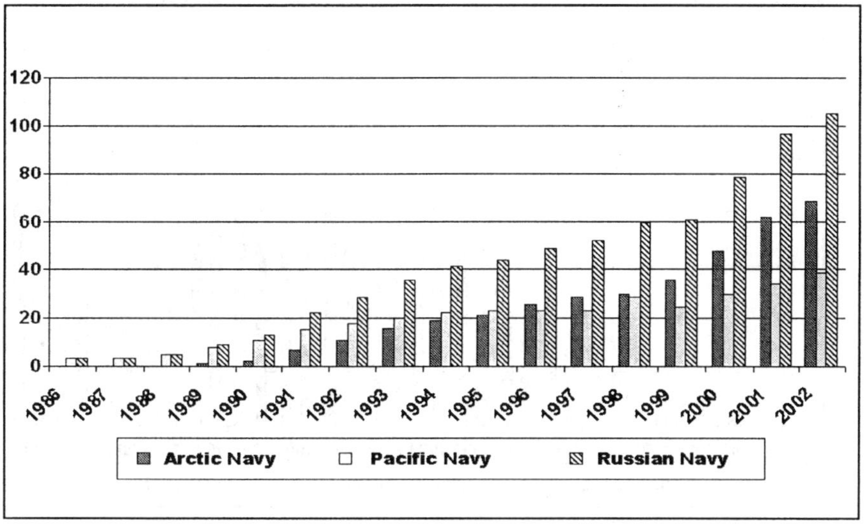

Figure 8. Dynamics of the Decommissioned NS De-fueling

Figure 9 shows the number of NSs housing SNF in their cores. One can see that after the year 2000 the pace of NS decommissioning remains behind that of NS utilization and, thus, the number of NSs stored with fuel gradually decreases.

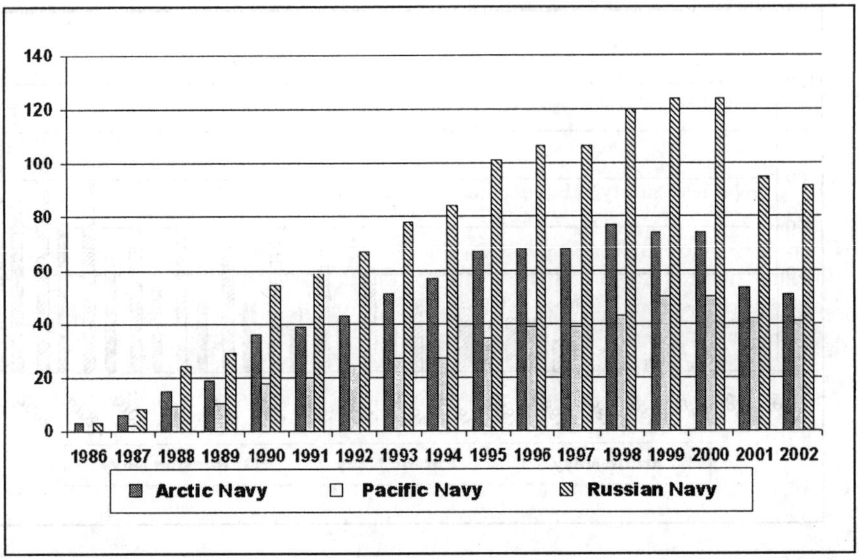

Figure 9. Number of NSs and Reactor Units Housing SNF in Their Cores

However, because of restricted capacities of the relevant infrastructure to perform SNF unloading, transportation and reprocessing, the NS de-fueling process can last for another 8-12 years even at the current actual pace.

Production Association (PA) "Mayak" is the only Russian enterprise which collects & reproceses the Navy SNF. The actual production capacity of the enterprise allows for reprocessing SNF of 10 NSs per year. Two special trains comprising 4 railcars each perform SNF transportation. To accelerate the pace of SNF removal from regions, work on constructing a buffer site of 120-150 container (cask) capacity is being carried out at PA "Mayak" to store TK-18 & TUK-108/1-type casks with SNF and also empty casks. The construction of container ships as well as of TUK-108/1-type two-purpose metal-concrete containers (adapted to the transport-technological flowsheet of the actual TK-18 casks) will also make a contribution to the acceleration of the SNF removal process. In 2002 some casks of TUK-108/1-type are to be constructed. Another 60 similar-type casks are to be built in the near future using funds of the international assistance programs.

Thus, to ensure safety, the following two tasks need urgent solution:

1. Constructing new engineering facilities in order to accelerate the pace of the decommissioned NS de-fueling (actually 170 NS reactors are waiting for de-fueling).
2. Accelerating the pace of SNF removal to PA "Mayak" for reprocessing (at present about 31500 SFAs are in waiting mode) and, simultaneously, implementing two-purpose metal-concrete casks into practice.

Besides, issues of managing damaged (off-test) SFAs, which cannot be actually reprocessed at PA "Mayak", as well as of highly-enriched SFAs of liquid-metal coolant NSs are equally important. So far, neither procedures nor techniques are available to solve these problems.

Utilization of these NSs with damaged Power Reactor Facilities (PRFs) and their SNF handling represents a special and a rather important task.

2. Nuclear safety related risks

From the nuclear safety standpoint NSs housing nuclear fuel present the greatest potential hazard. So far, 90 NSs are not de-fueled yet including 70 NSs in the Northwest region and 20 NSs in the Pacific region, their integral radiation potential being estimated at ~ 270 million Ci.

In most cases, at the decommissioned NSs, the major portion of core energy resource is spent. With consideration for the accumulation in fuel of important fission-fragment activity, the consequences of a hypothetical nuclear accident can reach levels 5-6 of the IAEA International Nuclear Event Scale (INES).

If one took into account the consequences of the Chazhma-Bay accident, which, occurred in 1985 with non-irradiated fuel (immediately after "new" fuel loading into NS core), those at a core with spent fuel would be much larger. In such a situation the release of long-lived radionuclides could be 3 to 4 orders of magnitude larger than in case of "non-irradiated fuel" accident. Such an accident could result in the contamination of both the water area and coastal territories, and its scale would exceed by many times the environmental consequences of the Chazhma-Bay accident. Nuclear & radiation risks are further aggravated due to the unsatisfactory technical condition of the decommissioned NSs under waterborne storage.

Data on technical status of the decommissioned NSs are generalized in Table 1.

In most cases the decommissioned NSs are unprepared for long-term safe afloat storage from both the technical and organizational viewpoint because of a failure to execute some necessary actions related to equipment dismantling, strong hull pressurization and NS preparation for long-term waterborne storage. Actually the system of NS day-to-day operation does not function, only <25% men of NS crews being available.

The technical condition of the first-generation NSs is particularly poor. Most of their mechanisms & equipment including such essential systems as sprinklers and water discharge installations have already spent their resources and became worn out due to long inaction. Corrosion of the light hull reached a dangerous level, and some driving ballast tanks became unsealed. Because of the above reasons hauling of the first-generation NSs from waterborne storage areas to utilization centers will be impossible within a few years unless special transport docks are used.

Taken together, the above circumstances create a threat of safe NS waterborne storage and necessitate urgent elaboration and realization of measures aimed at ensuring safe storage of the decommissioned NSs.

TABLE 1. Actual Technical Status of the Decommissioned Russian NSs

№	Technical Status	NS generation	Number	Comments
1	Stored with fuel in reactor cores	I	20	
		II	70	
2	Strong hull is depressurized	I	4	Two NSs have depressurized primary circuit
		II	8	
3	Driving Ballast Tanks (DBTs) are depressurized	I	47	Including: -Depressurization of over 50% DBTs in 17 NSs - work is being performed on DBT filling with foamy polystyrene
		II	23	
4	Power Reactor Facility (PRF) primary circuit is depressurized	I	33	
		II	49	
5	Failures in high-pressure air system: - whole system inoperable - used with restrictions	I	15	
		II	12	
6	Failures in pumping and discharge systems	I	37	
		II	19	
7	Failures in sprinkler systems	I	29	
		II	11	
8	Failures in principal power mains (blackout)	I	15	
		II	9	
9	Damaged state of the reactor	I	3	

3. Risks due to potential terrorist attacks

In the course of both Russian nuclear vessel operation and the subsequent utilization operations large amounts of SNF, LRW & SRW were accumulated within the Northwest and the Far East regions (see Table 2).

The integral radiation potential of the accumulated activity makes up about $400 \cdot 10^7$ Ci that exceeds by tens of times the integral activity released during the Chernobyl accident ($\sim 20 \cdot 10^6$ Ci).

TABLE 2. Generalized Data on the Accumulated Activity in Russian Regions

Region	Spent Nuclear Fuel (Core + SFAs), Ci	LRW, Ci	SRW, Ci
Murmansk region	$22 \cdot 10^7$	12	350
Arkhangelsk region	$25 \cdot 10^6$	22	1100
Σ Northwest region	$32 \cdot 10^7$	**34**	**1450**
Primorskiy kray	$\sim 10^8$	260	260000
Kamchatka	$8 \cdot 10^7$	330	1600
Σ Far East region	$13 \cdot 10^7$	**590**	**261000**

From the standpoint of potential terrorist attacks, CSEs (which storage facilities house considerable amounts of SFA & high-level RWs) are most hazardous. An estimate of the consequences of hypothetical events after unauthorized impacts allows for drawing up a list of the most hazardous events:

- explosion due to impinging of an anti-vessel missile or of a civil/military plane (with ammunition aboard) on SNF dry-storage facility/high-level SRW repository;
- NMSV sinking accompanied with SNF collapse after explosion.

Below are presented the results of some estimates of the consequences of unauthorized impacts (missile impinging, explosion and fire) on the available SNF & SRW coastal storage facilities. The calculations were performed by specialists of the Nuclear Safety Institute (IBRAE) of the Russian Academy of Sciences.

For example in Table 3, calculated data of the radioecological consequences in the case of an anti-vessel missile impinging on the SNF storage facility in Sysoeva Bay are demonstrated.

In is worth noting that the consequences of a similar event at the SNF dry-storage blocks in Andreeva Bay would be much larger because of its poorer protection (the facility in Sysoeva Bay is covered with plates above, whereas at Andreeva Bay CSE one has a virtually open-air SNF storage site).

TABLE 3. Results of Calculating Potential Radioecological Consequences in a Case of Anti-vessel Missile Impinging on a SNF Storage Facility in Sysoeva Bay

Distance along the trace axis, km	Surface contamination, 10^7 Bq/m^2	Inhalation dose, mSv	10-day external exposure dose, mSv	One-year effective dose, mSv
2	5	26	7	256
3	4	20	5,2	190
5	2	12	2,8	102
6	1,8	10	2,5	91
8	1	5	1,5	55
10	0,7	3,5	0,8	29

In Table 4 calculations of the radiation consequences in the case of an anti-vessel missile impinging on the high-level SRW storage facility at Sysoeva Bay are presented.

TABLE 4. Radiation Consequences Due to an Explosion at High-Level SRW Storage Facility (Sysoeva Bay, Far East Region)

Distance along the trace axis, km	Surface contamination, 10^7 Bq/m^2	External exposure dose during 10 days, mSv	One-year effective dose, mSv
2	8	45	1643
3	4	24	876
4	3,2	18	657
5	3	16,5	602
6	2,6	15	548
8	1,6	9,5	347
10	1,2	7	256
12	0,8	4,7	172
14	0,56	4	146

It is worthy of notice that the radiation consequences due to impinging of a civil plane on a RW storage facility could be worse by an order of magnitude because of the large fuel mass housed on board the plane. Both the damage and radioactive release due to fuel detonation would surpass the consequences of a potential explosion of standard ammunition.

The performed assessments of the long-term consequences of explosions at SNF & SRW storage facilities have demonstrated that, as a result of radioactive releases, extended territories (tens of kilometers from the source term) could become uninhabitable, the lifelong population exposure dose being equal to 1 Sv.

In Table 5, an estimate of dose loads resulting from fire at an open-air SRW storage site in the Northwest regions is given.

TABLE 5. Estimates of Potential Dose Loads in Case of Fire at Open-Air SRW Storage Site within a CSE in the Northwest Region

Distance along the trace axis, km	Integral surface contamination, (Bq/m^2)	Total one-year dose, mSv
0,5	$5.6*10^5$	4,4
1	$2.8*10^5$	2,2
2	$1*10^5$	0,72
3	$5.4*10^4$	0,4

The calculated model took into account both re-suspension of the surface contamination and its spreading together with the smoke plum with consideration for average statistical weather conditions in the region.

4. Risks resulting from fissile material proliferation

Huge amounts of SNF accumulated during NS utilization operations necessitate a thorough analysis of the problem from the viewpoint of nuclear material nonproliferation. Particular topicality of such an analysis is due to many factors among which high enrichment of vessel reactor SNF should be mentioned first.

It is known that, in keeping with the IAEA classification, uranium enriched over 20% is to be considered as a "potentially usable material to make a nuclear device".

The first-generation NSs with water-moderated reactors used ^{235}U of about 21% enrichment as their nuclear fuel; however the subsequent-generation NSs (equipped with the same type reactors) employed nuclear fuel of ~40% enrichment.

In addition to NS with water-moderated reactors, 8 NSs of the Arctic Navy were equipped with Pb-Bi liquid-metal coolant reactors. Their fuel composition comprised an intermetallic compound (UBe) with ^{235}U enrichment up to 90%. Every such-type reactor contained about 200 kg of ^{235}U.

Taking into account a relatively low fuel burnup fraction, the enrichment of considerable amounts of SNF can appreciably exceed the IAEA threshold level, and, thus, such a fuel can be theoretically used to create nuclear devices.

After a long-term SNF storage, its activity considerably decreases, and SFA handling operations can start. This thesis can be illustrated by an example of evaluating calculations.

The dose rate at 0.5-m distance from the very center of an "average statistical" NS SFA after 20-year storage is about 120 R/h and this allows an operator to perform relatively safe operations of separating the active, i.e. fuel-containing, part (its length is slightly less than one-half of the whole SFA length) from the upper guiding tube. The uranium mass within one SFA makes up about 1.4 kg, whereas the mass of SFA itself equals about 15 kg.

If a lead container 5-cm thick were arranged around the fuel-containing SFA part (slightly above 1-m length), the dose rate at some tens of centimeters from the container would decrease to few R/h, the weight of such a container being about 230 kg.

It is worth noting that some SNF storage facilities house SFAs removed from reactors 20 and more years ago.

The storage period for several hundreds of damaged SFAs of water-moderated reactors is to be even longer. So far, PA "Mayak" does not accept these SFAs for, to perform their handling & reprocessing, special procedures & techniques are needed.

A similar situation characterizes the fuel of liquid-metal coolant reactors: to manage such fuel, no appropriate technology and production basis have been developed yet.

The problem topicality increases further since physical protection of SNF storage facilities cannot be ensured at the needed level in all cases due to their remoteness and dispersed location.

5. International Cooperation on NS Utilization-Related Problems

The international cooperation concerning utilization of military units & installations created during the Cold War started in 1992 with the Umbrella Agreement between the Russian Federation and the United States of America on the Cooperative Threat Reduction Program (CTR Program). Next were concluded: -the Agreement between the Governments of the Russian Federation and Japan, 1993; the Agreement between the Ministries of Defense of Russia and Norway, 1995; the Declaration of Three Countries (Russia, the USA and Norway) on the Environment Protection in the Arctic, 1996, and other joint resolutions. In the context of their development some important international cooperation programs dealing with issues of NS utilization and environment protection in the Arctic region were agreed upon.

Among them, the Arctic Military Environmental Cooperation Program (AMEC Program) should be mentioned first. At present proposals are made to prepare a similar-type program regarding the Far East region.

In addition to AMEC Program, some other cooperative efforts are being carried out, such as: AMAP, ANVAP, YASAP Programs, etc. Their step-by-step realization will make it possible to resolve many issues related to environmental safety, in particular, due to NS utilization. By now we managed to reach many good results thanks to the implementation of different projects within the framework of these programs. Among them, the following achievements should be mentioned first:

- efficient equipment is being delivered to "Zvezda" and "Zvezdochka" shipyards to perform NS metal cutting;
- 4 special railcars were produced for SNF cask transportation;
- environmental rehabilitation of the contaminated territories in Andreeva Bay has started;
- TUK–type casks were designed and many TUK casks are being made to ensure temporary storage and transportation of SNF unloaded from the decommissioned NSs;
- Two facilities have been constructed and commissioned in Severodvinsk (the Northwest region) and Bolshoy Kamen' (Far East region) to process LRW;
- Sites for TUK temporary storage prior to transportation to PA "Mayak" are under construction.

However, despite important results, much work still must be done. Thus, in the nearest future one needs:

- initiate design and construction of coastal sites for RC long-term storage within both the Northwest region & the Far East region of Russia;
- assess the radioecological situation and perform rehabilitation of the Andreeva Bay CSE. Remove SNF from the CSE;
- manufacture the third special train for SNF transportation;
- develop basic principles for handling SNF from liquid-metal coolant reactors;
- resolve a number of practical issues related to NMSV utilization.

Along with practical issues, many important scientific problems remain still unresolved, in particular:

- scientific substantiation of priorities when performing multi-purpose utilization of nuclear powered vessels;
- studying the consequences of incidents/accidents in the course of NS utilization with an emphasis on their minimization;
- analysis of the vulnerability of radiation-dangerous facilities involved into the utilization process from the viewpoint of potential terrorist activities;
- forming adequate public opinion in regions concerned with NS utilization & RW management and many other tasks.

The actual level of cooperative efforts taken together with hazards issuing from the whole complex of nuclear vessel utilization activities requires new forms and further development of the international cooperation.

6. Conclusion

The most radical lines for reducing the different-nature risks related to the process of NS utilization in the Russian Federation are: -accelerating the pace of the decommissioned NS de-fueling, -subsequent placing of SNF into metal-concrete casks and -SNF transportation to PA "Mayak" for reprocessing.

Despite a complicated economic situation, Russia intensifies its efforts to promote utilization of all-type NSs. However under the current actual utilization rates the major portion of work can only be completed in 8-12 years, unless additional financing is assigned. In such conditions the international cooperation, which has already proved its efficiency and profitability, takes on special significance. At the same time one should be aware that the actual level of the international cooperation complies neither with the scale of the whole utilization problem nor with its importance in the present-day context of nuclear & radioecological safety.

OVERVIEW OF THE STATUS AND ISSUES RELATED TO THE DECOMMISSIONING OF NUCLEAR VESSELS IN RUSSIA

V. AKHUNOV
Chief of the Department of Environment and Decommissioning of Nuclear Installations, Minatom
Moscow, Russia

The process of Nuclear Submarine (NS) decommissioning at the Russian Navy was initiated by the termination of the lifetimes of many NSs began in the middle 80s of the last century and proceeded in keeping with the relevant international obligations of the Russian Federation (RF).

Before 1998 Russian shipyards utilized 3 to 4 NSs per year at most. As a result, many decommissioned NSs were accumulated within waterborne storage areas.

Thus, by the end of 1997 about 120 NSs were stored in a pending-state including about 110 NSs with reactor cores. Because of their poor technical condition, the driving ballast tanks of some of the decommissioned NSs became depressurized resulting in an increase in the probability of NS sinking.

In keeping with the RF National Safety Concept, an acceleration of the NS utilization process is one of the high-priority tasks of Russia regarding environmental protection.

Due to an urgent need for a solution of the problems related to the decommissioned NSs and environmental rehabilitation of the Russian Navy radiation-dangerous facilities, the RF Government designated Minatom of Russia as "The State Customer-Coordinator of the Work on NS Multi-purpose Utilization" and dismissed the RF Ministry of Defense from these unnatural functions.

Thus, in keeping with the RF Government Regulation of May 28, 1998, the RF Minatom, as the State Customer-Coordinator, ensures:

- Multi-purpose utilization of: decommissioned NSs, Nuclear Surface Vessels (NSVs) and Nuclear Maintenance Support Vessels (NMSVs). The latter category comprises: Floating Servicing Vessels (FSVs), special tankers and floating tanks to store Radioactive Wastes (RWs);
- Use of the funds (obtained when selling nuclear vessel utilization products) to finance multi-purpose NS utilization and its related work;
- Environmental rehabilitation of the RF Ministry of Defense (Navy) facilities involved in the process of temporary storage of Spent Nuclear Fuel (SNF) and of Liquid & Solid Radioactive Wastes (LRWs & SRWs).

To accelerate multi-purpose NS utilization, Minatom with the participation of interested ministries and departments, has drawn up the following basic organizational and technical documents:

- "Schedule of Urgent Work to be Realized in Support of SNF Unloading from Reactors of Decommissioned NSs and for Purposes of Reducing the Environmental Risk at Coastal Servicing Enterprises (CSEs)";
- "The Concept of Multi-purpose NS Utilization" (approved and accepted for implementation in keeping with a special order of the Russian Government);

A.A. Sarkisov and L.G. LeSage (eds.),
Remaining Issues in the Decommissioning of Nuclear Powered Vessels, 27–34.

- “Project of Federal Target Program for Multi-purpose NS Utilization for the Period 2000 through 2010”; and
- Joint Resolutions of the RF Minatom, RF Ministry of Defense and Rossudostroenie "On Measures of Ensuring SNF unloading from Reactors of NSs to Be Utilized" (are formalized once a year).

1. Actual Status of Resolving NS Utilization Issues

Since 1999, when the RF Minatom was designated the State Customer-Coordinator in order to ensure acceleration of the NS utilization process, much work has been performed.

1.1. ACCELERATION OF SNF UNLOADING OPERATIONS FROM NS REACTORS

The performed analysis has demonstrated that the lack of infrastructure capacities to perform SNF unloading, temporary storage, transportation and reprocessing is the principal factor hampering the whole NS utilization process. At the beginning of 1999 only three FSVs to unload SNF from NS reactors were available in the Russian Navy. Therefore, during 1994-1998 only 4 NS/year were defueled on average.

The facilities to temporarily store SNF issued from NS reactors were full and/or in poor technical condition.

SNF was transported for reprocessing in a special train comprising 4 transportation railcars. On average SNF of 4 to 5 NSs was transported per year.

Thus, the RF Minatom concentrated its efforts on ensuring: -most urgent defueling of the decommissioned NSs and –SNF safe management.

To ensure acceleration of SNF unloading rates:

- all available FSVs and SNF unloading equipment were overhauled, and supplementary sets of equipment were made; and
- a flowsheet for SNF unloading using servicing vessels of the “Murmansk Sea Navigation Enterprise” Public Corporation was implemented.

In 2002 SNF unloading operations are to begin at coastal unloading complexes of both “Zvezdochka” and “Zvezda” plants (it should be especially stressed that these complexes were constructed using US assistance funds).

To ensure safe temporary storage conditions, the mode of SNF dry-container storage is being developed. For these purposes:

- 48 metal-concrete new-type containers (casks) to store & transport SNF were designed, fabricated and prepared;
- a contract for making two additional container lots (25+35) using funds of the US assistance programs was signed;
- sites ensuring SNF temporary container storage (30 container each of the capacity) at Far East Plant “Zvezda” (Primorskij Kraj) and SMBE “Zvezdochka” (Severodvinsk-town) were completed;
- a site for SNF container reloading at SRE “Atomflot” in Murmansk is under construction;
- an understanding was reached on the construction of a buffer SNF container storage facility at PA “Mayak” using US assistance funds.

To accelerate the pace of SNF transportation to PA "Mayak":

- the second SNF transporting train of 4 railcars was built and commissioned;
- an agreement was reached on the construction of the third SNF transporting train of 6 railcars using funds of the US assistance programs.

The already constructed facilities make it possible to unload SNF from reactors of over 20 NS/year and ensure safe SNF management.

In 2000 – 2001 the actual rate of SNF unloading operations was 18 NS reactor/year.

The dynamics of SNF unloading from reactors of NSs to be utilized is illustrated in Table 1.

TABLE 1. Dynamics of SNF Unloading from NS Reactors

Year	1994	1995	1996	1997	1998	1999	2000	2001	2002
NS number	2	5	4	3	4	8	18	18	18 (planned)

The achieved SNF unloading pace (18 NS reactor/year) is optimal for it is coordinated with the capacities of the available industrial infrastructure on ensuring SNF storage, transportation and reprocessing.

Thus, the actions undertaken by the RF Minatom created a favorable background for completing SNF unloading operations from reactors of all decommissioned NSs in 2007.

1.2. REACTOR UNIT FORMING

Further acceleration of the SNF unloading process is hampered by the actual (rather low) pace of the subsequent NS utilization stages, i.e. cutting out of reactor compartments and their preparation for temporary waterborne storage.

Though 95 to 98 % of all NS radionuclides are concentrated in SNF, the residual activity of reactor facilities is also a radiation hazard and, thus, appropriate management is necessary. After SNF unloading, one needs to continue the NS utilization process in order to free up the waterside of the involved enterprises for the next (coming for defueling) NSs.

The dynamics of Reactor Unit (RU) forming during NS utilization is illustrated in Table 2.

TABLE 2. Dynamics of Reactor Unit Forming

Year	1994	1995	1996	1997	1998	1999	2000	2001	2021
Formed RUs	4	5	6	5	2	6	16	8	17 (planned)

Actually the utilization of Russian NSs is carried out at 8 ship-repair & shipbuilding enterprises under the jurisdiction of Rossudostroenie and the Russian Navy, which perform NS-cutting and RU-forming operations. These are:

- "Nerpa", "Poliarninskiy Shipyard " and "Sevmorput'" enterprises in Murmansk region;

- "Zvezdochka" and "Sevmachpredprijatie" enterprises in Arkhangelsk region;
- Far East Plant "Zvezda" and "Chazhminskij SY" in Primorskij kraj;
- "Viliuchinskij SY" in Kamchatka region.

The actual procedures & techniques used for NS utilization consist of: -first, cutting out of the reactor facility as a "Three-Compartment Unit" (TCU) comprising the reactor compartment plus two adjacent compartments to ensure buoyancy and -next, TCU sealing for subsequent temporary waterborne storage. Such a NS utilization technology is far from optimal because:

- TCU-forming operations are more expensive than those of One-Compartment Unit (OCU) creation because, in the first case, many openings need to be sealed;
- additional funds are required to ensure both the functioning of the RU temporary waterborne storage centers and for RU periodical docking and maintaining repair.

Thus, to ensure safe long-term storage, OCUs, which can be formed on the basis of TCUs are preferable.

Advanced NS utilization technologies provide for OCU forming and the ensuring of their subsequent long-term storage within coastal sites. But the implementation of this technology is hampered by a lack of appropriate process & transport facilities and coastal storage sites.

At present 2 feasibility studies aimed at creating 2 coastal complexes for long-term storage of OCUs have been developed: one complex is to be constructed in the Northwest region (Saida Bay, Murmansk region) and the other one in the Far East region (Razbojnik Bay, Primorskij kraj), their cost being estimated at 4 billion rubles.

There is no way of assigning budget funds to construct the above facilities in the immediate future, because this would result in a sharp decrease of SNF unloading & NS utilization paces. To find the needed resources, the RF Minatom carries on negotiations with potential foreign investors, such as the USA and Japan.

The NS hull cutting out and RU forming processes represent a rather protracted (up to 2 years) and expensive (up to 80% of all costs) NS utilization stage. Therefore, depending on annual financial resources, the number of formed RUs varies from year to year.

An analysis of the results of work performed during 1999 through 2001 demonstrates that under the condition of appropriate financing all decommissioned NSs will be utilized by 2010.

1.3. RW MANAGEMENT

The construction, operation and utilization of the Russian Navy nuclear vessels resulted in accumulation of considerable amounts of RWs within both the RF Northwest region and the Far East region.

To optimize RW handling operations the following actions were performed during 1999 through 2001:

To manage LRWs:

- thanks to funds of the US Cooperative Threat Reduction (CTR) Program stationary RW processing complexes at "Zvezda" and "Zvezdochka" plants were put into operation;
- a floating complex “Landysh” to process mixed-composition LRWs was commissioned at "Zvezda" plant thanks to funds provided by Japan;
- modernization of a facility to process LRWs at SRE “Atomflot” is being completed using funds Russia, Norway and the USA;
- 3 mobile LRW processing facilities were made and commissioned in the Murmansk region, Kamchatka region and Primorskij kraj.

As the above processing facilities are commissioned, all LRWs produced as a result of nuclear vessel operation and utilization will be processed & conditioned; consequently the amount of accumulated LRWs will gradually decrease.

The integral amounts of LRWs processed in 2001 in the Northwest and the Far East regions were 2000 m^3; 4200 m^3 of LRWs are to be processed in 2002.

The actual pace makes it possible to process all LRWs produced during NS utilization operations and reduce volumes of already accumulated LRWs by 2500-3000 m^3 per year.

To manage SRWs:

- SRW process and storage complexes were constructed and commissioned at the "Zvezda" and "Zvezdochka" enterprises;
- a design has been developed and delivery of equipment has started for a SRW conditioning center at "Poliarninskij SY" using funds of the AMEC Program (USA, Norway);
- a detailed design to construct an experimental-industrial facility for disposal of low-level and medium level RWs (produced during nuclear vessel operation and utilization in the Northwest region) in the South Isle of the Novaya Zemlia archipelago has been developed. By now a Statement on Intentions to construct the facility has been argreed upon, and the project documentation is forwarded for public examination and environmental examination.

So far, most of SRW is not processed and is stored within the available storage facilities.

Toxic chemical wastes produced during NS utilization operations are also stored without being processed.

Both stationary and mobile LRW & SRW processing complexes constructed during the last few years have created guaranteed conditions for the ultimate termination of marine disposal of NS-origin RWs.

In Russia this circumstance has created a favorable background for joining the LC.51 (16) correction to the London Convention of 1972 on the prevention of Marine Pollution by Dumping of Wastes and Other Matter.

1.4. NS WATERBORNE STORAGE AND ENSURING SAFETY

The process for decommissioned NSs to transfer to enterprises executing NS utilization operations started in reality in 1999 after the entering into force of many documents (standard, legal and organizational) regulating the NS transfer procedure – acceptance, organizing their safe handling by civil crews and ensuring safety of utilization operations.

Altogether during 1999-2001 about 40 NSs were accepted by the concerned enterprises, NS maintenance costs being financed in keeping with the contracts made with the RF Minatom.

The collected experience has demonstrated the inexpediency of urgent transfer of all decommissioned NSs under the jurisdiction of the above enterprises because of the following reasons:

- lack of the needed watersides to arrange all decommissioned NSs;
- lack of civil crew personnel needed to maintain NSs because of a rather protracted educational cycle;
- social problems in the NS basing areas (housing problem for civil crews in NS basing areas, their forced separation of families, appreciable rise in price of NS handling operations because of high travel expenses for such crews, etc.).

In addition, in keeping with their contracts with Minatom, the shipyards ensure durability, floodability, explosion-proof properties, fire-, nuclear and radiation safety of all NSs decommissioned from he Russian Navy independent of their owner and location (i.e. at enterprises or at Navy basing sites).

Principal indices of NS utilization activities performed by the RF Minatom are illustrated by data of Table 3.

TABLE 3. Principal Results of NS Utilization Related Work Performed by the RF Minatom

Number of NSs	During 13 years (January1,1986 –December 31,1998)	During 3 years (January 1,1999 – December 31, 2001	Total (as of January 1, 2002)
NS decommissioned and to be utilized	about 160	about 30	about 200
SNF unloaded from NS eactors	53	44	97
NSs utilized with RU forming	38	30	68
NSs in pending state, total	120	-	122
NSs in pending state with SNF	105	-	93

2. Rehabilitation of RW & SNF Storage Facilities

At the present time 4 Coastal Servicing Enterprises (CSEs): 2 CSEs in the Murmansk region (in Andreeva Bay and Gremikha-settlement), a CSE in Sysoeva Bay, Primorskij krai, and a CSE in Krasheninnikov Bay, Kamchatka region, are decommissioned from the Russian Navy.

In 2000, in keeping with the RF Government decree, State Unitary Enterprises "SevRAO" (Murmansk) and "Dal'RAO" (Vladivostok) were created by Minatom, which accepted CSEs of the Russian Arctic & Pacific Navy to ensure their operation and perform rehabilitation of CSE sites and of RW & SNF storage facilities.

Within CSE storage facilities and sites SNF & RW are accumulated and stored.

The lifetime of basic CSE installations is used up, and a part of them is in an emergency state. There are soil contamination spots for which decontamination is needed.

Actually, rehabilitation procedures are being developed at radiation-dangerous installations of all the above CSEs. In particular:

- a deepened site for low-level SRW storage of 4500 m^3- capacity was commissioned;
- bulky SRWs of 180 m^3 -volume were fragmented;
- 800 t of SRWs were collected, packed, removed from open-air sites and placed into storage facilities;
- 2 LRW-storing tanks were dewatered and decommissioned;
- in 2001 project documentation for constructing installations to ensure physical protection of nuclear & radiation facilities at "SevRAO" and "Dal'RAO" was completed; in 2002 work on the construction of these systems is to begin;
- work has started to restore communications and create appropriate infrastructure for ensuring a safe working environment for the involved personnel.

Since December 27, 2001 "Dal'RAO" ensures single-handed protection of its nuclear- and radiation facilities (i.e. without resort to the Navy elements). So it will also be for "SevRAO" after the second quarter of 2002.

3. Actual Status of NMSV Utilization Operations

In total 5 FSVs, 10 special tankers and 25 floating tanks housing Spent Fuel Assemblies (SFAs), LRW and SRW are decommissioned at the Arctic Navy & the Pacific Navy and are to be utilized.

In the Northwest region the following are to be utilized: Floating Shops (FShs) PM-50, special tankers TNT-8, TNT-19, TNT-25 and TNT-29.

In the Primorskij kraj: FShs PM-80, PM-125, PM-133, special tankers TNT-16, TNT-17 and THT-27.

In the Kamchatka region: FShs PM-32, special tankers TNT-23 and TNT-42.

In addition, four Floating Control Dosimetric Stations (FCDS-4, FCDS -9, FCDS – 14 and FCDS-49) are located at Poliarninskij SY. Their lifetime is up, all FCDS are charged-off and are to be utilized. FCDS-4 is to be utilized in 2002. Procedures & techniques to utilize FCDS-14 have been developed.

At present SFA unloading from PM-80 is nearing completion, preparative operations for PM-80 utilization have started.

SFA unloading from damaged facilities of PM -32 is to start in 2002.

In 2001 a radiation-technical survey of PM -50, PM -125 and PM -133 was performed; primary data to develop procedures for their utilization were prepared. In 2002 utilization of PM-50 is to begin.

In 2001 TNT-29 utilization procedures & techniques were developed, and practical operations are to start in 2002.

Principal procedures & techniques to utilize floating tanks (FT-50 design) were also developed in 2001.

Thus, in 2002 the RF Minatom begins practical work on NMSV utilization: during the year three NMSVs (PM-50, TNT-29 and FCDS-4) are to be utilized.

4. International Assistance

In 2001 international cooperation and gratis aid to solve problems of NS multi-purpose utilization proceeded in keeping with the agreements in force.

Under the SOAE Program the USA carried out: -financing of 9 strategic-missile NS utilization operations ($50.3 mil.), the construction of 2 coastal SNF unloading complexes ($23.5 mil.) and -building of RW-handling complexes at enterprises "Zvezda" and "Zvezdochka" ($45.2 mil.). The contracts were made directly with the above enterprises, and they are to be completed in the second half of the year 2002.

Within the framework of the tri-lateral agreement between Ministries of Defense of Russia, Norway and the USA (AMEC Program) both the certification and fabrication of 100 containers to store SRWs at SMBE "Zvezdochka" were funded ($142 000).

In keeping with an intergovernmental agreement between Russia and Japan, a floating LRW processing complex was completed ($35 mil.). Contracts for implementing 2 new projects and proposals for future cooperation (15 new projects) are being prepared.

5. Conclusions

- As a whole, the tasks placed on the RF Minatom by the Decree of May 28, 1998 of the RF Government are in progress;
- The tasks addressed to Minatom by the RF President in his appeal to the Federal Assembly "On Basis of Home and Foreign Policy of the Russian Federation in 2001" to Minatom have been resolved;
- Coastal Servicing Enterprises of the RF Ministry of Defense have been transferred under the jurisdiction of the RF Minatom;
- Attendant operations at 20 NSs, and SNF unloading operations at 18 NSs have been performed;
- The needed financing to perform multi-purpose NS utilization in 2002 - 2005 is estimated at a level of 2.5 billion ruble/year at least. If appropriate financing is provided, basic problems related to NS utilization will be resolved during 2007 - 2010.

OVERVIEW OF STATUS AND ISSUES RELATED TO THE DECOMMISSIONING OF NUCLEAR VESSELS IN FRANCE

A.TOURNYOL DU CLOS
Director, Advisor to the President
CEA, France

Abstract

France has built a small fleet of nuclear powered vessels: 14 submarines and 1 aircraft carrier (the latter being equipped with 2 nuclear plants); 2 land-based reactors have also been built specifically for the purposes of research, development and training related to nuclear propulsion.

The first 4 submarines, and the eldest land-based reactor, have already been decommissioned and are, at present, at various stages of dismantling. On the other hand, there is a steady flow of new constructions to replace older plants.

A policy of dismantlement, able to accommodate existing and future plants, is hence needed.

Due to the limited space available in France, and to the existence of a strong civilian nuclear program (for electricity production), the French government has chosen not to have a specific permanent repository for decommissioned naval nuclear plants, but to turn those plants into pieces acceptable (size and radioactivity) by the civilian national radioactive waste storage organization. This applies also, as far as possible, to the spent fuel.

To achieve this goal, and address the relevant problems, the French safety and industrial organizations have been adapted.

Introduction

Since the beginning of the sixties, France has developed, slowly but steadily, a small fleet of nuclear powered vessels.

This development can be divided in two phases:

- phase 1, from the mid-sixties to the mid-nineties, building up the fleet,
- phase 2, from the mid-nineties to now on, putting into active service a new generation of ships.

The figures are given in table 1 below.

A.A. Sarkisov and L.G. LeSage (eds.),
Remaining Issues in the Decommissioning of Nuclear Powered Vessels, 35–38.

Table 1. France nuclear powered fleet

1st generation	2nd generation
2 land-based prototypes	1 land-based prototype
6 SSBN "Le Redoutable" class	4 SSBN "Le Triomphant" class
6 SSN "Rubis" class	6 SSN "Barracuda" class
	1 CVN "Charles de Gaulle"

Insofar as the ships of the 2nd generation are being built, the older ones are decommissioned and enter the dismantling process. The average rate is presently one submarine decommissioned every two or three years ; this is not a very heavy rate, but altogether it asks for a policy of dismantlement, able to accommodate existing and future nuclear plants.

Dismantlement policy

From the beginning, French Defence Ministers decided to stick, as much as possible, to the organization and principles put into place for the needs of the civilian nuclear program.

This means that dismantling will be conducted accordingly to the standardized levels of IAEA, i.e.

- level 1: spent fuel and radioactive liquids are removed;
- level 2: all movable equipment is withdrawn, confined part of the plant is sealed and reduced to a minimum;
- level 3: all radioactive material is removed, decontamination is pursued until no further control is necessary.

Coming back to nuclear submarines, level 1 is easily achieved, as it is not very different from the plant situation during ship overhaul or major refits.

To achieve level 2, we separate the reactor compartment from the rest of the ship, we seal it and store it on a ground facility located inside Cherbourg Naval Dockyard. The rest of the ship is decontaminated, controlled and sent for scrap like any conventional submarine[1]. The reactor compartment will stay in this intermediate storage facility for roughly 15 years ; a duration calculated to allow enough time for short lives corrosion products to disappear, and hence reduce the radioactive dose to workers during the next phase.

After this 15 years period, work will be resumed on the reactor compartment in order to achieve level 3. At this time, all remaining pipes, structures, and equipments will be cut into pieces, conditioned and sent to ANDRA for definitive storage.
ANDRA is the French national agency qualified for long term storage of radioactive waste.

[1] With the notable exception of Le Redoutable, which has been turned into a museum, and is now part of "La cité de la mer " in Cherbourg.

Status of nuclear fleet

Today, 4 submarines of the first generation of SSBN have been decommissioned; table 2 below, summarize the present status of those submarines.

Table 2. Status of French decommissioned submarines

IAEA level	Characterization	Responsibility	Submarine
Level 1	Cold shut down without fuel	DCN Cherbourg	Le Terrible Le Foudroyant
Level 2	Reactor compartment removed	DCN Cherbourg	Le Redoutable Le Tonnant (2002)
Level 3	Dismantlement completed	ANDRA	

The industrial issues related to the dismantling of French nuclear submarines, the specific dispositions for spent fuel and radioactive waste, will be addressed later in this workshop.

Nuclear safety organization

Let me now say some words about the nuclear safety organization.
The French President has just signed (February 2002) some decrees modifying substantially the nuclear safety organization.

The idea was to create a new institute, totally independent from all previous organizations, industrial or public, working in the nuclear field. This institute called "Institute for Radiological Protection and Nuclear Safety (IRSN)" will carry out research, analysis and work within the fields of nuclear safety, protection against ionising rays, control and protection of nuclear materials and protection against acts of malevolence.

IRSN is a public establishment of an industrial and commercial nature (EPIC), under the joint authority of the Ministers of Defence, Environment, Industry, Research and Health. It groups more than 1500 experts and researchers; it will take the place of the former IPSN (Institute for Nuclear Protection and Safety) and the former OPRI (Office for Protection against Ionising Rays).

Depending on the subject, IRSN will send its reports and analysis either to the Minister of Industry or to the Minister of Defence, who will give when required the authorizations, take decisions of a regulatory nature and realize on site inspections.

The schema below shows the French new organization for nuclear safety and radioprotection.

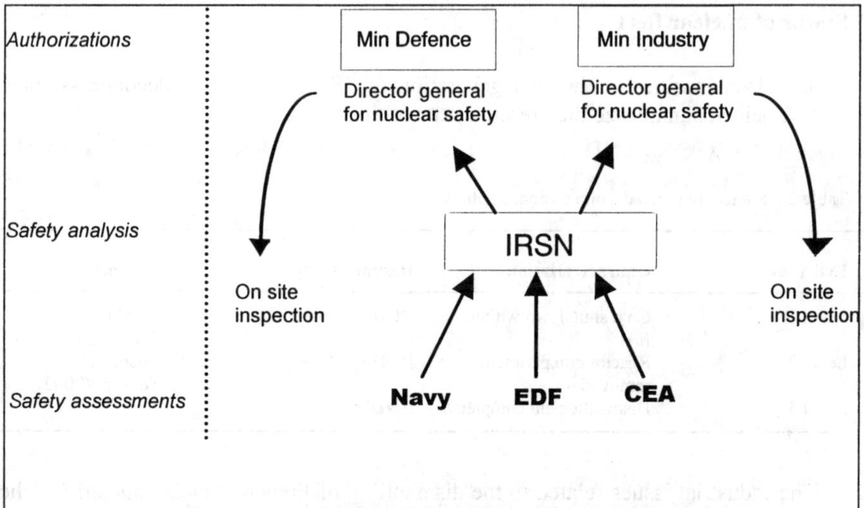

FRENCH NUCLEAR SAFETY ORGANIZATION

Conclusion

Due to the limited space available in France, and to the existence of a strong civilian nuclear program (for electricity production), the French government has chosen not to have a specific permanent repository for decommissioned naval nuclear plants, but to turn those plants into pieces acceptable (size and radioactivity) by the civilian national radioactive waste storage organization. This applies also, as far as possible, to the spent fuel.

This decision puts some constraints on the Navy but has two main advantages:

- First, it is cost-effective; we do not have to create or manage any permanent specific facility for waste coming from nuclear naval plants.
- Secondly, it proves to public opinion that the Navy has nothing to hide as far as protection of the environment is concerned.

INACTIVATION AND RECYCLING OF NUCLEAR VESSELS IN THE USA OVERVIEW AND STATUS

MALCOLM MACKINNON III
RADM, USN (Ret.)

1. Introduction

I traveled to Moscow seven years ago to participate in the 1995 workshop. The Cold War hadn't been over for very long. I came with some uneasiness although I knew the subject of what to do with inactivated nuclear vessels was an important issue for both the Russian Federation and the United States. My uneasiness was quickly put to rest and the tone of the Workshop can best be illustrated by an inscription from the author in a book presented to me. The book was authored by Admiral, retired, Nicolas Mormul.

The inscription indicated that "The time has come at last when the enemies gather at the table to decide how to dispose of their weapons." Well said words, clearly reflecting the feelings at that Workshop, feelings that continue with me to this present day.

I spent almost the entire Cold War in the uniform of the U. S. Navy, serving in destroyers and submarines and then becoming an Engineering Duty Officer. My subsequent duties involved the design, construction, and repair of nuclear submarines and aircraft carriers. When I retired in 1990, we were just beginning to realize that we had a major problem in what to do with our inactivated nuclear ships and just how to do it.

Our paper given here in 1995 covered the rudiments of the problem and what we were doing about it.

My purpose today is to review the process currently employed in the U. S. to decommission, inactivate, defuel, and dispose of/recycle our nuclear vessels. It has changed very little in the past seven years. In addition I shall briefly discuss the status of the situation today.

2. Review of U. S. Processes

The following comprise the processes that the United States Navy utilizes to inactivate and recycle its submarines:

- Removal from service: The ship normally steams to Puget Sound Naval Shipyard under its own power. There it is removed from active service by means of a Decommissioning Ceremony. All weapons and explosive devices were removed prior to arrival at the Shipyard. The ship is moved into drydock.

- Defueling: The reactor is shut down and defueled in a planned sequence.

A.A. Sarkisov and L.G. LeSage (eds.),
Remaining Issues in the Decommissioning of Nuclear Powered Vessels, 39–41.

- Movement of fuel: The fuel elements are transferred to specially designed containers for transport by rail to storage at the Idaho National Engineering and Environmental Laboratory.

- Preparation for disposal and recycling: All expendable materials are removed. Classified and sensitive materials are removed. The main storage battery is disconnected and removed. All fluids are removed and their tanks and systems flushed and dried as necessary and all radioactive systems are sealed. Electrical systems are disconnected. Equipment removal begins. The ship is moved into a specially configured drydock where final dismantlement takes place.

- Reactor Compartment removal and disposal: The ship is drydocked such that the Reactor Compartment is positioned over a specially designed cradle. All interferences are removed and the Reactor Compartment is cut free, a meter or so forward and aft of its shielded bulkheads. The compartment is then slid free from the rest of the ship. Additional sealing bulkheads are welded in place at each end. The compartment and cradle are then jacked up and slid onto a barge that had been drydocked with the ship. The barge is then undocked, leaving behind the two parts of the submarine, which had been moved into the dry portion of the dock. The barge is then towed out of Puget Sound, down the coast and up the Columbia River to the burial site at Hanford, Washington.

- Ship dismantlement, recycling, and disposal: The remaining two pieces of the submarine are then stripped of all equipment and interferences and cut into smaller pieces and removed from the drydock by cranes. Waste materials are carefully handled and the toxic and hazardous wastes are segregated for special handling. The handling and disposition of this hazardous and toxic waste material presents a problem equal in magnitude to that posed by the nuclear and radioactive material. The ferrous and non-ferrous metals are cut to sizes best suited for transportation and recycling. Equipment that can be reused is disposed of separately.

I will now briefly discuss some of the issues we face in taking care of toxic and hazardous wastes.

3. Handling of Hazardous and Toxic Wastes

The hazardous and toxic materials identified on board our nuclear vessels are:

- Polychlorinated Biphenyls (PCBs)
- Asbestos
- Heavy Metals: Lead, Mercury, Cadmium
- Organotin
- Ethylene Glycol
- Halogenated Fluorocarbons

Of all the non-nuclear waste materials the polychlorinated biphenyls (PCBs) present the largest problem. Regulations require its treatment as toxic waste when existing in concentrations greater than 50 parts per million. There are complex issues regarding identification, testing to determine concentration, handling and disposal. All hazardous material requires special handling and disposal and can include the use of protective clothing.

PCBs are found in many areas including electrical cable insulation, gasket material, paints, adhesives, as well as in liquid form in electrical transformers and capacitors. Asbestos is found in pipe and ventilation insulation, in gaskets and packing, in acoustic insulation and in deck tiles. Lead is found in ballast weight, paint, batteries, gaskets and plumbing systems. Mercury is found in gauges and instruments, and cadmium in plated fasteners.

Organotins are found in anti-fouling paints. Ethylene glycol is found in gauges, cooling pump fluids, and in air conditioning and refrigeration systems. Halogenated fluorocarbons are found in air conditioning and refrigeration systems.

4. Status as of December 2001

The following is the status of inactivation and recycling of nuclear ships in the United States as of December 2001. Data is from open sources.

- Reactor Compartments in storage: 102
- Submarines Recycled/Scrapped: 81
- Nuclear Cruisers Recycled/Scrapped: 1
- Pending: 33 Submarines, 8 Nuclear Cruisers (all defueled)

PERSPECTIVES ON RISKS ASSOCIATED WITH NUCLEAR VESSELS

POVL L. ØLGAARD
Department of Radiation Research
Risø National Laboratory
DK-4000
Roskilde, Denmark

1. Introduction

The record of the operation of nuclear vessels demonstrates that accidents have unfortunately occurred in the past (see e.g. [1]), and there seems no reason to believe that such accidents may not occur also in the future. Nuclear vessels are primarily naval vessels, in particular submarines. Such vessels must be able to work under wartime conditions where risks have to be taken, and this fact may be reflected in the training of the crews and in the safety culture of nuclear fleets. This may be the explanation why more accidents have occurred with naval reactors than with civilian power reactors.

Most of the risks, which apply to operational nuclear vessels, do also apply to nuclear vessels during decommissioning. If the reactor core is not properly cooled for an extended period after final shutdown, a loss-of-cooling accident (LOCA) may be the result. As long as the fuel remains in the core and in particular during defuelling when control rods and fuel elements are moved criticality accidents may occur. The handling of the spent fuel after defuelling may also result in accidents. During towing or floating storage of a decommissioned nuclear vessel the vessel may sink and if the vessel is not recovered this will ultimately lead to contamination of the sea. Fortunately it seems that so far there have been no serious accidents during the decommissioning of nuclear vessels, but this fact should not lead to complacency.

It may be claimed that if people follow the rules and regulations there will be no accidents. This is not necessarily the case. For example, if a major leak develops in the primary cooling system due to corrosion, and the emergency core cooling system does not have sufficient capacity to cool the reactor, the crew, working according to the rules, may not be able to prevent a LOCA. But it is even more important to accept that people do not always follow the rules, either due to ignorance or to carelessness, and that this fact has to be taken into account when planning safety procedures. From [2] and [3] it is seen that most criticality accidents could have been avoided if the personnel involved has followed the rules. They did not.

A.A. Sarkisov and L.G. LeSage (eds.),
Remaining Issues in the Decommissioning of Nuclear Powered Vessels, 43–47.

2. The Physics behind Nuclear Accidents

To appreciate the risk of nuclear accidents it is important to understand the physics behind them. Nuclear fission, which is the basic process of energy production in all nuclear reactors, is inherently coupled to two physical phenomena, the nuclear chain reaction and the production of radioactive nuclides, and one or both of these two phenomena are fundamental to all nuclear accidents.

During shutdown the reactor is strongly subcritical and during normal operation the fission rate of the core is controlled in such a way that the reactor is just critical and the power production stable. However, changes of the core configuration, e.g. in connection with refuelling or defuelling may unintentionally increase the reactivity and make the reactor prompt critical. This will cause the power level to increase so rapidly and violently that the fuel will be damaged and radioactive nuclides in the fuel released. The result is a criticality accident.

Once the control rods are inserted into an operating reactor, the chain reaction and the energy release due to fission is stopped. However, the energy production of the core is not. During operation of reactors large amounts of radionuclides are produced by fission (fission products) or by neutron capture (activation products). The decay of these radionuclides will continue to produce significant amounts of power after the reactor shutdown. After extended operation the power level will immediately after shutdown drop to about 6% of the level before shutdown and then gradually decrease. After 24 hours the power level has dropped to about 0.5%. This decay heat has to be removed. If this is not done, the temperature of the fuel will increase and the fuel may ultimately melt, causing a core meltdown and a release of radionuclides from the fuel. The result is a LOCA.

Should a nuclear vessel sink and not be salvaged, the radionuclides produced during operation, whether fission products or activation products, will at some point in time start be released due to corrosion of the seawater.

The storage and handling of the spent fuel after refuelling or defuelling may also result in nuclear accidents. Criticality accidents may occur due to in-leakage of water in a dry storage facility. Another possibility is the drainage of water from a storage pool due to a leak. Hereby the water shield is removed, and the result is a radiation accident.

The types of nuclear accidents discussed above are all inherently coupled to the use of nuclear reactors and apply to operating as well as decommissioned nuclear vessels, at least as long as the fuel has not been removed. Proper safety measures have to be taken to avoid that such accidents occur.

3. Loss-of Cooling Accidents

As mentioned in section 2 LOCA type of accidents may occur if the reactor fuel is not properly cooled. Such accidents may occur in operating reactors if e.g. a major leak develops in the primary cooling system. In such cases the reactor will immediately be shut down, but the decay heat will still have to be removed. A LOCA may also occur in shutdown reactors, in particular shortly after reactor shutdown when the decay heat is still significant, e.g. if the coolant is drained from the core due to operator error. Modern reactors are usually provided with an emergency core cooling system (ECCS), which will provide the necessary cooling during a LOCA, but this is not necessarily the case for old reactors. Depending on the reactor design natural circulation cooling of the reactor may provide sufficient cooling after hours or days after shutdown. When a naval reactor has been shut down for three years or more a LOCA will not occur, even if the reactor tank is totally drained because of the drop in the decay heat [4].

LOCA's have occurred in naval reactors. According to [1] five LOCA's have occurred in first generation Soviet nuclear submarines, all while the reactors were operating at power. The only LOCA reported in [1], which did not involve an operating reactor is the LOCA at the icebreaker NS Lenin. The accident occurred shortly after the reactors had been shut down for refuelling, and the reason was presumably that the central part of the core of one of the three reactors was drained due to an operator error. This was possible at the first Lenin reactors because the inlet coolant tube was situated at the bottom of the reactor tank. According to [4] this design is not used in any Russian submarine reactor. Here both the inlet and outlet coolant tubes are situated above the reactor core so that the core can not be drained. Early Soviet naval reactors were not provided with an efficient ECCS.

It seems as if the reactor compartment and the submarine hull acts as a rather efficient containment in case of a LOCA. During the June 1989 LOCA, which involved a first generation Soviet nuclear submarine the Norwegians took water samples close to the submarine and found very low levels of radioactivity. So while the interior of the submarine was contaminated with activity released from the damaged fuel, little activity escaped to the environment.

From this information it seems reasonable to conclude that there is little risk of LOCA's in decommissioned nuclear submarines, provided forced cooling is provided during the first period after reactor shutdown. After that natural circulation should be able to take over and after a couple of years the decay heat is reduced to such low levels that even if all water is removed from the reactor tank, the fuel will not overheat. Further, even if a LOCA should occur during the first period after shutdown, the activity released to the environment should be quite limited, but the dismantling of the submarine will of course be much more complicated.

4. Criticality Accidents

Criticality accidents may in principle occur at any time while there is fuel in the reactor. However, experience shows that this type of accidents does not occur during normal operation or when the reactors are in stable, shutdown condition. They occur primarily during reactor shutdown when control rods or fuel elements are moved, i.e. the reactivity is changing, and the nuclear instrumentation is not working. Under these circumstances the personnel has no indication of how close to criticality the reactor is.

In [1] five cases of criticality accidents in Soviet nuclear submarines are reported. Two of these involved refuelling while the remaining three occurred during shipyard work. In all cases the reactors were shut down, and it seems also that the nuclear instrumentation was not working, and that the personnel did not follow the existing rules and regulations.

Fuelling accidents (re- or defuelling) differ from other criticality accidents since during fuelling the hull above the reactor has been opened and there is free access to the environment for the radionuclides released during the accident. Fortunately, both refuelling accidents occurred after the new fuel had been inserted in the reactors. Thus the activity of the core was limited to that of the radionuclides produced during the power excursion. In the case of a defuelling accident with a decommissioned submarine with spent fuel in the core, the activity release, in particular of fission products with long half-lives such as ^{137}Cs and ^{90}Sr, will be significantly higher.

To prevent criticality accidents during the defuelling of decommissioned reactors a number of safety measures may be introduced. [4] outlines a number of such measures. All cables with electric power supply to the control rod drive mechanisms are cut, one meter length of the cables is removed and the cable ends are isolated. The control rod drives are blocked by stoppers and weldings with the rods in their lowest position. The electricity supply to the control panel is cut and the room locked. These measures will ensure that criticality accidents will not occur during floating storage of the decommissioned submarines.

Defuelling is only carried out three or more years after the final shutdown of the reactor. Before the defuelling is started, the water is drained from the reactor tank, the primary circuits, other circuits and from storage tanks and the piping between these parts of the power plant are blocked. The reason for the drainage is that a drained reactor can not go critical, even if all control rods are removed.

The reactor tank lid is removed and replaced by a refuelling flask which can be placed over any of the fuel element positions and with which the fuel elements can be removed. Part of the flask also acts as a radiation shield.

This procedure should ensure a safe defuelling, if all measures are carried out correctly. However, a major weakness in the procedure is that the personnel performing the work has no indication of how close to criticality the reactor is at any given time. If e.g. the water of the reactor tank has only been partly removed and still covers the core, the check of the water level of the tank is not carried out correctly or not at all, and the control rods are removed together with the tank lid, the reactor may well go critical. Therefore, the drainage prior to defuelling should only be carried out after a strong neutron source and a thermal neutron detector are placed in or close to the core so that the personnel will during the full defuelling process have a signal indicating how close the reactor is to criticality. The signal should be displayed visibly as well as audibly. Without such a warning system there will always be a non-negligible risk of a criticality accident. Many if not most of the criticality accidents of the past could have been avoided if proper instrumentation had been in place.

For operational as well as for decommissioned nuclear submarines the most damaging of all types of accidents is a re/defuelling accident, since because of the direct access to the environment the activity released during the accident will contaminate not only the submarine but also its surroundings. If there is a major town near the place of the accident, the consequences for the local population may be very unpleasant.

5. Sinking Accidents

When floating storage for many years is used for decommissioned nuclear submarines there will always exist the risk that the submarines will sink due to e.g. corrosion. If the sinking occurs during floating storage at a naval base it should be fairly easy to salvage the submarine so that the contamination of the seawater is limited, even if the submarine has not yet been defuelled.

If the submarine sinks at sea while being towed from one base to another salvage will of course be much more difficult, in particular if it sinks at great depth. However, measurements of the activity release from sunken nuclear submarines with fuel in the reactors and with nuclear weapons indicate that the activity release is very slow and limited so, depending on the position of the sunken submarine, it may not be worthwhile to salvage it.

6. Spent Fuel Accidents

Once the spent fuel has been removed from the reactor it has to be stored in water pools or dry storage shielding casks until it can be sent to reprocessing or disposal. During this storage period it must be ensured that the fuel is provided with the necessary radiation shielding and that the fuel is stored in a safe geometry.

In the case of wet storage, e.g. in a storage pool, it must be ensured that should a leakage develop in the pool, the water level can be maintained, so that the radiation shield, the water, does not disappear, before the fuel is moved. Further, the water of the pool has to be cleaned continuously so that the contamination of the surrounding areas by the leaking water is kept to a minimum. Finally the fuel geometry has to be safe also when the fuel is moved in the pool.

In the case of dry storage in shielding casks it is important to ensure that the geometry used will remain safe even if water enters the cask, e.g. during flooding so that criticality accidents can not occur. According to [5, 6] a special problem exists in connection with the storage of spent fuel from naval reactors with liquid metal cooling. Here water penetration may cause criticality.

7. Conclusion

The operation as well as the decommissioning of nuclear submarines involve the risk of nuclear accidents. The accident with largest consequences for the environment is a criticality accident during defuelling. For decommissioned submarines the reactors contain substantial amounts of radionuclides with long half-lives and during defuelling the reactor compartment is open to the environment. Therefore the defuelling should be carried out in such a way that the personnel are warned of an approaching criticality accident.

There are other nuclear accident risks that should also be prepared for such as loss-of cooling accidents, sinking of submarines and spent fuel accidents.

8. References

1. Olgaard, P. L., Accidents in Nuclear Ships. Accidents in Nuclear Ships. NKS/RAK-2(96)TR-C3, Dec. 1996.
2. McLaughlin, T. et al., A Review of Criticality Accidents. 2000 Revision. LA-13638, 2000.
3. Olgaard, P. L., Criticality Accidents. Risø-I-1511(EN), Dec. 1999.
4. Elatomtsev, V. V. et al., Nuclear Safety Assessment of Stored Afloat Non-Defuelled Decommissioned Nuclear Submarines. Final Report. Kurchatov Institute, Institute of Nuclear Reactors. No 31/8010, 1997.
5. Pamkravtov, D. et al., The Safety Assessment During the Long-Term Storage of Spent Fuel of Nuclear Submarines with Heavy Liquid Metal Coolant. In: Ecological Problems in Nuclear Submarines Decommissioning. International Seminar, July 4-9, 2001, Severodvinsk, Russia. Abstracts Volume, p. 123-124.
6. Bugreev, M. et al., Step-by Step Project on Problem for Long-Term Storage of Spent Fuel of Nuclear Submarines with Heavy Liquid Metal Coolant. In: Ecological Problems in Nuclear Submarines Decommissioning. International Seminar, July 4-9, 2001, Severodvinsk, Russia. Abstracts Volume, p. 125-126.

ON RISK IN DECOMMISSIONING OF NUCLEAR SUBMARINES (NS)

O. KOVALEVICH, V. NIKITIN, V. SHEMPELEV, A. SHULGIN
Nuclear & Radiation Safety Research and Technical Center,
Gosatomnadzor,
Moscow, Russia

1. Risk all over the world and in RF

The Sixth All-Russian Scientific and Practical Conference "Accident Risk Management" [1] was held in 2001 in Moscow. The objective of the conference was to review accidents in detail and to develop recommendations on the main lines and priorities of scientific & technological policy and practical activities in Russia in the field of natural and man-induced risk management. Theoretically such a gloomy picture was painted that it could be doubted, whether separate regions or civilization as a whole could survive in the XXI century, if individual countries and the whole mankind do not take decisive measures to mitigate adverse man-induced impacts on the environment.

The issue of concern, NS utilization risk, is just a component of this holistic problem and it should be mentioned, that it doesn't play a governing part. Solution of this problem is however inseparably linked with other risk problems (economical, methodical and mainly with those related to management and disposal of radwaste in nuclear energy use).

Since financial resources of the state are limited, a definite strategy is required to find the way out of the crisis and to search priority ways and effective solutions for nuclear deterrent management.

It seems to be expedient, based on a resolution of the Security Council of the RF, to highlight the following directions of activity related to risk management and to fix liability for development of these directions:

- development and adoption of the Federal Law on the state concept of risk management, which would establish a list of binding "risk" indicators (State Duma of RF, Government of RF);
- development and introduction of generally accepted concepts of risk and its derivatives, relevant general methods, criteria and assessments (Russian Academy of Science);
- relevant branch programmes on risk assessment and management (NS utilization included);
- combining of the branch results of risk assessment and management into unique national programme.

Before we speak about "risk", it should be agreed what we mean. In paper [2] more than ten examples are given of how the term "risk" is used in different activities and some proposals for "risk" definition are provided. Finally, in different papers there is a widespread notion that the basic risk component is the probability of some negative event (or it is called a "threat") and associated damage (loss).

A.A. Sarkisov and L.G. LeSage (eds.),
Remaining Issues in the Decommissioning of Nuclear Powered Vessels, 49–61.

In the case of variety of such events, risk is considered as product of probabilities and damage integrated or summed. There is one more aspect of a multidimensional "risk" notion. It is the preparedness of people to perceive risk they are exposed to, and this aspect could be of critical importance [3]. In paper [4] an effort was made to analyze typical types of activities associated with risk as this enables one to distinguish three different groups of activities, which are often defined by the same term "risk".

Task 1. There is one or some initiating events resulting in an accident (emergency situation - ES) with consequences, which are certainly considered as severe. Evaluation of consequences in natural or monetary indicators is either obvious or rather time-consuming, but it is not made, as it is not the task.

"Probability" of such an event is considered as "risk", all efforts are made to reduce probability (i.e. to exclude the event).

This "risk" evaluation scheme is used, for example, for risk-analysis of strength and reliability of equipment and structures at potentially hazardous facilities, when failure of equipment and structures operated under load is an event resulting in considerable damage. The same scheme is used to accomplish tasks associated with "risk" of an aircraft crash, train collision, shipwreck and so on.

Task 2. One certain or several initiating events, which result in definite adverse consequences, are postulated. This task is to mitigate consequences of these specific events, while the probability of the event is not defined. "Risk" indicator is the scale of "consequences" for specified events. Assessment of the "risk" of ionizing radiation belongs to such type of task, as far as radiation dose is postulated and its impact on the human beings is defined. This scheme of "risk" assessment is used, for example, in tasks associated with the survival of the car driver and passengers in the case of a head-on collision, survival of the submarine crew when the submarine is submerged, i.e. in all tasks related to survival of the human beings in extreme cases.

Task 3. For each facility there are a certain number of events and corresponding adverse consequences. Different events occur with various probability (frequency) and the corresponding consequences differ in scale (in natural or monetary indicators). For a rationally organized society it is necessary to take into account both "frequent" and "rare" events [4]. "Risk" from such a facility, taking into account these two constituents, is a product (sum of products) of event probabilites (frequency) and all possible consequences. More specific aspects of the specified task can be considered with more detailed information about "risk" probability indicators and special assumptions upgrading “frequent” and “rare” events. In assessment of the risk the of NS decommissioning we are confronted with this task.

2. The State of risk problem in NS decommissioning

The risk problem for NS decommissioning shall to this extent fall into each of the above-stated lines of state policy in risk management (see section 1).

We have been facing this problem for about 10 years and a large number of papers and issues of various levels, which touch upon different scientific and engineering aspects, were dedicated thereto (White Book - 1993, White Book - 2000. Conference Proceeding, March 1995, Severodvinsk, June 1995. Moscow, 1997, Moscow, 2001. Severodvinsk. Bellona's Reports. ISTC project 101, many articles).

The engineering problem of NS utilization already finds its practical solution in the works of specific design institutes and organizations of the Russian Shipbuilding Company [5].

We do not make it the aim of this paper to solve some particular scientific or engineering task which is a part of the problem. The objective is to try to conceive, if it is possible, to assess the NS decommissioning risk taking into account the state-of-the-art.

Under "risk" notion we will hereinafter mean and sometimes use the idea "global risk", having in mind the risk for all stages of NS decommissioning considered below from removing of the NS from the Navy up to spent fuel loading into containers for transportation to the radio-chemical plants and storage of reactor compartments in the ship-way of the long-term storage facility. All other aspects are outside the sphere of concern.

Risk assessment in NS decommissioning is characterized with the following specific aspects:

- lack of standardized methods, criteria and regulatory basis, which, as it was mentioned before, is typical throughout the country;
- now we are just at the initial stage of the process, which could last, taking into account the current work pace, for tens or more years; new engineering problems can emerge and it is not yet clear how to assess them;
- the overall NS utilization system is not yet clear, as regards the time and the scope. The time period and the scope of NS utilization (including ships and vessels) are not definite. We are not able to speak about global risk associated with the overall system in future due to the reasons above or because we do not know how the adverse effects will show themselves over time;
- in NS cutting at a wharf or on shore, it is very important, from the point of view of the personnel and population safety, which chemical toxic substances are used (we will not consider the issue);
- an important place in the process risk assessment is occupied by submersion of the Navy NSs and their components, i.e. ingress of radioactive substances into sea and ocean water; consequences of such an ingress are less known yet.

As an example of probable adverse consequences of submersion refer to paper [6], which describes a hydraulic method of safe neutralization of toxic spills, in water areas. The matter concerns submersion in the Baltic See at the depths of 200-700 m of about 300 thousand tons of ship and vessel fragments with toxic substances and explosives in 1941-1945. “The problem is being discussed, no concrete work, however, is being carried out. There are only observations. A scientific basis has not been created for solution of the problem, which could pose a threat for national safety of RF” (in quotes a statement from paper [6] is presented). TV information about the presence of tizit in the Baltic Sea fish is the first sign of the problem.

Authors of the report propose a scheme for removal of hazardous substances from the sunken ships (using a jet device) and their further safe disposal. Will such activities appear to be useful to some extent for our case? We at least have to think about such approaches.

Under the circumstances, the global results being obtained now present to a greater extent risk assessment for those living now, but not for the future generations.

3. The present state of the NS decommissioning

About 190 NS have been removed from service by the Navy at the current time, while spent fuel (SF) has been unloaded from about 60 NS. The SF unloading rate did not exceed 6 cores per year (but in 2000-2001 it was increased up to 18-21), the utilization rate is about 6 NS/year. Under the actual state of affairs it becomes inevitable, that a large number of NS are forced to be afloat for a long period (20-25 years), including those with cores not unloaded.

At present about 40 reactor compartments are being temporary stored afloat in the Northern region, they are mainly of the 3-section variant; there are 19 such compartments in the Pacific region also.

It should be emphasized that utilization of NSs by a 3-section variant is a forced temporary measure and it doesn't solve the task for the long-term. In the future we'll have to transport 3-section and multi-section reactor compartments to plants of the State Russian Nuclear Ship-Building Center (GRTSAS) to remove adjacent-to-reactor compartments before long-term storage of reactor compartments themselves on ashore sites.

A special hazard is presented by the NSs of the first generation, which were handed over by industry to the Navy 30-40 years ago and are located at distant bases; the cores were not unloaded from most of the NSs. For some NSs the driving balance tanks are not tight and filled with water. Reserve buoyancy is significantly decreased. NS transportation to the industrial plants without use of floating docks is not safe and represents one more potential source of contamination of the water area.

4. Indicators of risk assessment at the stages of NS decommissioning and utilization

If it is possible to assess global risk associated with NS utilization, we have to be clear which parameters are available, which parameters we are not well known and which are not known at all. The adverse impact of radiation on human beings and the environment could be via air and contamination of coastal territories, as well as by ingress of radioactive substances into sea and ocean water.

Table 1 presents the main stages of the technological cycle of NS utilization currently being used; in Table 2 we would like to analyze hazard sources, which are present at different stages of the technological cycle. "Dash" in a square in the Table 2 means that the indicator is not related to this stage; a "question mark" means that up to now no information is available. The data presented have been taken from paper [7].

TABLE 1. The Main Stages of Technological Process of NS Utilization

No	
1.	Removal from the Navy and keeping till transfer to the GRTSAS
2.	NS transportation by sea to the GRTSAS plant
3.	Unloading of spent fuel assemblies
4.	Unloading of damaged fuel assemblies
5.	Temporary storage on site, transportation of SF, damaged included
6.	Cutting off and formation of reactor compartments (RC) modules
7.	Transportation of the RC modules to the site of temporary storage
8.	Storage of the RC module in the temporary storage facility and transportation of the RC module to the GRTSAS plant for cutting
9.	RC cutting out of the module, RC handling before placing in the ship-way in the long-term storage facility (LTSF)
10.	RC LTSF operation and long-term storage of RC
11.	Storage of SF in ashore storage facilities
12	Unloading of SF from ashore storage facilities and its transportation

TABLE 2. Main Risk Indicators at Different Stages of NS Utilization

	Stages of utilization Indicators	1	II	III	IV	V	VI	VII	VIII	IX	X	XI	XII
1	2	3	4	5	6	7	8	9	10	11	12	13	14
1	Number of facilities n_i (items)	93	71	93	6	0	120	120	59	0	0	5	5
2	Activity at the facility $Q_i°$ (Ci)	$2{\cdot}10^{6(1)}$	$2{\cdot}10^{6(1)}$	$2{\cdot}10^{6(1)}$	$4.10^{8(2)}$	$7{\cdot}10^{4(3)}$	$7{\cdot}10^{4(3)}$	$7{\cdot}10^{4(3)}$	$7{\cdot}10^{4(3)}$	$7{\cdot}10^{4(3)}$	$7{\cdot}10^{4(3)}$	$2{,}7{\cdot}10^5$	$2.7{\cdot}10^5$
3	Average duration of the stage T_i (year)	12	0,03	0,05	0,25	4	3	0,05	5	0,5	70	35	10
4	Probability of facility submersion in keeping or utilization, $p_{3\,I}$	$7.3{\cdot}10^{-4}$	10^{-3}	$7.3{\cdot}10^{-10}$	$7.3{\cdot}10^{-4}$	0	0	$7.3{\cdot}10^{-4}$	$7.3{\cdot}10^{-4}$	$6{\cdot}10^{-3}$	0	0	0
5	Probability of activity release into environ-ment (criticality) $P^{1\,(4)}$	$4{\cdot}10^{-21}$		$4{\cdot}10^{-21}$	$4{\cdot}10^{-21}$	0	0	0	0	0	0	?	

TABLE 2. Main Risk Indicators at Different Stages of NS Utilization (Con't)

6	Amount of activity released from the facility in case of criticality Q^1_i (Ci)	$3,4 \cdot 10^5$	0	$3,4 \cdot 10^5$	$1,4 \cdot 10^8$	—	—	—	—	—	—	—	—
7	Risk of activity release into the environment $R^0_i = Q^1_i * P^1_i \eta i$ $\eta_i = \{n_i . m_i ...$	$1 \cdot 10^2$	$1 \cdot 10^2$	$1,3 \cdot 10^{-13}$	$9,6 \cdot 10^{-12}$	—	—	~	—	—	—	«-	--
8	The number of fuel assemblies in reactors and storage facilities	$4,3 \cdot 10^4$	$3,2 \cdot 10^4$	$4,3 \cdot 10^4$	840	0	0	0	0	0	0	$7 \cdot 10^4$	$7 \cdot 10^4$
9	Cases of facility submersion	2	—	—	—	—	—	—	—	—	—	—	—

TABLE 2. Main Risk Indicators at Different Stages of NS Utilization (Con't)

10	Expected time of activity release from submerged facilities into water (year)	?											
11	Risk of activity release from submerged facilities $R^2_i=Q^2_i*P_{3i}$ ηi	$1{,}8 \cdot 10^5$	$1 \cdot 10^5$	$1{,}8 \cdot 10^5$	$1{,}8 \cdot 10^6$	0	0	$6{,}3 \cdot 10^3$	$6{,}3 \cdot 10^3$	$3 \cdot 10^{10}$	0		

(1) Average activity at one facility

(2) Average activity at facility in accident

(3) Average activity at facility after core unloading

(4) Probability of radioactive substances release into environment in facility keeping or utilization

As it has been already mentioned, the objective of this paper is not to obtain definite risk values for NS decommissioning stages, but to have a complete picture of the solution of scientific, engineering and practical problems, taken into account in global risk assessment.

It is obvious that we are still far from getting even assessed risk values for the overall complex problem. For some questions specialists have, at least, evaluation answers. Others are far from being solved. In particular, the development of long-term physical-chemical and biological processes during the presence of radioactive substances in sea water has not been solved.

Another important aspect, which has to be taken into account, is that we do not review the process of SF transportation to the radio-chemical plant, its treatment and, which is the main thing, the follow-on disposal of high-level radwaste. This is a problem for mankind, which arises from the use of nuclear energy. In the future it may give more concern than NS decommissioning and utilization, including the removal of spent fuel.

5. Criticality accidents in SF unloading

During NS operation there were three criticality accidents. Two of the accidents included K-11 in 1965 at Severodvinsk and in 1985 at seaside, Chazhma bay. Both accidents had severe radiation consequences [8].

Though the accidents mentioned were outside the technological process of NS utilization, SF unloading is provided for in the process of utilization. Adequate attention should be paid thereto. In our opinion we choose the way mentioned in Task 1 (see Section 1 of the report), i.e. take measures to prevent such situations. To avoid criticality the reactor cover is pulled up after the water has been removed, which makes criticality improbable.

6. The problem of risk assessment in submersion of NS or its components

It is obvious from the Table 2 that at many stages there is a probability for submersion of the NS or its components and release of radioactivity into the water. It is still a problem to define the consequences and to assess the risk. Some specialists believe, that submersion of large amounts of activity (core, for example) while surrounding strength barriers (reactor vessel, submarine hull and others) are maintained, enables long-term isolation of the activity which is inside and prevents its migration into sea and ocean water areas. Hopes are also rested upon the large volumes of the World's Oceans and the strong dilution of the activity. We cannot agree with the above statement. Any material in the marine environment is exposed to impacts, it changes its behavior and sooner or later radioactivity will ingress into the water. This is confirmed by the example with conventional ammunition in the Baltic Sea.

Remains of ships sunk tens and hundreds years ago, which we sometimes see on TV, give us a clear idea about the processes under water. It appears that "somnolent" information on TV about first-class arrangement of monitoring around

"Komsomolets" and other similar enclaves should not be misleading. Monitoring gives an insight of to-day's state and cannot forecast the future.

In our opinion submersion of facilities with radioactive substances is not in compliance with the statement presented in the report of U.N.O Commission headed by Mr. Harlem Brundtland, that "Imperative of our civilization shall be to provide future generations with starting conditions, comparable with those, which living generations have" and with the accepted world idea of "steady position".

The Law "On Environmental Protection" has prohibited release of radwaste into seas and oceans. The "White Book – 2000"(contains data) indicates how much radwaste was released into the World's Ocean and who has done this. It is known, the number of NSs with operating cores that had accidents and are now on the sea bottom. There were cases of submersion of NSs being kept at the wharf. Accidents in operational NSs or accidents in NSs removed from service can be repeated. Thus, the fact of accidental submersion of large amounts of radioactive substances (occurred in the past and not excluded in future) is still true. The question is, how to take into account this fact in assessment of prolonged-in-time risk. To believe that there will be no harm from submerging of radioactive substances even in the far future attempts to overlook the problem. What war the London convention adopted for; why are national and international funds spent for adequate solution of the problem; what is the answer to the public?

What is the way out? In our opinion, in the first place, efforts should be concentrated on Task 1 (see section 1 of the report), i.e. to aim basic efforts at excluding the of submersion of radioactive facilities by taking engineering and administrative measures at all stages of utilization.

At the same time it is necessary to continue intensive studies of consequences caused by ingress of radioactive substances into World's Oceans. Investigations [8], which have been carried out up to now, either operate on release of small amounts of radioactivity, or state the current situation and do not provide a long-term forecast. Only if the above factors are taken into account, can we assess the global risk in full scope.

One of the first directions of the investigation seems to be obtaining information about the tightness of closed typical sealed reservoirs with radioactive substances inside. Wide experience has been accumulated in ship building on studying of corrosion of different materials in sea water. At the same time we would like to have data not only on surface corrosion of structures with release of induced activity into water; we also need definition of at least two typical time points (time point of loss of tightness by considered structure and time point of release of radioactive substances into surrounding water) for some specific selected structures (ship hull, containment, protection enclosure, reactor vessel, tanks with LRW and so on). Similar experiments could be performed at simulators in water area of plants with required simulation of the base and connecting materials, temperature, stresses and other parameters. What these data are required for?

For one thing, they will enable defining beforehand, how much time is left before depressurization and release of activity from closed reservoirs, to calculate the amount of released radioactive substances into the water area and possibility to take some protective actions (like those described in paper [6]). Simulation and similarity of the processes will probably enable reducing the time for the experiments.

Studies in the second direction, which were performed and are being performed by researchers, do not yet provide all information enabling evaluation of the global risk. A necessary and important task would be studying of nuclide migration though sea and ocean water areas, its impact on the sea biota and via biota on the human beings. It would be beneficial to have a regulatory document of the NRB type, which would establish permissible concentrations of hazardous nuclides and radioactive substances in seawater (local and global), permissible concentrations in sea products and so on. As we know, such proposals have already been made [8]. Only after this information is available and relevant monitoring is provided, we would be able to evaluate NS utilization risk rather reliably. The impact on human beings would be considered as a consequence and this impact would be surely not momentary, but prolonged in time (which is less hazardous) and will strongly depend on geography.

These scientific problems would be, in our opinion, the main problems for to-day and could be the subject for further international co-operation.

7. The problems associated with auxiliary ships and shore infrastructures

Removal of NSs from the Navy resulted in more than 300 cores (or more that 70 thousand fuel assemblies), more than 14 thousand m^3 of LRW and more than 26 thousand m^3 of SRW being stored for a long time at overfilled facilities of the Navy, both on shore and on special ships. Practically no vacant reservoirs are available. Total activity of SRW and LRW accumulated during operation of NSs exceeded 270 thousand Ci. There is direct evidence of the non-satisfactory technical state of a number of ashore storage facilities and exhausted capacities of the Navy floating and ashore infrastructures intended for SF, LRW and SRW [9].

Increasing the number of NSs waiting for their turn for utilization causes involvement of a large number of personnel, requires the use of new areas of land and sea water for keeping NSs before storage, for long-term storage of NSs removed from the Navy and for generated RW management.

The threat of radioactive contamination of the environment, the location of sites for keeping NSs removed from the Navy relatively close to settlements, gives some social slant to this problem and requires selecting such technologies, which would minimize expected adverse consequences.

Studying of the risk presented by these facilities is a part of the overall task.

8. On Regulatory documents

The purpose of the regulatory documents of different hierarchical levels is to regulate activities to prevent release of radioactive substances outside established boundaries, which may violate the safety limits.

Different stages of the technological process in utilization of NSs removed from the Navy are in this regard of different importance. This issue was discussed in paper [5]. We came to the conclusion, that for many stages, in particular for NS cutting up ashore, available regulatory documentation on decommissioning of nuclear facilities, radwaste management, transportation of spent nuclear fuel and some other issues could be used for the process of NS utilization. Some corrections may be required.

Limits, imposed by the regulatory documentation on utilization the process, result in a negligibly small reduction of risk from many utilization processes. It may be confirmed by information presented in papers [10,11], where small dose loads are shown in most technological processes during NS cutting ashore.

At the same time, Table 2 contains many "question marks". They mean, that no relevant regulatory documentation is available and that we have very little scientific information to establish safety requirements.

First of all, this is a problem of accidental submersion of NSs or compartments of NSs removed from the Navy. It seems, efforts should be aimed in the direction of Task 1 (section 1 of the article), i.e. try to minimize probability of submersion of such facilities.

At the same time, as far as accidental submersion of radioactive facilities has occured and taking into account a large number of already submerged facilities, the problem of release of radioactive substances from the sunk facility, migration of radioactive substances in the seas and their impact on biota - is a problem to be investigated and regulated.

9. Role of the Public

An approach to making decisions by responsible people, and by the society, in conditions of risk plays an important role in the solution of the NS utilization problem [12]. The stages of discussion of this procedure and interaction between the society and authorities are the subjects for wide discussion, including different political, ethical and moral aspects.

In the assessment of current approaches and risk problems, an impression is created in the West, that the "risk" notion is being transformed not only into some quantitative value (probability, consequences and others), but also into some qualitative multiple-factor complex, where we have to solve many different problems. Quantitative risk indicators are necessary, but according to experience, they are often far from being adequate. In this case the work may not achieve its goal, as it may not be understood by the public. To convince the public of the safety of NS utilization - is an important task of national and international scale.

Risk prevention and the need to protect against its consequences require clear scientific, engineering and other assessments, approaches. These should be formulated, so all the participants have the same understanding of the issues. This structure of assessment of approaches causes different variations of the questions:

- technical - if variants of technologies have been developed, which are capable of preventing a risk or mitigating its consequences;
- scientific - what we know about the risk and what we need to forecast the probability of an event and its consequences judging by their assessment?
- economical - which funds could be allocated for complete risk prevention or its reduction to a negligibly small values?
- insurance - what types of risk could be indemnified ?
- political and ethical - which are the largest concerns of the society? Which type of risk is considered acceptable?

- legal - what is the liability of the officials for failure to prevent risk and failure to take measures to protect from risk?
- social and political - what process of risk acceptance should be used? Who should be responsible for making decisions on behalf of the population?

Any approach, which limits consideration of any of thee aspects, will inevitably lead to adverse results and, finally, to failure.

10. Conclusions

Trying to answer the question, raised in the very beginning of the report, if we are able to assess global risk (see our explanation of the word "global" in this paper) in NS utilization, we would not be able to give an affirmative answer to the question. There are too many unclear issues. We define risk as the sum of products of event probability and its consequences. The situation will be even more complicated, if the public starts asking questions, mentioned in section 9. We have already mentioned in the beginning, the specific problems we have. The main one - at what time period we want to give an answer-in 10, 100, 1000 years. Depending on this, we determine a range of scientific, engineering and social problems, which are to be solved.

11. References

1. Risk Management of Emergency Situations (2001) *Sixth All-Russian Scientific &Practical Conference.* Ministry of Emergency Situations of Russia, Ministry of Industry and Science of Russia, Russian Academy of Science.
2. Safety of Russia. Legal, Social & Economic, Science & Technical Aspects. (1999) *Volume «Dictionary of Terms and Definitions»,* **2,** MGF «Znanie».
3. Kovalevich, O.M. (2001) Risk Notion and Its Derivatives. *Safety Problems in Emergency Situations, VINITI,* **1**.
4. Shevelev, Y.B. (1988) *Management of Population Security (Safety) in USSR,* preprint of Kurchatov IAE, IAE - 4731/3.
5. Kovalevich, O.M., Gavrilov, S.D., Egorov, Y.V. et al. (2001) Regulatory Basis for Ensuring Nuclear, Radiation and Ecological Safety in NS Utilization. State and Perspectives, *Proceeding of International Scientific Workshop.*
6. Smoldyrev, A.Y. (1999) Hydraulic Method of Safe Disposal of Toxic Pits in Water Areas. *Safety Problems in Emergency Situations, VNITI,* **9**.
7. Problems of Decommissioning and Utilization of Nuclear Submarines. Report by "Bellona" Association, (1996), in A. A. Sarkissov (ed), *Materials of International Scientific Workshop RAS/NATO, Moscow,* KomTeh, Moscow.
8. Dovgusha, V., Tikhonov, M. (2002) On Potential Nuclear and Radiation Hazard in Complex Utilization of NS and Nuclear-Powered Ships, *Bulletin on Nuclear Power Engineering* **2**, p. 56.
9. *Safety Analysis and Assessment of Radiation Risk for Population of Severodvinsk in Potentially Hazardous Nuclear Activities at PO «Sever» (NPO «Zvezdochka)* (1994) International Fund of Rehabilitation of Territories Exposed to Radioactive Impacts, Severodvinsk.
10. *Analysis of Consequences of Potential Radiation Accidents in Utilization of NSBM (nuclear submarine with ballistic missiles) of «Typhoon» Type, Assessment of Radiation Risk* (1999) I. I. Afrikantov DDOMB Nizhny Novgorod.
11. Kovalevich, O.M. Presentation at VAZDOR.

DECOMMISSIONING AND DISMANTLING OF FRENCH NUCLEAR SUBMARINES: INDUSTRAL ISSUES

BERNARD ROBIN
Technicatome (France)

AMAURY DE BUZONNIERE
Technicatome (France)

The paper aims at presenting the industrial issues of the decommissioning and dismantling of French Nuclear Submarines and will introduce a short film of DGA/DCN which shows the dismantling operations carried out on the Le Redoutable, first French nuclear submarine.

France has elaborated a specific regulation for the decommissioning of nuclear propulsion reactors.

This specific procedure notably in terms of safety inspection is based on the IAEA decommissioning levels 1, 2, 3.

The responsibilities are shared among four organizations who are :

- the CEA (Atomic Energy Commission) which delegates to Technicatome, prime contractor of nuclear propulsion reactors, the technical design and specification of dismantling works,
- the DGA (General Direction of Armament) which delegates to DCN (Direction for Naval Construction) of Cherbourg the execution of works.

Concerning the studies, Technicatome scope of work consists in realising the decommissioning and dismantling dossiers with a specific approach for each of the nuclear submarines due to the fact that there are some differences among the submarines of a same generic type and also considering the operation of the submarine itself.

So, these studies aim to define:

- the reactor status at the various decommissioning levels,
- works and supplies needed to reach the required decommissioning level,
- management strategy for waste produced during dismantling operations,
- radioprotection and containment to be assured during the dismantling operation and interim storage phases.

The main technical options for decommissioned reactors compartments at level 2 are the following:

- maintenance of 2 containment barriers periodically controlled between the environment and activated surfaces.
- the first barrier is the primary circuit fitted with sealed shutters in place of removed equipment,
- the second barrier consists of the walls of the reactors compartment itself which are made airtight before separation of the reactor compartment,

A.A. Sarkisov and L.G. LeSage (eds.),
Remaining Issues in the Decommissioning of Nuclear Powered Vessels, 63–64.

To comply with these principles, the following rules are applied:

- limitation of corrosion risks in the circuits. For that purpose, the circuits are emptied and maintained in dry air as well as the reactor compartment itself,
- limitation of fire risks by removal of any permanent electricity supply,
- definition of radioprotection zones allowing taking the safety measures for workers and also to predetermine the potential activity of waste,
- Protection of workers from irradiation notably by putting protections around steam generators and primary circuits and keeping the secondary circuits of steam generators in water up to the end of the dismantling operations and drying them only at the end,
- protection of workers and the reactor compartment from contamination due to a dynamic containment,
- radioprotection control for workers and rooms,
- waste reduction and management according to French regulations. In relation with radioprotection zones, waste are sorted according to their nature and activity, controlled, treated and stored before evacuation.

The monitoring of the reactor compartment isolated in level 2 and stored in a special hall in the Cherbourg shipyard is performed as follows:

- every month, without entrance into the reactor compartment, control of the RC atmosphere (temperature and hygrometry),
- every three months also without entrance into the RC, control of RC atmosphere is completed by a radioactivity measurement.
- once a year, the control is performed in two phases :
  - before entrance in reactor compartment, control of RC atmosphere as above mentioned,
  - with entrance in RC :
    - for the first barrier: control of hygrometry and air tightness, radiological cartography in RC,
    - for the second barrier: visual inspection, air tightness control and drying of the RC if required

In conclusion:

- The experience gained whilst dismantling " Le Redoutable " to decommissioning level 2 has enabled a reference system to be drawn up for operation, dosimetry, produced wastes, costs and related schedule.
- Optimization of dismantling operations have been possible for the 3 additional submarines with a reduction of dosimetry and costs.
- Inspections of the reactor compartment of Le Redoutable stored at decommissioning level 2 have shown since 1993 that radioactivity has been correctly contained without any specific constraint (no corrosion detected).
- The shift to the decommissioning level 3 after decay storage is being studied by TECHNICATOME. The dismantling of the Le Redoutable NPP can be anticipated as it is now being decided by EdF.

INTERNATIONAL COOPERATION IN NUCLEAR VESSEL DECOMMISSIONING AND UTILIZATION

BILATERAL COOPERATION RELATED TO THE DECOMMISSIONING OF NUCLEAR VESSELS BETWEEN THE USA AND THE RUSSIAN FEDERATION – STATUS AND ISSUES

CDR MARK A. BAKER
United States Navy

1. Purpose

This paper provides a summary discussion of the efforts undertaken by the U.S. Defense Threat Reduction Agency through the Cooperative Threat Reduction Program with the Russian Federation to eliminate Submarine Ballistic Missile Launchers and the dismantling of the associated Strategic Ballistic Missile Submarines (SSBNs).

2. Program Basis

In 1991, recognizing the significant changes taking place in Russia and the opportunity to establish improved relations in a long term program to assist the Russian Federation (RF) and other countries of the Former Soviet Union (FSU) in preventing the proliferation of weapons of mass destruction, including related technology and information, Senator Sam Nunn and Senator Richard Lugar conceived and established the legislation required to create The Cooperative Threat Reduction (CTR) Program. Through this program, the countries of Belarus, Ukraine, and Kahzakstan were provided assistance and became non-nuclear states. To establish the program in Russia, an "umbrella agreement," officially entitled "Agreement Between the United States of America and the Russian Federation Concerning the Safe and Secure Transportation, Storage and Destruction of Weapons and the Prevention of Weapons Proliferation dated 17 June 1992; as extended on 16 June 1999", was signed by the Presidents of the United States of America and Russia to provide the indemnification and tax/duties exemptions needed for the United States to legally perform such work. To implement the program to reduce the Strategic Offensive Arms in Russia, an agreement entitled, "Agreement Between the Department of Defense of the United States and the Committee for Defense Industry of the Russian Federation Concerning Cooperation in the Elimination of Strategic Offensive Arms dated 26 August 1993; as amended 3 April 1995, 19 June 1995, 27 May 1996, 11 April 1997, 11 February 1998, 9 June 1998, 16 August 1999 and 8 August 2000 (SOAE Agreement)" was signed by the Secretary of Defense of the United States and the Minster of the Committee for Defense Industry for the Russian Federation. Under these agreements, the CTR program successfully assisted the RF in reducing its nuclear weapons stockpile to less than the START I Treaty limit of 4900 attributable warheads by eliminating long range bombers, intercontinental ballistic missiles and silos, and submarine launched missiles and the

A.A. Sarkisov and L.G. LeSage (eds.),
Remaining Issues in the Decommissioning of Nuclear Powered Vessels, 67–71.

associated strategic ballistic missile submarines (SSBNs). The CTR Program continues under these agreements and is projected to continue through 2007.

3. Strategic Offensive Arms Elimination Program Objectives

In accordance with Strategic Arms Reduction Treaty (START I) Conversion or Elimination Protocols, the SOAE Program plans to provide assistance in equipment, direct funding of work, and infrastructure improvements to eliminate approximately:

- 1025 Intercontinental Ballistic Missiles (ICBMs)
- 565 ICBM Silos
- 208 ICBM Mobile Launchers
- 131 Strategic Bombers
- 713 Nuclear Air-to-Surface Missiles (NASMs)
- 677 Submarine Launched Ballistic Missiles (SLBMs)
- 612 SLBM Launchers
- 41 Strategic Ballistic Missile Submarines (SSBNs).

Specifically related to expediting the elimination of SLBM launchers under Annex C of the SOAE Agreement, and dismantlement of associated SSBNs the CTR program has provided equipment and funded contracts for infrastructure improvements, elimination of SLBM launchers and the dismantlement of SSBNs. The U.S. program objective to eliminate a total of 677 SLBM launchers and dismantle the associated 41 SSBNs is based on recent statements from the Russian Federation indicating plans to reduce their nuclear weapons to a level of approximately 1500 warheads. The SSBN Dismantlement Program also provides for environmentally compliant processing of low level radioactive waste (LLRW) and dry storage of spent nuclear fuel (SNF). A summary of SSBNs to be dismantled under the CTR Program includes: two Yankee Class, 34 Delta Class, and five Typhoon Class Russian Strategic SSBNs. The elimination work is being accomplished at five START-designated SLBM launcher elimination facilities: the State Machine Building Enterprise (Zvezdochka) and the Northern Machine Building Enterprise (SevMash) at Severodvinsk; the Far East Shipyard (Zvezda) at Bolshoi Kamen near Vladivostok; Ship Repair Facility 49 at Vilyuchinsk near Petropavlovsk; and, the Ship Repair Yard (Nerpa) at Snezhnogorsk near Murmansk.

4. SSBN Dismantlement Program Breakdown by Project Area

4.1 SSBN ELIMINATION:

The SLBM Elimination and associated SSBN Dismantlement Program began in 1993 with equipment purchases and deliveries. Following this demonstration of good faith, the RF permitted visits to proposed dismantlement sites and discussed potential dismantlement bottlenecks. Infrastructure and additional equipment were purchased. In 1997, it became clear that the RF did not intend to devote sufficient funds to eliminate the SLBM launchers and dismantle the associated SSBNs at rates necessary to meet the START I Treaty requirements. The CTR Program developed and implemented contracts for the SSBN dismantlement preparation, SLBM launcher elimination and SSBN dismantlement. The CTR Program successfully completed negotiation of the first contract to eliminate a Delta Class SSBN at Zvezdochka in March 1998 (Contract completed in August 1998). The program supplied equipment that supported the completed in August 1998). The program supplied equipment that supported the elimination of five SSBNs prior to direct contracting for SSBN dismantlement. Since August of 1998, CTR has contracted for dismantlement of five SSBNs at Nerpa (four completed, one in progress), five SSBNs at Zvezdochka (all completed), ten SSBNs at Zvezda (eight completed, two in progress), one in progress at Sevmash (Typhoon 712), and one in progress at Ship Repair Facility 49 (SSBN 367). This brings the total to 27 SSBNs eliminated (396 launchers eliminated which represents 816 attributable warheads deactivated) with support from the CTR Program.

(Note: SSBN 381 dismantlement in progress at Zvezdochka Shipyard was funded by the Russian Federation and is using U.S. provided and maintained equipment.)

To date, the CTR Program has funded $205,000,000 in SSBN Dismantlement, $38,800,000 for LLRW Facilities, and $18,300,000 for SNF disposition and has $18,000,000 in pending contracts for SLBM launcher elimination and SSBN Dismantlement efforts.

4.2 INFRASTRUCTURE AND EQUIPMENT PROVIDED

The CTR Program has funded $38,500,000 in equipment and infrastructure improvements at the Nerpa, Zvezdochka, and Zvezda Shipyards. Infrastructure improvements included concrete pads for cutting areas, lighting towers, low pressure air systems, oxy-acetylene systems, and electrical power distribution upgrades. Major equipment provided includes Harris Metal Cutting Shears and metal bailers, Triple S Dynamics cable cutting equipment, and a variety of cranes, trucks, cutting tools, and non-standard equipment needed for the dismantlement work.

4.3 LOW LEVEL RADIOACTIVE WASTE

Zvezdochka Engineering Enterprise: CTR Program funded the construction of the environmentally compliant LLRW volume reduction facilities at the Zvezdochka Engineering Enterprise Shipyard at Severodvinsk. The work at Zvezdochka included the construction of solid waste treatment facility, including decontamination equipment and a low force compactor, a liquid waste treatment system utilizing sorbent technology, and a laundry water treatment system. The facility was commissioned on 18 October 2000.

Zvezda Far East Shipyard: CTR Program funded the construction of LLRW volume reduction facilities at the Zvezda Far Eastern Shipyard at Bolshoi Kamen. The work at Zvezda included the construction of a solid waste treatment facility, laundry water treatment system, and a solid waste storage facility. The LLRW facility was commissioned on 2 August 2001. Low Level Liquid Waste will be processed at the Zvezda Shipyard on board a processing barge (Landysh or Suzeron) provided by the Japanese Government.

4.4 SPENT NUCLEAR FUEL (SNF)

SNF Defueling Vessels ($8.7M): The project provided interim maintenance to three Russian spent naval fuel (SNF) defueling vessels, pending completion of construction of On-Shore Defueling Facilities at Zvezdochka and Zvezda.

SNF On-Shore Defueling Facilities ($24.6M plus additional funds pending contract awards): These facilities at the Zvezda and Zvezdochka Shipyards are designed to remove shipyard dependency on the SNF defueling vessels and provide transient storage for SNF with supporting rail heads to service the storage areas. Contracts for the construction of these facilities were signed in late 1998. The facilities are expected to be in operation by the end of 2002.

SNF Transportation Railcars (TK-VG-18-2) ($2M): The U.S. has contracted with Atomspetstrans to manufacture and deliver six TK-VG-18-2 SNF Transportation Railcars. These cars will each carry two TUK-18 or TUK-108/1 SNF Transportation and Storage Casks and will be able to reach locations that the TK-VG-18A railcars cannot. These newly designed railcars are approximately 5 meters shorter than the older version and have substantially reduced per axle load. The railcars are expected to be placed in operation in mid-2003.

SNF Dry Storage Facilities (total cost to be determined): In addition to the transient storage pads provided as part of the On-Shore Defueling Facilities, CTR is pursuing the remodeling of Building 301 at Mayak Production Association in Ozersk for Dry Storage of SNF using TUK-108/1 SNF Transportation and Storage Casks. This facility is expected to be able to store 150 or more casks. The completion of the facility is planned for December 2005. Additionally, a storage pad funded by Norway and U.S. CTR funds with a capacity of 19 casks is being constructed at RTP Atomflot in Murmansk under the Arctic Military Environmental Cooperation (AMEC) Program.

SNF Transportation and Storage Casks (TUK 108/1) ($9.6M to date, additional funding pending future orders): The TUK-108/1 SNF Transportation and Storage cask was developed under the AMEC Program using Norwegian, Russian, and U.S. CTR funds. CTR has established a very flexible contract with Sevmash Shipyard for the serial production of TUK-108/1 SNF Transportation and Storage Casks. The first order for 25 casks to be manufactured and shipped to the Far East was placed in November 2001. The production capacity of Sevmash for these casks is expected to be about 300 per year once they have completed line certification and the first article testing. The Ministry of Russia for Atomic Energy is analyzing the need for these casks to ensure the casks are not over produced and thereby create an additional problem.

5. Current Issues and Future Challenges

Overall, the SSBN Dismantlement Program has been highly successful. The cooperative attitude demonstrated by the former Ministry of Economics, the Ship Building Agency, and the shipyards and their ability to coordinate issues with the MOD and MinAtom has been key to this success.

For planning purposes, the Defense Threat Reduction Agency, CTR Program Directorate (DTRA/CT) has developed the following contracting schedule for SSBN dismantlement. In FY 2002, contract negotiations are planned to support dismantling one Delta at Zvezda and one Typhoon at SevMash. In FY 2003, contract negotiations are planned to support dismantling one Delta at Zvezda and one Delta at Zvezdochka. In FY 2004, contract negotiations are planned to support dismantling one Delta at Zvezda, and one Typhoon at SevMash. In FY 2005, contract negotiations are planned to support dismantling one Typhoon at SevMash and one Delta at Zvezda. In FY 2006, contract negotiations are planned to support dismantling one Typhoon at SevMash, one Delta at Zvezdochka, and one Delta at Zvezda. In FY 2007, contract negotiations are planned to support dismantling one Delta at Zvezda and one Delta at Zvezdochka. It should be noted, however, that this plan has not been confirmed by the Russian Federation as the reorganization of the Russian Nuclear Force Structure remains a very complex problem for the Russian MOD.

The remaining issues in the SSBN Dismantlement Program include:

- the remaining dismantlement schedule;
- completion of current projects;
- disposition of equipment provided to shipyards (maintenance).

Remaining issues for Russian Federation as a result of SSBN Dismantlement:

- ultimate disposal of floating reactor compartment units containing high level radioactive waste;
- long term storage of solid low level radioactive wastes;
- implementation of the long term storage or reprocessing of the SNF.

MILITARY ENVIRONMENTAL COOPERATION ON RADIOACTIVE AND NON-RADIOACTIVE WASTE IN RUSSIA'S NORTHWEST AND FAR EAST

DIETER RUDOLPH
U.S. Department of Defense

INGJERD KROKEN
Norwegian Ministry of Defense

EDUARD LATYSHEV
Russian Federation Ministry of Defense

Abstract

The Arctic Military Environmental Cooperation (AMEC) Program is a cooperative effort between the Ministries of Defense of Norway and Russia and the Department of Defense of the United States. The goal of the program is to mitigate the impact of military operations in the fragile Arctic environment. This paper discusses the history of the Program, current efforts, accomplishments and challenges. It provides an overview of the AMEC projects addressing five radioactive waste program areas: spent nuclear fuel, liquid waste, solid waste volume reduction, solid waste storage technologies, radiation monitoring and personnel safety.

1. History of the Program

This program is a Norwegian initiative that grew out of the acknowledgement in the late 1980's of the cross-border nature of environmental problems, especially with respect to radioactive waste. The changed political climate in the early 1990's opened the way for bilateral cooperation between Norway and Russia in the field of nuclear safety. While the main emphasis in the early years was placed on data collection and analysis, Norwegian efforts soon were focused on eliminating bottlenecks in the established Russian infrastructure, specifically as to the transport and handling of spent nuclear fuel and the radioactive waste associated with the dismantling of Russian nuclear submarines.

In 1995, Norway approached the United States and Russia with proposals for a discussion of defense-related environmental problems in the Arctic region. This resulted in a joint declaration, signed in September 1996 by the Defence Ministers of Norway and Russia and the United States Secretary of Defense, on Arctic Military Environmental Cooperation (AMEC) under which the three nations' armed forces will work together to ensure that their military activities do not harm the Arctic environment, especially with regard to contamination by radioactive materials. This cooperation is intended to develop

A.A. Sarkisov and L.G. LeSage (eds.),
Remaining Issues in the Decommissioning of Nuclear Powered Vessels, 73–78.

better methods and infrastructure for the processing of radioactive waste from military sources, the handling of spent nuclear fuel and radioactive waste from decommissioned submarines and the assessment of environmental risks associated with military activities. The list of collaborative projects includes, for example, the development of a 40-ton container and associated storage platform for spent nuclear fuel, monitoring systems for radioactive waste and mobile units for the collection of both solid and complex liquid radioactive waste.

For Norway, the AMEC cooperation forms an integral part of the overall cooperation with Russia on matters of nuclear safety as detailed in the "Nuclear Plan of Action", which allocates $70M towards addressing nuclear safety problems. For the U.S., the program is executed as an International Environmental Cooperation Program. What makes this project so unique is that it is a "military-to-military" cooperation project focusing on the task of solving the serious environmental problems stemming from military activities in Northwest Russia.

It is important to note that although this is a military-to-military programme led in the case of the United States by the Department of Defense, the Department of Energy, the Environmental Protection Agency and the Department of State all participate. Most of the United States funding, amounting to some $24 million dollars since the start of the program, is provided by the Department of Defense, the Department of Energy has contributed some $2 million dollars to date. In Norway the Ministry of Foreign Affairs provides the funding for the Ministry of Defence (since AMEC is an integral part of the Plan of Action for nuclear safety issues and the environment). To date Norway has provided about $4.2 million dollars.

The Russian Federation funding is provided by the Ministry of Defence and the Ministry of Atomic Energy and includes "in kind contributions".

2. Current Efforts

AMEC is a forum for dialogue and joint activities among military and environmental officials from Norway, Russia and the United States. AMEC is led by designated senior officials-- the AMEC Principals from the United States Department of Defense and the Defence Ministries of Norway and Russia. A Steering Group, assisted by the Technical Guidance Group consisting of experts in the field, identifies and prioritizes military environmental projects in the Arctic region.

Projects address five radioactive waste program areas: spent nuclear fuel, liquid waste, solid waste volume reduction, solid waste storage technologies radiation monitoring and personnel safety; and two non-radioactive waste program areas: remediation technologies, clean ship technologies. As a result of the limited AMEC budget, prototype development is the goal. Mass production of the prototype will have to be accomplished using Russian, International or other U.S. funds. This is the approach used for the spent nuclear fuel project. The U.S. will purchase this cask for the spent nuclear fuel from the Ballistic Submarines, while Russia will purchase these casks for the fuel from the General Purpose Nuclear submarines.

2.1 NAVAL SPENT NUCLEAR FUEL MANAGEMENT

Two Projects are currently underway – the development of a prototype cask for storage and transport of spent nuclear fuel from submarines and a transshipment pad at RTP

ATOMFLOT in Murmansk. This pad will be used for temporarily storing casks as they are unloaded from the service ships, and on loaded onto railroad cars. The prototype cask was completed in October 1999, and was certified for storage and transport by the military certification authority. Currently, certification by the Russian civilian certification authority is being considered. The pad construction is complete and the pad should be in operation by the end of the year.

2.2 NAVAL LIQUID RADIOACTIVE WASTE TREATMENT

The goal of this project is to develop a specialized mobile liquid waste processing facility to support the dismantlement of nuclear submarines. This project will be part of an integrated radioactive waste pretreatment facility at Polyarninsky Shipyard in the Murmansk region and will be included in Phase II of the radioactive waste pretreatment facility. Preliminary design discussions have been conducted.

2.3 TECHNOLOGIES FOR SOLID RADIOACTIVE WASTE VOLUME REDUCTION

The goal of this program area is to develop an integrated treatment system necessary to process solid radioactive waste. This is the core facility of the radioactive waste pretreatment facility at Polyarninsky Shipyard and should be in place by the end of 2002. A phased approach is used. The first phase, an initial Technology Assessment is completed. Phase two, the design and construction of a Mobile Pretreatment Facility, is in progress.

2.4 SOLID RADIOACTIVE WASTE STORAGE TECHNOLOGIES

This program area which addresses the Russian shortage of interim storage for naval solid radioactive waste and includes a one-year test of a coating material to seal surfaces and facilitates decontamination, was successfully completed in August 1999. Results showed that this coating material could be applied in the harsh arctic climate and that contamination could be removed in one decontamination cycle without any harmful effects on the coating. Four hundred steel radioactive waste containers for the transport and storage of solid radioactive waste have been produced and are now being tested by the Russian Navy. Future plans include demonstrating the applicability of interim storage facilities to support the solid radioactive waste pretreatment facility at Polyarninsky Shipyard.

2.5 RADIATION MONITORING, PERSONNEL AND ENVIRONMENTAL SAFETY

This program area enhances monitoring, radiation dosimetry and safety procedures and techniques in the Russian military particularly as they pertain to nuclear submarine dismantlement. Shipyard workers and nearby civilian populations are concerned about radiation exposure resulting from submarine decommissioning and dismantlement activities. An initial project to provide equipment and training to improve efficiency of measures taken to protect the health of workers, the public, and the environment has been completed. Another project - the PICASSO system includes the application of radiological sensing equipment interfaced with computerized graphic displays that are tailored to specific radiological waste storage and submarine dismantlement sites. The system automates and greatly enhances the ability for local and centralized monitoring of radiological hazardous material. PICASSO will be integrated with the interim Spent Nuclear Fuel storage pad at RTP Atomflot in Murmansk and the radioactive waste pretreatment facility in Polyarninsky Shipyard

2.6 TECHNOLOGIES FOR THE REMEDIATION OF HAZARDOUS WASTE SITES ON ARCTIC MILITARY BASES

This Norway/Russia project focuses on the selection of proper and effective technologies for dealing with military hazardous material spills in the Arctic environment. Phase 1 of the project is a study, analysis and data presentation of a military fuel storage facility in the Arkhangelsk region in Northwest Russia and is completed. Phase 2 of the project involves actual remediation of the area studied, where the Russian party will select an appropriate technology required for decontamination of the area.

2.7 REVIEW AND IMPLEMENTATION OF "CLEAN SHIP" TECHNOLOGIES

This program area examines technologies for the collection and comprehensive processing of "non radioactive" naval ship waste. The initial phase of the project provided an assessment of the waste problem on board Russian naval vessels. This study quantified and characterized the wastewater discharges – about 250,000 gallons daily - from the Russian Northern Fleet ships. The next phase is the demonstration of "Clean Ship" technology plans to evaluate an Oily Wastewater treatment system aboard a Russian naval ship.

3. Accomplishments

There are direct and indirect accomplishments. Perhaps the most significant accomplishment is indirect – providing the Russian Navy and Ministry of Defense with the building blocks for a comprehensive radioactive waste processing facility at Polyarninsky Shipyard in the Murmansk region. This facility will have the capability to process both solid and liquid radioactive waste generated from the dismantlement of nuclear submarines as well as civilian generated radioactive waste. The plan for this facility was approved by Admiral Kuroyedov on 20 March 2002.

The direct accomplishments are listed by program area. In the area of Spent Nuclear Fuel Management, the prototype container for naval spent fuel has been developed and is certified for military use. Certification by civilian authorities is the next step and will allow the use of this cask for the transport and storage of civilian fuel from nuclear power plants. The United States has contracted with SEVMASH in Severodvinsk for the mass production of this cask. The cask will be used for the transport and interim storage of spent nuclear fuel from Russian Ballistic Nuclear Submarines that will be dismantled as part of the Cooperative Threat Reduction (CTR) Program. Russia has started the mass production and use of this cask for the spent nuclear fuel from the General-Purpose nuclear submarines. The interim storage pad for spent nuclear fuel at RTP Atomflot has been completed and should be in use by the end of the year. The use of this pad will reduce spent nuclear fuel shipment time from 3 months to 3 weeks, and the pad and cask together have helped eliminate the spent nuclear fuel bottleneck.

Russian radioactive waste technologies have been improved through the completion of the surface coating demonstration at RTP ATOMFLOT and the manufacturing of U.S. and Russian made steel radioactive waste transport/storage containers. The radioactive waste pretreatment facility at Polyarninsky shipyard will be the first Russian Navy comprehensive radioactive waste treatment facility for both solid and liquid waste.

Radiation monitoring, personnel and environmental safety have been improved through the transfer of radiation monitoring equipment and the training of Russian Naval radiation experts in both Norway and the U.S. A system for monitoring the local radio-ecological conditions will be included at the RTP Atomflot spent-nuclear-fuel defueling facility and at the Polyarninsky waste treatment facility.

4. Challenges

There are a variety of challenges facing the AMEC Program. This includes the lack of a separate legal agreement and the proposed termination date of September 2002 for U.S. participation.

4.1 LACK OF A LEGAL AGREEMENT

Three years ago, a separate legal agreement was proposed to cover Norwegian, Russian and U.S. AMEC activities – the "Trilateral Legal Agreement". The key stumbling blocks are: liability, access, taxes and status of personnel. A proposal to use the language from the Cooperative Threat Reduction (CTR) Agreement that covers these issues was recently rejected by Russia since the CTR Agreement has not been ratified by the Russian Duma.

The Multilateral Nuclear Environmental Program with Russia (MNEPR) Framework Agreement seems to have resolved these issues. The problem is that the U.S. has not agreed to the MNEPR liability language since it has stricter liability requirements than the donor countries.

Both Norway and the U.S. have used existing Bilateral Legal Agreements to provide the required legal coverage to implement the AMEC projects. The U.S. uses the CTR Agreement, which means that the U.S. can only participate financially in those projects directly related to Ballistic Submarine dismantlement.

Norway participates financially in those radioactive waste projects included in the Framework Agreement. New projects can only be included by mutual agreement of both parties communicated via formal diplomatic channels.

The U.S. has decided that AMEC cannot use the CTR Agreement after September 30th. This means that if there is no alternative legal coverage, the U.S. cannot continue its active participation in this program.

4.2 CONTINUED U.S. PARTICIPATION

In 1999 Congress provided the U.S. AMEC Program with it's own funding allocation and requested that the Department of Defense provide project details and to propose a program end date. The Department decided to use September 30, 2002 since that was the date funding for the AMEC stopped. Based on the success of the program, the U.S. is prepared to continue, provided that it will use its own legal agreement.

5. Conclusion

Although the AMEC Program operates on a relatively small budget – typically about $4-5M per year – it has made a significant difference in addressing both radioactive and non-radioactive issues in the fragile Arctic environment. Not only has AMEC developed a spirit of cooperation and trust between the militaries and other agencies of the three nations, but it also provides viable technical solutions for those concerned with preserving the environment.

NON-PROLIFERATION AND OTHER SECURITY-RELATED ISSUES ASSOCIATED WITH THE DISMANTLING OF NUCLEAR VESSELS IN NORTH-WEST RUSSIA

O. REISTAD and A. SOERLIE
Department of Radiation Protection and Nuclear Safety
Norwegian Radiation Protection Authority
Grini Naeringspark 13
N-1332 Oesteraas, Norway

1. Introduction

Nuclear safety has been in the focus for international cooperation on Russian general purpose nuclear submarine dismantling. Even if the presence of security-related issues has been admitted, the threshold for discussing these issues openly outside the scope of disarmament has been high. This paper suggests, on the basis of technical analysis, site descriptions of Russian naval sites and previous experience in dealing with similar issues for Russian civilian ships, that security-related threats should come more in the forefront when considering submarine dismantlement. In addition, the results of the Russian/ Swedish/ Norwegian governmental cooperation on similar issues are presented.

No effort has been made in this paper to define the term 'a security-related issue'. INFCIRC 225/ Rev. 4, as published of IAEA, covers deliberate acts that could endanger social stability, public health or safety [1]. In this paper, deliberate actions to increase the risk of radioactive releases and proliferation of nuclear materials have a natural focus. Relevant examples on such deliberate actions are the desperate attempt of a Russian soldier to sabotage a vessel before committing suicide, and the different thefts of fresh nuclear material from the naval facilities in Murmansk and Zapadnaya Litsa. This approach does not suggest a new, formal ranking of the overall importance of the different risks, but reflects, by the authors' opinions, a need for discussing a complementary side of the overall risk picture compared to the continued focus on environmental risks.

2. The core of the issue: proliferation

The crucial question is; in what respect does dismantling of a Russian nuclear submarine fleet represent a proliferation risk? Several hypotheses have been put forward, such as for example in [3], but without any immediate response from either the Russian or other

A.A. Sarkisov and L.G. LeSage (eds.),
Remaining Issues in the Decommissioning of Nuclear Powered Vessels, 79–90.

governments or institutions concerned with the overall safety when dismantling Russian vessels. So what are the issues?

2.1. SPENT NAVAL FUEL – PROPERTIES AND BARRIERS

All spent nuclear reactor fuel is a proliferation risk. The most protected group of materials as defined by Convention on Physical Protection of Nuclear Materials (CPPNM), category 1, includes Plutonium (Pu) and Highly Enriched Uranium (HEU) irradiated in a reactor but with radiation level equal to or less than $1\cdot10^5$ Ci/ kg Pu, HEU or U-233 at one meter unshielded [1]. The presence of one of these types of material requires for example restricted access, determination of people's trustworthiness, constant surveillance and response forces. IAEA has earlier taken an initiative for evaluating specifically the radiation levels for nuclear reactor spent fuel to examine whether there should be a separate category of conversion time for direct-use nuclear material contained in such fuel wherein the total activity was equal to or greater than above-mentioned limit [4]. The U. S. National Academy of Sciences' Committee on International Security and Arms Control (CISAC) has introduced the "spent-fuel standard" regarding the intrinsic properties of the fuel as a criterion for judging the adequacy of resistance to theft and proliferation produced by a disposition option [5]. Even if disposition options are not the focus at this stage in this work, the methodology and results when assessing different intrinsic properties of the spent nuclear fuel are relevant when considering other types of spent fuel, as for example spent naval fuel from the dismantling of nuclear vessels. This includes barriers for diversion, including radiation levels, detection and possible use in weapons.

TABLE 1. Ranking of the proliferation risk associated with Russian spent marine fuel

Priority No.	Type(s) of material	Reactor and fuel characteristics	Type of vessel (as indicated in available sources):	Cooling time:
1	HEU	Initial enrichment > 70%, low-, medium and fully burned fuel, high initial amount of U-235	Icebreakers, submarines of third/ subseq. generations	Any cooling time
2	Pu	Initial enrichment below 20%, low- and medium burned fuel	Submarines of first and second generation	Any cooling time
3	Pu	21%> Initial enrichment > 30%, low- and medium burned fuel	Submarines of third/ subseq. generations	Any cooling time
4	HEU	20%> Initial enrichment > 80%, high initial amount of U-235, low- and medium burned	Submarines of third/ subseq. generations	Any cooling time
5	Pu	30%> Initial enrichment > 80%, low- medium and fully burned fuel	Icebreakers, submarines of third/ subseq. generations	Any cooling time

Preliminary analysis made on this basis as presented in [6, 7], as displayed in table 1, shows that the content of highly enriched uranium (HEU) should be the major concern, especially spent fuel with particular high initial enrichment and low-burn-up. More general security concerns are related to the removal of spent fuel removed from the vessels with little or no indication of operation history and isotopic concentrations. Regarding the specific barriers, the results show that after more than 30 years storage after decommissioning, the fuel assemblies cannot be regarded as being self-protected, and this limit might be closer to a handful of years rather than decades [6]. Combined with small fuel dimensions, the threshold for non-authorized uses is considerably lower for marine spent fuel than for commercial spent reactor fuel with average burn-up histories.

2.2. PROLIFERATION OF CONCEPTS AND TECHNOLOGY

Considering the technology component of the proliferation risk, the overall security concern is associated with the proliferation of nuclear submarine concepts and technology as an important factor for the survivability and operational capability of nuclear weapons. Only Nuclear Weapon States have deployed nuclear submarines, however, India, a de-facto nuclear weapon state, is building a nuclear submarine. Several other countries have considered the option, however, so far they have dismissed their plans as too costly [8]. One exception is India, which again has hired a Russian submarine. Another concern is the spread of weapons systems; primarily suitable for missiles delivery, and parts thereof as stated in the different lists specifying types of technology listed in the Zangger Trigger List and of the Nuclear Suppliers Group. A specific example on how a group of North-Korean citizens attempted to gain access to naval yards engaged in the dismantling of SSN's has been presented in [3]. Similar attempts at sites in Northwest-Russia have yet to be registered.

3. Environmental security issues

In order to facilitate a more systematic approach, the scenarios relevant for the dismantling of nuclear vessels might be divided in the following topics:

- Non-defueled vessels;
- Spent fuel storage facilities;
- Reactor compartments and waste;

The possibility of releases from spent marine fuel, reactor compartments and radioactive waste has not been properly addressed, probably due to the lack of information on the physical lay-out of the spent fuel, the reactor compartments, radioactive waste and their storage conditions, and the relative lower risk associated with this material compared to the spent fuel stored in the decommissioned vessels awaiting further dismantling. The focus is put on the presence of spent fuel due to the fact that the main part, 90-99%, of the activity rests with the fuel. The impact assessments have concentrated on scenarios including incidents involving spent nuclear fuel either as part of sunken submarines or reactor compartments with possible effects on the marine environment as in [9], or decommissioned, non-defuelled vessels [10]. As a result, the current focus has been put on safety issues and not on threats or releases as a result of deliberate acts or acts of terrorism. Recent publications mention terrorism as a possible reason for specific initiating events [11], however, others haven't included this at all as a significant initiating event [12].

An evaluation of the intentional increase of risk introduces a broader scope in the accident analysis. For example, sabotage and terrorist attacks might be performed by internal and/ or external groups. In the event of internal actions, the persons involved are expected to have a good knowledge of the site and the systems and have access to protected areas. A related subject to the security of decommissioned nuclear submarines is the security of the nuclear icebreakers in homeport, in the open sea or in other ports, for which scenarios have been developed in a Scandinavian-Russian cooperation to identify a design-basis threat for the protective systems. The protection system and response forces have been designed accordingly to the identified threat.

Some information on the conditions of the spent fuel and radioactive waste have been released as the Russian pledge for international assistance to upgrade the infrastructure at former naval military sites, for example in Gremikha and Andrejeva Bay, has grown stronger. This information, however, represents far too little of the information necessary to complete proper risk assessments, including other initiating events issues than those related to safety. As the international assistance continues and increases, there is a need for transparent assessments of risks based on open source information on these issues in order to minimize the risk for all types of accidents and incidents which could affect the environment and the regional and social stability and the international support for accelerating the solution of the problems.

4. Risk reduction and barriers for incidents

As stated in this paper, important scenarios have not been adequately assessed in the efforts to map the risks associated with the dismantling of Russian submarines. This part will address the barriers for such incidents to happen, however, with a continued focus on what has been called security-related threats in the introduction. Important elements in this picture are the physical elements around the material and the facilities, often described as material control, protection and accountancy. In this listing, also the regulatory part and the overall policy issue will be mentioned. First, however, this paper will touch upon the soft core of the issue; further international assessments to address all possible threats systematically and prioritise the available resources.

4.1. THE SOFT CORE: FURTHER RISK ASSESSMENTS

Several studies focusing on Russian marine reactors have used the safety report for the cargo ship *Sevmorput* as the basis for their models of Russian marine reactors [13]. This report constitutes the most comprehensive report on the details of Russian marine reactors. However, this study does not contain details on the reactor and fuel configuration. In addition, no analysis has been performed of the relevance of these data when considering Russian military reactors. Another relevant example on studies of Russian marine reactors is [9]. This is one of the most substantial sources on technical data on older design of Russian military vessels. This report also contains a detailed description of the reactor in the icebreaker Lenin. Most of the other relevant publications have only focused on how the different data are important in a policy context, and have not made any scientific analysis to consider their relevance and accurateness.

In order to address the risk in more detail, the relevant Russian authorities have to initiate an evaluation on how to address the risks and the response of spent marine fuel without compromising what traditionally have been considered military secrets. The issue has to be raised in order to assure the necessary level of transparency in the efforts to increase the rate of dismantling, taking all relevant factors into account, including the international need for access and independent analysis as a prerequisite for assistance. The current Russian attitude has already delayed international efforts to address the potential proliferation risk associated with the spent marine fuel, and remedial efforts at former Russian operative naval bases such as Gremikha and Andrejeva Bay. In other contexts, however, also involving naval fuel, the issue of increased transparency has been emphasised [13].

4.2. THE PHYSICAL LAYER: OVERSIGHT AND PROTECTIVE SYSTEMS

Both for proliferation of nuclear materials and possible deliberate acts to release radioactive material, oversight and adequate protection are of vital importance. Today, the responsibility for the sites and the spent nuclear fuel has been transferred from the Ministry of Defence to Minatom. In order to fulfil the tasks Minatom has set up two organisations, SevRao and DelRao in 2000. The current situation is that there has been no substantial upgrading of the physical protection and accountancy systems at sites storing spent naval fuel in Northwest-Russia.

The general problems affecting the physical protection layer in Russia have been extensively discussed, especially in the US which operates the largest programs in this area. The conclusions in for example in [14, 15, 16] state that there is a need for improved long-term efforts especially when it comes to accountancy in the Russian-American cooperation. The lack of sustainability, i.e. the indigenous ability to run, operate and maintain the installations after the equipment has been installed and the system commissioned, has also been specifically addressed in the US [15]. This is especially important when it comes to the dismantling of Russian submarines since this activity will continue for many years to solve the current backlog. The lack of adequate systems at sites in Northwest-Russia has been addressed in bilateral meetings between Russian, Swedish and Norwegian authorities [17].

4.3. THE SOCIETAL LAYER: REGULATOR'S ROLE AND SOCIAL ISSUES

The surrounding social environment in which the dismantling process will take place influences the overall risk. Several incidents of diversion of nuclear material have occurred in Northwest-Russia. For the people involved, the future prospects for the Russian navy and the local sites might be considered to have played an important role. Northwest-Russia is experiencing extensive changes in the military presence. This is especially relevant for cities and villages close to naval sites with decommissioned vessels, spent naval fuel or radioactive waste. This development will probably continue for several years ahead. To a certain extent this influences the overall policy. For example, the economic incentives are important when the Russian government decides to allow import of large amounts of foreign spent fuel.

There has been confusion on how Russia plans to regulate the area of decommissioning nuclear submarines. The responsibility for the decommissioned submarines was transferred to Minatom in 1998, along with the responsibility for the waste and the spent fuel. The regulatory authority, Gosatomnadzor (GAN), understood the development concerning decommissioned submarines in such a way that the organization obtained a extensive regulatory responsibility in this area, however, this was not in line with the opinions in Minatom. While the Ministry of Defence has agreed to the fact that this should be kept outside its military responsibility and under civilian control, the role of the civilian nuclear regulatory authority Gosatomnadzor and other similar agencies seems unclear, at least when it comes to inspecting the sites. The preliminary solution is that GAN is not conducting inspections at military installations. If military material is present at civilian facilities, the facility itself is responsible for making a plan for how and when inspections might be conducted [18]. Lately, the military GAN inspection authority has been responsible for inspections and licensing procedures at military sites including also sites transferred to the responsibility of Minatom. GAN, and other Russian authorities such as, for example, the health inspectorate Medbioextrem, will at least be responsible for developing and setting requirements. Due to the volatile relationship between the Russian operator of these sites and the regulatory authority, the actual situation as of today might be changed.

5. Assistance programs

Considering the different initiatives to address the proliferation issues and other security related threats associated with the dismantling of Russian strategic nuclear submarines, the number of actors is limited. The different actors involved are also considering the issue from several different angles. No country has launched a specific initiative to address the proliferation risk of the spent fuel, however, several countries are about to offer assistance on issues like material accountancy, physical protection and regulatory development.

Another type of assistance also relevant for reducing security-related threats, are the assistance programs aiming at reducing the overall environmental risk associated with the dismantling program. A brief description might include actors like Norway, Sweden and EU in particular. Especially Norway has engaged itself in this work. Among the most relevant projects are the cleaning-up of Andrejeva Bay, including only physical protection at this stage, construction of a ship for transport of spent fuel and radioactive waste from naval sites to railway lines, and, already completed, the construction of four railway cars for transport of spent fuel. Also the EU is about to launch a Northern Dimension Environmental Program under the auspices of EBRD, which will include a so-called nuclear window. At this stage, this seems to be in the order of 300 million Euro, however, the details of this initiative have not been made public yet. Other, more general initiatives such as the EU's TACIS program include assistance both to the national operator of installations with fissile material and to the regulatory authorities; however, this type of general assistance will not be described in this paper.

5.1. NORDIC AND BRISTISH SECURITY-RELATED ASSISTANCE

Norwegian and Swedish authorities have cooperated on reducing the proliferation risks from Russian fresh and spent marine nuclear fuel at the Murmansk Shipping Company (MSCo) for several years. UK has also participated in the assistance efforts and is likely to continue the cooperation with the Scandinavian countries when it comes to efforts at former military naval sites in Northwest-Russia. This is, however, dependant on the signing of bilateral or multilateral agreement covering, for example, tax exemption and the necessary liability clauses.

The most substantial project on the ground has been to strengthen the control and physical protection of fissile material from the Russian fleet of nuclear-powered civilian vessels operated by Murmansk Shipping Company (MSCo) to prevent unlawful diversion, radiological consequences and sabotage. The stand alone physical protection system on the nuclear-powered icebreaker *Sovjetsky Sojuz* consists of both organisational measures such as rules, schedules, programs, personnel training procedures and maintenance measures, and mechanical and tele-technical protection equipment such as reinforced doors, locks, sensors, alarm system and closed circuit television systems. Hardware and support were also parts of the delivery. The Swedish Nuclear Inspectorate (SKI) has, together with British authorities, the Department of Trade and Industry, completed a similar project at the cargo ship *Sevmorput*. As of this year the physical protection system on two additional icebreakers, *Yamal* and *Russia*, will be upgraded.

These experiences might be valuable for the implementation of similar measures. Possible areas include the State System for Accountancy and Control in order to give the supervisory authority a full overview of the material at all stages during transportation, unloading, handling, storage etc. of spent nuclear fuel. At different stages the responsible operating organisation might change, and the full overview is important so nothing "falls in between rocks". The shipyards, where the dismantling activities actually take place, must have a reliable registration system that is compatible with the GAN system.

Another important co-operation area is the support and joint project activities with the GAN. Here, the Nordic countries have assisted in four main areas:

- Development of laws, regulations and requirements: As of December 2001, 15 regulations were developed and 9 put into force by GAN;

- Technical assistance: equipment, Cerenkov viewing devices (CVD), computer hardware and software, development of computer programmes, and other physical protection related equipments. This includes a custom-made electronic accountancy system for fresh and spent fuel installed at MSCo in February 1998;
- Training: administrative aspects, management, quality assurance and technical. A concrete programme has been the training of GAN inspectors in handling and use of CVD, in Sweden. These inspectors will then have the responsibility for the training of other GAN employees in Russia;
- Physical protection: upgrading and installation of PP systems at concrete objects and sites;

In addition, the SKI has completed several projects to improve the general competence of the regulatory authority in the field of nonproliferation both in Russia and the other states in the former Soviet Union [19].

5.2. US ASSISTANCE PROGRAMS AND INITIATIVES

The Cooperative Threat Reduction (CTR) Program, a targeted assistance program for completing the implementation of START I in Russia in order to secure the continuation of the disarmament work and the prevention of the spread of nuclear weapons will soon be completed. The CTR Program has decided to allow their equipment to be used for the dismantlement of vessels other than the SSBNs now that these have been taken care of. However, there are gaps in the support of the operation of this equipment that are important to address. The dismantling of SSNs and the strategic cruise-missile nuclear submarines, has not been the subject of any U.S initiatives due to the non-strategic nature of the weapons and the difficulty of conveying to Congress the proliferation threat posed by the spent fuel. An environmental threat has not been sufficient. However, in order to explore the possibility, the US Congress' took an initiative in a bill called "Cross Border Cooperation and Environmental safety in Northern Europe Act of 2000", requesting an evaluation of the environmental risks associated with the dismantling of SSN's. The result indicated no shift in the U.S. policy on this issue [19]. However, the U.S. had earlier attempted to check out the non-proliferation relevance of some projects which were assisting Russia in the dismantling of SSN's. These included, for example, the railway cars for spent fuel financed as part of the Norwegian assistance program for nuclear issues in Russia.

Another U.S. actor has been the MPC & A Program administrated by the US DoE. Among its goals has been to improve US national security by reducing the risk for proliferation of fissile material in Russia. However, spent Russian naval nuclear fuel has been included in the MPC and A Program. GAN was included in the program because several civilian institutes with weapons material were identified, and consistent regulations were important for efficient implementation at each site. The DoE cooperation with GAN has, until recently, been focusing on development of regulatory documents, design of a federal material control and accountancy system and equipment and training for inspectors.

6. Conclusions and recommendations

The current ambiguity is only giving away valuable time to a possible proliferator, and making it more difficult to achieve international funding and assistance. There are no legal obligations for Russia as a nuclear weapon state (NWS) for making declarations or allowing international cóntrol of any type of nuclear material beyond which is given in its national legislation regarding the production, use and storage of nuclear materials.

Spent nuclear fuel represent a proliferation risk as important as the fresh fuel, dependant on the amount and quality of the fissile material in the spent fuel and its radiation levels. Some of the vessels are known to have used fuel initially enriched to 90% and the spent fuel will subsequently contain large amounts of HEU. There may be much material of high proliferation risk in the back end of the Russian submarine cycle. The fuel itself, however, will often be protected by the reactor pressure vessel, the compartment and the hull. The materials most easy available are probably at the different bases. At the same time, there is little or no protection at the sites, there is no central or regional accountancy system for spent naval fuel, and there is no central policy describing how to avoid proliferation of spent fuel and submarine technology. Together with a continuous struggle between several Russian nuclear authorities on administrative and regulatory issues regarding spent military naval fuel described in this paper, the national ability to account for and control spent fuel of any enrichment and radiation level is questionable and a proliferation concern.

Despite numerous attempts from others, including the US authorities, very little information on the vessels, their fuel and their protective systems has been released, either as part of an on-going cooperation or publicly. In order to have a more precise description of the situation on the ground, and to initiate an international cooperation to solve the current problems, the Russian authorities must take an initiative to address the transparency problem. This should be possible without compromising their need for secrecy. In order to turn this situation around as fast as possible and establish a transparent, reliable and efficient nonproliferation regime for the Russian submarine dismantling process, the following measures must be initiated and implemented as soon as possible:

- Facilitate continued use of dismantling technology after the specified number of SSBN's has dismantled; The US has supported Russia with important equipment that must be kept at the dismantling sites in Russia to facilitate an efficient process in the future. This is an urgent needed measure as the current work to complete the provisions in START II is drawing to a conclusion.

- Conduct an international assessment: Initiate and contribute resources to an international assessment of the material attractiveness of Russian spent naval fuel. The US must play an important role in this work and should provide experience and the preliminary results from the cooperation with Russia describing the proliferation risk associated with the dismantling of SSGN's and SSN's.

- Use an integrated approach addressing both environmental and security needs: Today the co-ordination of environmental and security upgrading measures is left to bilateral meetings between the different actors, or to the different Russian local actors. For full transparency, better co-ordination and increased effectiveness of ongoing activities, Russian authorities should take the lead in getting an integrated approach underway.

- Establish plan for improving MPC & A for spent naval fuel, including regulatory issues: Today there are either no systems or inadequate systems. Russian authorities have to establish plans for improving MPC & A for spent naval fuel. Accountancy systems have to be in operation at all dismantling sites and within all responsible Russian organisations. Compatibility between the systems should be achieved. It will be important with parallel activities. Laws and regulations should be developed at the same time as establishment of accountancy systems and implementation of technical measures.

Based on the Scandinavian engagement in both environmental and security issues in the Northwest-Russia and their overall regional competence and the US efforts for strategic disarmament in the same area, contact and cooperation between representatives from these countries might facilitate insight and creative solutions for future efforts.

7. References

1. IAEA (2000) Guidance and considerations for the implementation of INFCIRC/225/Rev. 4, The Physical Protection of Nuclear Material and Nuclear Facilities, IAEA-TECDOC-987 (Rev 1), International Atomic Energy Agency, Vienna.
2. PIR Center, *Yaderny Control*, , spring 1996, p. 16.
3. Moltz, C. and Robinson, T. (1999) Dismantling Russia's Nuclear Subs: New Challenge to Nonproliferation, *Arms Control Today*, June 1999, pp. 10-15.
4. Bragin, V. and Ong, L. (1992) Radiation Levels Associated With Nuclear Reactor Spent Fuel, STR-287, International Atomic Energy Agency, Vienna.
5. National Academy of Sciences (2000) The Spent-Fuel Standard for Disposition of Excess Weapon Plutonium – Application to Current DOE Options (prepublication copy), National Academy Press, Washington, DC.

6. Reistad, O. and Gussgard, K. (2000) An estimate of the amounts of ^{235}U, ^{239}PU and the material attractiveness in naval irradiated nuclear fuel from the first and second generation of Russian submarines, 41st INMM Meeting in New Orleans, International Nuclear Materials Management, 2000.
7. O. Reistad, K. Gussgard (2001) Russian spent marine fuel as a global security risk. Conference proceedings (in press). International Conference on Security of Nuclear Material – Measures to Prevent, Intercept and Respond to illicit Uses of Nuclear Material and Radioactive Sources, International Atomic Energy Agency (IAEA), Stockholm , 7-11 May 2001.
8. Chunyan, M. (2000) The Review of Submarine Proliferation (unpublished – received September 2000).
9. IAEA (1997) International Arctic Seas Assessment Project - Radionuclide release from marine reactors dumped in the Kara Sea. IAEA-TECDOC- 837, International Atomic Energy Agency, Vienna.
10. North Atlantic Treaty Organisation (1998) Cross-Border Environmental Problems Emanating from Defence-Related Installations and Activities. Pilot Study - Phase II: 1995-98 Environmental Risk Assessment for Two Defence-Related Problems. North Atlantic Treaty Organization, Report No. 227, Oslo.
11. Lisovsky, I. V. (2000) The Radioecological Risk of Decommissioning of Nuclear Submarines – Possible Accidents and Normal Conditions, paper P-5-323 in International Radiation Protection Association International Congress 10, Hiroshima, 2000 Harmonization of radiation, human life and the ecosystem : Proceedings : IRPA – 10, Hiroshima, CD-Rom.
12. Ølgaard, P. (2001) NKS-SBA-1 Status Report: The Potential Risks from Russian Nuclear Ships, Nordic Nuclear Safety Research, Roskilde.
13. Safety report of Sevmorput - *Information of Safety of Icebreaker - Transport lighter-container-ship with nuclear propulsion plant Sevmorput.* Approved by the Register of Shipping of the USSR (unpublished – received 1992, available copies at Norwegian Radiation Protection Authority (www.nrpa.no)).
14. Doyle, J (1998) Improving Nuclear Materials Security In the Former Soviet Union: Next Steps for the MPC&A Program, *Arms Control Today*, March 1998, pp. 12-18.
15. Potter, W. and Wehling, F. (2000) Sustainability: A Vital Component of Nuclear Material Security in Russia, *The Nonproliferation Review*, 7 (Spring 2000) (http://cns.miis.edu/pubs/npr/vol07/71/potweh71.htm).
16. Bunn, Matthew (2000) The Next Wave: Urgently Needed New Steps to Control Warheads and Fissile Material. Carnegie Endowment for International Peace and Harvard Project on Managing the Atom, Washington DC.

17. Protocol from the 8th coordination meeting between Gosatomnadzor, Swedish Nuclear Inspectorate and Norwegian Radiation Protection Authority in St. Petersburg, January 2002 (unpublished – received January 2000).
18. Centre for Nonproliferation Studies (2000) Study of Minatom's organizational structure, (draft report), Monterey, US (unpublished – received February 2000).
19. US State Department (2001) Environmental Security Threats From Decommissioned Russian Marine Nuclear Reactors and Associated Spent Nuclear Fuel, Radioactive Waste, and Contamination - Report from US State Department (submitted to the U.S. Congress Pursuant to U.S. Public Law 106-255), October 2001.

EXPERIENCE OF INTERNATIONAL COOPERATION AND UNRESOLVED ISSUES OF UTILIZING NUCLEAR SUBMARINES AND NUCLEAR MAINTENANCE SUPPORT VESSELS IN THE FAR EAST REGION OF RUSSIA

YURY SHULGAN
Far East Plant (Zvezda)
Bolshoy Kamen', Russia

The Far East Plant (FEP) "Zvezda" began Nuclear Submarine (NS) utilization (i.e. decommissioning and distmantlement) operations in 1989. During the following 22 years 25 NSs including 20 strategic missile NSs that makes up 90% of all already utilized NSs owned by the Pacific Navy were utilized. Because of this about 1/3 of the decommissioned Pacific NSs no longer constitute nuclear, radiation or environmental threat.

These activities required considerable efforts and important financing. Since 1994 work aimed at upgrading the infrastructure of NS hull dismantling operations has been performed at FEP "Zvezda" thanks mostly to funds of the US Cooperative Threat Reduction (CTR) Program. As a result of these activities a unique complex was built at the Russian Far East. The complex includes:

1. A guillotine of 2000-t capacity to perform cutting of large- and medium-width vessel structures including NS strong hull.
2. A concrete pad and a frame crane to ensure the guillotine operation.
3. Excavating machines with mounted cutting tools for medium-width vessel structure cutting.
4. A packing press to perform pressing & packaging of thin-sheet vessel structures.
5. A transporter and casks for scrap metal loading and transportation.
6. Gas-cutting tools and equipment to perform gas-, plasma- and gasoline cutting.
7. Equipment and tools to ensure mechanical air cutting.
8. Frame cranes and truck cranes of 32-t through 80-t capacity.
9. Different-type trucks & trailers.
10. A facility to process NS dismantled electric cables.

All this equipment taken together with vessel elevators and slips of the plant make it possible to utilize over four NSs per year.

It is widely known that initially NS utilization operations were accompanied by low-level Radioactive Waste (RW) disposal into the sea. This was due to lack of an appropriate infrastructure in the Russian Far East able to ensure RW temporary storage, treatment and disposal. Actually the needed infrastructure to handle RWs

A.A. Sarkisov and L.G. LeSage (eds.),
Remaining Issues in the Decommissioning of Nuclear Powered Vessels, 91–95.

and, then, store temporarily the derived Solid RW (SRW) is constructed. It is worthy of notice that all these RW managing facilities are located at FEP "Zvezda".

The considered RW handling infrastructure consists of:

1. Floating complex "Suzeron" to process low-level Liquid Radioactive Wastes (LRW) provided by the Japanese Government within the framework of the Program of Eliminating Nuclear Weapons to Be Reduced in the Russian Federation. The complex processes low-level (i.e. up to $1 \cdot 10^{-5}$ Ci/l) LRW of integral concentrations up to 20000 mg/l. These LRWs comprise: -up to 17000 mg/l of chlorides, -up to 1000 mg/l of nitrates, -up to 500 mg/l of oxalates, -up to 200 mg/l of sulfates, phosphates and hydrazine hydrates, -up to 150 mg/l of ammonia and -up to 5 mg/l of mineral oils.

The Environmental Legislation actually in force in the Russian Federation (RF) prohibits waste dumping into the aquatic environment which is polluted/contaminated over the limits fixed by The Hygienic Standards of Hazardous Substance Admissible Concentrations in Water Medium. Thanks to LRW processing at the "Suzeron" floating complex, the following characteristics for treated/decontaminated wastewaters are reached:

a) Radionuclide concentrations are below the intervention levels for drinking water (i.e. admissible specific activities) fixed in Russian "Radiation Safety Standards (RSS)-99":
   - for 6 radionuclides by 3 orders of magnitude;
   - for 4 radionuclides by 2 orders of magnitude;
   - for 5 radionuclides by an order of magnitude.

b) Hazardous chemical admixture concentrations are below the admissible limits determined by The Hygienic Standards for Fish-Industry Reservoirs (such as, e.g. Ussury Bay of the Japan Sea) and below the admissible limits fixed by the State Drinking Water Standard:
   - for 2 components by 5 orders of magnitude;
   - for 1 component by 3 to 4 orders of magnitude;
   - for 4 components by 1 to 2 orders of magnitude.

The complex capacity exceeds 2 000 m^3 of LRW/year.

2. Another new RW handling facility is a SRW repository of 1500 m^3 capacity to store solidified concentrates after LRW processing at the "Suzeron" floating complex as well as conditioned LRW produced when laundering working clothes and processing SRW.

3. The third new installation is a low-level LRW & SRW processing complex that comprises workcells for laundry LRW decontamination and those for SRW sorting, cutting and pressing.

Both facilities possess decontamination centers, laboratories, systems of power supply, physical protection, radiation & technological control and environmental monitoring. Note that these installations were constructed using funds of the US CTR Program.

All the above-listed facilities taken together with the existing infrastructure of RW control, collection, interim storage and transfer for processing (laboratories, coastal and floating tanks, pipeline systems and special trucks) constitute a unique RW managing complex in the Russian Far East. FEP "Zvezda" is the only plant that ensures LRW decontamination/processing down to levels, which satisfy the most strict waste disposal standards, and the only complex having licenses of the RF Gosatomnadzor. It is also worth noting that the laboratories of FEP "Zvezda" are accredited in the RF State Standard Committee.

Actually Spent Nuclear Fuel (SNF) is unloaded from NS reactors by the Russian Navy units using a Floating Servicing Facility (FSF), and the whole SNF handling flowsheet represents a rather sophisticated multi-step procedure. Thus, first, SNF is unloaded from NS reactors and put into FSF storage tanks, then, when full, FSF moors to a pier within the naval base where empty Transportation Casks (TC) of TK-18-type are already available. Next, these TCs are submerged by a floating crane into a special area within FSF wherein SNF is loaded into TCs. Thereafter full TCs are unloaded by the same floating crane and placed at a trailer moving to the Coastal Servicing Enterprise (CSE) owned by RF Ministry of Atomic Energy (Minatom). It is here that packages containing SNF are taken out and transported to a coastal SNF repository. As a special SNF transportation train arrives, SNF packages are removed from the coastal repository, placed again into TCs and, finally, transported by trailers to a special site of their loading into the train. When the procedure is completed, the train departs for the Ural Region at PA "Mayak". It is obvious that every SNF handling operation is potentially hazardous from the radiation contamination standpoint. Moreover, such a SNF management flowsheet necessitates much time and restrains the whole NS utilization process.

Actually using funds of the US CTR Program the construction of a coastal facility is being completed within the "Zvezda" site to unload SNF from reactors of NSs to be utilized. The coastal facility consists of the following installations:

- a sector of SNF loading into TCs,
- two interim-storage repositories for empty & full TCs of 30-unit capacity,
- a site for TCs loading into a special SNF transporting train,
- rail tracks for the train makeup and
- a railway section up to the railroad station at Bolshoy Kamen' owned by the Far East Railroad Enterprise of the RF Ministry of Communications.

By now the facility installations are completed. When the facility is commissioned, all potential radiation-dangerous SNF handling operations and the period of time from SNF unloading from NS reactors to its loading into a special train transporting SNF for PA "Mayak" will be considerably reduced. Thus, SNF will be unloaded from NS reactors by a gantry crane and loaded immediately into TCs within the loading sector at the coastal facility. Using electric trolleys TCs will be removed from the loading sector building and reloaded to a trailer by truck cranes, and, next, reloaded once more by the same crane into a TC interim storage repository. Upon SNF transporting train arrival, TCs will be loaded directly into the train railcars by truck cranes and transported from FEP "Zvezda" to PA "Mayak".

Thus, thanks to international cooperation and financial support provided by the USA and Japan, a multipurpose NS utilization infrastructure comprising the 3 following unique complexes has been created at FEP "Zvezda":

1. A complex to perform SNF unloading, interim storage and sending for reprocessing.
2. A complex for RW collection, temporary storage, transportation & processing and also for temporary storage of the derived SRW.
3. A complex of vessel salvaging and NS hull dismantling & cutting.

The created infrastructure taken together with the accumulated technological experience of NS utilization imparts the key function in multi-purpose utilization of the Pacific Navy NS to the FEP "Zvezda" in the Russian Far East.

As it was mentioned above, 20 NSs (out of 25 units already utilized by FEP "Zvezda") were equipped with ballistic missiles and were to be utilized within the framework of the CTR Program in keeping with the Strategic Offensive Arms Elimination (SOAE) Agreement and the Strategic Arms Reduction Treaty (START). These NSs, which had relatively short operational periods (about 25 years), were transferred to utilization without long-term waterborne storage. As a consequence, they possessed operable systems of the above-water floodability, fire- and explosion safety, nuclear & radiation safety and a satisfactory hull technical status. The utilization process for already decommissioned Strategic Ballistic Missile NSs (SBMNS) is to be completed in 2002 or 2003.

Further lowering/elimination of potential nuclear and environmental hazards in the Pacific area depends on utilization (down to scrap metal) of the decommissioned multi-purpose first-generation NSs. NSs of this type were built 30 through 40 or even more years ago. By now over 20 decommissioned multi-purpose NSs containing SNF are stored afloat over 10 years without docking. All the considered NSs have unsealed ballast tanks, low buoyancy and, thus, cannot be transported to SNF unloading and utilization sites unless special vehicles are used. Moreover, if unloading SNF in afloat conditions, the design requirements to lurch & trim of such NSs and their Power Reactor Facilities (PRF) are not observed. This means that because of the poor technical status of most such-type NSs neither nuclear nor radiation safety is ensured when performing potentially dangerous (from the standpoint of nuclear & radiation safety) SNF unloading operations and their related actions.

In this context a set of upgrading proposals was developed by FEP "Zvezda" leaders and forwarded to RF Minatom (appointed a customer-coordinator for multi-purpose NS utilization in Russia); thereafter these proposals are to be presented to the Japanese Government. They include:

- salvaging and hauling of multi-purpose NSs to FEP "Zvezda" using special transport-transfer floating docks;
- unloading SNF from the concerned NS reactors after NS docking and using the available coastal complex facilities; and
- moving NSs to the plant slips to perform hull-cutting operations (down to scrap metal).

Actually, in addition to NSs of poor technical status, for which SNF cannot be unloaded, the Pacific Navy has to handle three NSs containing damaged PRF. In this case, because of extremely high dose rates within the reactor compartments of two such NSs, normal unloading of nuclear fuel is impossible: e.g., when staying only about 4 minutes in any reactor compartment room, one can get maximum permissible annual exposure dose.

To utilize these NSs, a detail design known by the name "Sarcophagus" was developed by Central Design Office (the considered NS designer). Its main points consisted in reactor compartment cutting out and, next, placing into a floating sarcophagus made of SBMNS hulls. However extremely high dose loads for the involved personnel did not comply with the so-called "well-grounded exposure" principle. In keeping with this principle any activity related to a radiation source handling is inadmissible if the risk of harm due to the radiation exposure exceeds potential benefit for the society.

Actually no work is underway on the "Sarcophagus" Project, and to solve issues of utilizing NSs with damaged PRF, technical proposals "Fakel" ("Flare") and "Ukrytie" ("Shelter") were developed. Their main point consists in NS salvaging and, next, placing in a dry site/floating dock followed by insulation of the whole NS hull without cutting out the reactor compartments with damaged PRF. Unfortunately, at present, these technical proposals are not developed further.

When considering NS utilization issues, one also has to consider the decommissioned Nuclear Maintenance Support Vessels (NMSV) that performed NS nuclear fuel loading/unloading operations. This category of nuclear vessels comprises: -Technical Pouring Tankers (TPT) that received, stored temporarily and transported LRW from NSs and -Floating Servicing Facilities (FSF) which performed NS nuclear fuel replacement.

The procedure of NMSV utilization is principally different from that of NSs (except for NSs with damaged PRF) because of the following reasons. In the NS case the ionizing radiation sources are confined within the reactor compartments, whereas in the case of NMSV the whole vessel represents a source of ionizing radiation with the only exception being living & service rooms. It is significant that not only such NMSV structures that can be potentially decontaminated should be considered as SRWs. Pipelines, fittings, pumps and other equipment of sophisticated configurations, inner voids and other large structures also fall into this category, and both procedures & techniques for their decontamination necessitate further development.

One should be aware that in all the above-considered cases one has to deal with an increasing, with time, probability of initiating accidents with environmental consequences due to failures in ionizing radiation source management. And it is only an efficient international cooperation that is able to eliminate such a threat.

INTERNATIONAL CO-OPERATION FOR NUCLEAR SUBMARINE DISPOSAL: CURRENT RESULTS AND FUTURE PROSPECTS

N. KALISTRATOV
Zvezdochka Shipyard

Zvezdochka (Federal State Unitary Enterprise – Zvezdochka State Engineering Enterprise) of Severodvinsk, a part of the Russian State Center for Nuclear Shipbuilding, stands among Russia's major businesses involved in the disposal of strategic nuclear-powered submarines.

Indeed, Zvezdochka has mastered actually all submarine disposal process options, with all the operations performed in conformance to the basic process developed by Onega (State Unitary Enterprise – Onega Research, Development Technological Bureau) – the lead institution for Submarine disposal process designing integral to Zvezdochka.

At this time, Zvezdochka has implemented the following:

- missile compartments have been cut out in 29 nuclear powered submarines (NPSs);
- 16 submarines have been disposed.

In order to enable comprehensive disposal of nuclear submarines Zvezdochka is provided with the required structures and facilities, including covered slipways, a docking basin and special purpose quay with a 475 m long moorage wall.

Furthermore, relevant infrastructure facilities are in place to make possible the works that involve handling, storing or conversion of radioactive waste, including a special purpose tanker ship, three on-shore liquid radioactive waste storage facilities (1,200 m^3), permanent storage facilities and temporary sites for solid radioactive waste storing.

Measures have been taken at Zvezdochka to establish the infrastructure required to enable unloading of spent nuclear fuel, specifically:

- all-purpose test-beds have been constructed and commissioned for post-overhaul testing of reloading equipment on submarines of all generations;
- the reloading facilities OK-300PB #.5 and OK-300PBM have been overhauled;
- PM-125 and PM-63 Tenders have been overhauled.

At the same time, however, there are certain problems specific to every shipbuilding business in Russia. A removal of those problems would substantially improve the efficiency of submarine disposal operations and save national budget expenditure.

A decisive factor in the implementation of the nuclear submarine disposal program is the assistance rendered by the United States in financing the operations related to disposal of strategic nuclear submarines decommissioned by the Russian Navy and establishment of the infrastructure that would ensure nuclear and/or radiation safety of the process. The assistance is provided under the Russian-US Intergovernmental Agreement for Cooperative Threat Reduction (Nunn-Lugar Program).

A.A. Sarkisov and L.G. LeSage (eds.),
Remaining Issues in the Decommissioning of Nuclear Powered Vessels, 97–101.

More specifically, the following was made possible owing to the US-supplied funds:

- 5 strategic nuclear-powered submarines were disposed (including two Delta-1, one Delta-2 and two Delta-3 Submarines);
- a workshop with an automated cable-processing line was constructed;
- three special-purpose sites were constructed to enable cutting submarine hull structures, including a Harris press concrete-bed site with access and sub-crane ways, centralized oxygen, propane-butane distribution and relevant load-lifting equipment;
- a low-radiation waste processing plant was constructed at Facility # 160/161;
- process scaffolding and containers for metal scrap and dismantled equipment were manufactured;
- Navy-owned equipment providing for unloading of spent nuclear fuel from NPSs being disposed was overhauled;
- establishment of infrastructure required to support unloading of spent fuel from submarine reactors is nearing completion within the spent nuclear fuel unloading base and the spent fuel container storage and special train reloading site.

In addition to the above contracts, US companies supplied Zvezdochka with high efficiency equipment for hull cutting and re-processing of the products of disposal. The implementation of those contracts enabled Zvezdochka to actually establish a regional nuclear submarine disposal center offering the most comprehensive disposal cycle. Incidentally, the standard rated slipway period of six months has been reduced at Zvezdochka to as little as only 2.5-3 months. A spent nuclear fuel reloading and special train loading station (for trains bound for "Mayak") is currently operational within the water area managed by Zvezdochka at the White Sea Naval Base port. Maintenance and uninterrupted operation of the described infrastructure facilities were made possible due to financing gained through the mentioned contracts.

Contract # 4400019409 of November 18, 1999 with SAIC (Science Applications International Corporation) of the United States for designing and manufacturing of one hundred transportation containers made of steel for solid low radiation waste within the framework of the international AMEC (Arctic Military Environmental Cooperation) Program was completed by Zvezdochka on April 28, 2000.

Another Contract with SAIC Company (# 44000435416 of March 7, 2000) that involved certification of the mentioned lot of 100 containers was completed, as well.

Finally, one more contract was signed at the end January of 2001 with SAIC to manufacture 300 SRW transportation containers within an estimated contract term of eight months.

Norway's Kvaerner Maritime A.S. and Zvezdochka performed the following works under the Russian-Norwegian Co-operation Program:

1. Contract # 610/26-14 of May 6, 1998 "Agreement on Implemented Cooperation for Overhaul of LLRW Tanks, Pipelines at Zvezdochka of Severodvinsk", with the overall Contract value of 3 million USD. Zvezdochka performed NKr 10,056,019 worth of works under the Contract.

The following equipment or facilities were involved:

- the building of Facility # 159 (special chemical water treatment) was overhauled;
- the layout of the decontamination station was changed;

- tanks A-01 and AO-03 with integral measuring instrumentation equipment, piping and valves were overhauled.

The overhaul operations were completed on September 6, 1999 in compliance with the Contract.

2. Contract # 2398 of September 30, 1999 "Overhaul of Tanks A-02/1, A-02/2 at Liquid Radioactive Waste Storage Facility (Facility # 159), Zvezdochka".

A review of the national budget allocation progress proves the financing of submarine disposal operations is unlikely to improve in the coming years. By far more dramatic to the Russian Federation is the issue related to nuclear-powered multi-purpose submarines, especially considering their greater total numbers and earlier decommissioning dates. The Russian Navy has decommissioned over 80 NPSs (mainly, attack Subs) with SNF being still unloaded. While the technical condition of most submarines should be described as unsatisfactory, the condition of quite many is critical. The described situation poses apparent nuclear, radiation and environmental threats both to the Russian Federation and the Arctic nations. Adequate management of the problem, through bringing the decommissioned submarines to a nuclear safety standard followed by proper disposal involves complex technical solutions and immense costs unbearable by the state budget of this country at this time.

The US Administration is reviewing the issue of financing the program for nuclear attack submarine disposal. A matter of concern to the United States is that the nuclear submarines of that class may be sold to third countries or refurbished into carriers of strategic nuclear weapons. To stimulate an applicable positive decision on behalf of the United States the joint Russian-US commission is working to produce a relevant report to the US Congress. In our view, the implementation of the described program should be a matter of special attention of both the Ministry for Atomic Energy and the *Rossudostroyeniye* Shipbuilding Association of Russia.

The Russian-Norwegian Agreement dated May 26, 1998 provides for construction of a container carrier ship designed to transport spent nuclear fuel and/or radioactive waste produced by disposal of Russia's Nuclear Powered Submarines decommissioned by the Russian Navy in the northern regions of Russia.

Funding required for the container ship construction is supplied by Norway, Canada and the United Kingdom. The Norwegian party is currently selecting the vessel construction project management company, following which bids will be invited within an international container ship construction tender. Once commissioned, the new container carrier would contribute to better management of the environmental issues in the North-West of Europe.

Signing the contract with STORVIK Zvezdochka Norway A.S. including subcontracted works by Zvezdochka for construction, testing and commissioning of the Mobile Unit for Solid Radioactive Waste Treatment under the AMEC Program is scheduled for end-January of 2001. The target date for completion of the works is early 2002.

Standing among the most vital problems emerging from nuclear submarine disposal operations is the issue of handling submarine reactor compartments that require specified approaches and definition of certain process and/or technical solutions that would assure radiation and environmental safety at every phase of the reactor compartment handling operations.

Triple-compartment units stored afloat for fifty or more years are exposed to corrosion, ice or wind impacts and require continuous maintenance similar to any different floating structure. The transition to single-compartment disposal method would dramatically improve the environmental situation in the North-West of Europe. This would require construction of a temporary storage site for the reactor compartments for eventual transition a to single-compartment nuclear submarine disposal process and eliminate the need to store the reactor compartments in the coastal waters of the North-Western region, i.e., remove the situation fraught with unpredictable consequences.

As the proposals were reviewed with respect to the establishment of a temporary storage site for submarine reactor compartments, this enterprise outlined a project option related to construction of infrastructure facilities for temporary storage of reactor compartments.

Another serious problem faced by Zvezdochka today is the refurbishment of the combustible solid radioactive waste incinerator shop. The existing incinerator plant designed by the NIIKHIMMASH R&D Institute for Chemical Engineering of Yekaterinburg has been in operation since 1983.

The plant provides for solid waste volume reduction of 20-80 times during incineration. The refurbishment job statement implies maintenance of the furnace, smoke circuit, service systems and replacement of the control and measuring instrumentation.

In 1999-2000 Zvezdochka was involved in negotiations with the representatives of the Atomic Energy Commission of France on the Commission's joining the project. After principle consent was obtained, Technichatome Company was engaged to develop a feasibility study. Three out of eight phases of Zvezdochka/Technichatome's joint projects have been completed by today, and the feasibility study is to be finished this year.

Technichatome have proposed the following in addition to the original job statement:

- replace obsolete filters with state-of-the-art equipment;
- replace extraction fans;
- make ash collection process mechanical;
- provide ash cementing unit;
- include solid radioactive waste sorting phase;
- make loading phase automatic.

Technichatome intend to co-ordinate with the Atomic Energy Commission and recommend that the EC Commission provide the funding required for refurbishing the incineration plant. The Company requests relevant support from the Ministry of Atomic Energy.

On its own part, Zvezdochka requests that the Atomic Energy Commission's recommendations to the EC Commission regarding process refurbishment funding are supported, as well.

Certain problems related to radioactive waste handling require immediate solution if regional radiation and environmental safety is to be improved. The problems as follows:

1. Radioactive metal scrap re-melting shop

At least 400 tons of radioactive metal scrap impossible to deactivate are being stored at Zvezdochka. Bulky equipment items include pumps, pipes, reactor control and protection system actuators, etc. The amount of scrap may increase if light structures of submarine reactor compartments are dismantled during the pre-storage phase. A particularly notable increase is possible if it is required to dispose of Project 326M Tenders, special-purpose tanker ships of which the designed service life has expired.

In view of the apparent complexity of the site in question, it would be practical to try and locate the funding for the initial (project development) phase.

2. Deactivation shop reconstruction

The existing shop has been operated for over thirty-five years and needs overhaul and upgrade with due regard for the process of metal scrap pre-remelting deactivation to be integrated.

3. Solid radioactive waste storage facility (Facility # 162) rehabilitation

The described storage facility has been in operation since 1963. Over the years it was filled with non-containerized solid waste and requires adequate maintenance. Unloading the facility, reprocessing of the bulk of solid radioactive waste and repair of the facility building would improve storing conditions and vacate some 60 percent of the volume for future utilization.

Resolving the problems outlined above would ensure improved efficiency, higher rates of disposal of nuclear submarines and of radioactive waste to eliminate the risks of nuclear and/or radiological accidents in the Arctic region, lower mutual tension and contribute to the cause of peace and mutual understanding between our nations.

RADIOECOLOGY, RADIATION, AND SAFETY ISSUES

SYSTEM ANALYSIS OF ENVIRONMENTAL RISKS: BASIC PROBLEMS AND METHODS OF SOLUTION

L. BOLSHOV
Director, Nuclear Safety Institute (IBRAE) of the Russian Academy of Sciences
Moscow, Russia

R. AROUTIOUNIAN
First Deputy Director, Nuclear Safety Institute (IBRAE) of the Russian Academy of Sciences
Moscow, Russia

When considering environmental threats issuing from radiation-dangerous facilities (including the decommissioned NSs), most often the attention is focused on radiation risks. Almost no consideration is given to other hazards coming from the concerned or nearby facilities despite the fact that those threats can exceed by many times the radiation danger. Among such hazards chemical risks and also indirect threats, such as: ecological, psychological, at al. should be mentioned first.

One should also take into account an exaggeration of the radiation hazard by the population resulting from unjustified information quite often provided by the mass media.

1. Risks for the Population and Personnel under Normal Operation of Minatom Facilities

According to the data of the Russian Federation (RF) Ministry of Health, the technogenic-origin exposure resulting from both past and current operation of the Russian nuclear industry facilities is responsible for a negligibly low contribution to the integral population exposure throughout the RF territory (Table 1) [1].

A.A. Sarkisov and L.G. LeSage (eds.),
Remaining Issues in the Decommissioning of Nuclear Powered Vessels, 105–112.

TABLE 1. Structure of the Population Exposure in Some Russian Regions in 1998 [1]

Administrative region	Exposure from natural radiation sources, %	Medical exposure, %	Exposure from global fallout and past radiation accidents, %	Technogenic exposure from enterprises using ionizing radiation sources, %
Area affected by the Chernobyl Accident (1986)				
Bryansk region	51.9	37.3	10.8	0.01
Kaluga region	74.9	24	0.9	0.18
Orel region	64	32.7	3.3	0.03
PA "Mayak" area including territories affected by the Kyshtym Accident (1957)				
Sverdlovsk region	58.7	39.5	1.7	0.14
Cheliabinsk region	74.5	24.7	0.6	0.24
Area affected by nuclear weapons tests				
Altajskiy kray	81.9	17.8	0.29	0.01
Area around nuclear power plants in operation				
Voronezh region	62.4	36.9	0.6	0.11
Murmansk region	73.6	25.5	0.6	0.26
Smolensk region	58.5	39.8	1.7	0.04

It should be stressed that, even if applying the conservative Nonthreshold Concept regarding the radiation effects within the low-dose range, the radiation risk due to the use of nuclear technologies is negligibly low when comparing not only to chemical risks but also when considering the structure of the integral hypothetical radiation risk from all exposure factors.

Under normal operation of nuclear industry facilities, the population exposure gives appreciable contribution neither to the structure of the integral risks nor to that of the radiation risk. Even in the vicinity of the largest nuclear-technical cycle enterprises (including the contaminated territories of PA "Mayak" due to a radiation accident in 1957) technogenic-origin annual population exposure doses in 1993 through 1996 did not exceed 5% of the natural radiation background, the latter varying within 2–2,5 mSv/year (see Table 2). For comparison: radiation doses due to medical exposure make up 1 to 3 mSv/year. The technogenic-origin exposure doses recorded in the vicinity of more recent Minatom enterprises, such as: Mining Chemical Combine (MCC), Krasnoyarskiy kray, and Siberian Chemical Combine (SCC), Tomsk-town, are even lower.

TABLE 2. Annual Exposure Doses and Hypothetical Radiation Risks for the Population in the Vicinity of RF Minatom Enterprises in 1993–1996 [2]

Enterprise	Exposed population, thousand people	Annual effective dose, mSv/year			Life-long risk for the population
		External	Internal	Total	
PA «Mayak»	320	0.01	0.10	0.11	$5.5 \cdot 10^{-6}$
MCC	200	0.03	0.02	0.05	$2.5 \cdot 10^{-6}$
SCC	400	0.0004	0.005	0.0054	$3 \cdot 10^{-7}$

Under normal operation of the present-day nuclear technologies both the population and the personnel exposure doses are far below the levels at which reliable data on harmful health effects are identified (Figures 1 and 2) [3].

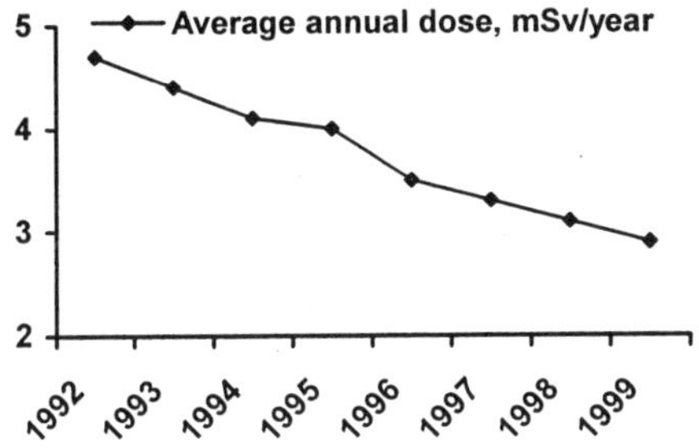

Figure 1. RF Minatom Personnel Average Annual Exposure Doses during 1992-1999 (Annual Admissible Limit for the Personnel – 50mSv; since 2000 - 20 mSv) [3]

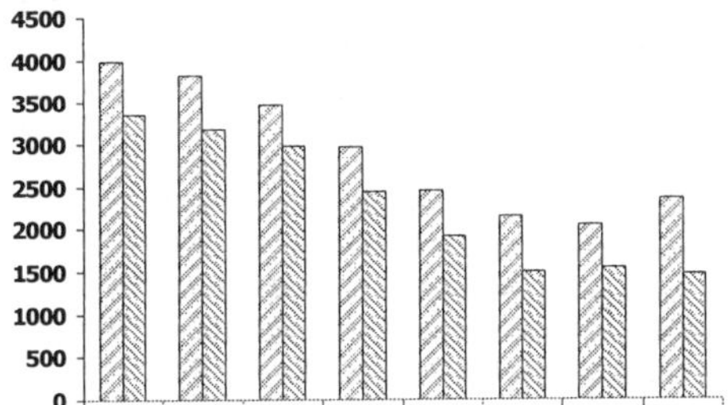

	1992	1993	1994	1995	1996	1997	1998	1999
Minatom	3980	3817	3472	2964	2455	2144	2050	2355
NPP	3345	3180	2979	2445	1911	1499	1546	1464

Figure 2. External Exposure of RF Minatom Personnel over 1,5 rem during 1992-1999 [3]

At the present time, when considering the prospects of the up-to-date power technologies, the environmental safety indices become most important. Comparative analyses of the integral damage for both the population health and the environment from different types of power generation were performed in thousands of investigations and published in many generalizing materials of authoritative international organizations.

Most detailed assessments of the damage per a unit of generated power throughout the power cycle were carried out with respect to the nuclear industry. Huge amounts of experimental and calculated data concerning the personnel & the population exposure throughout the nuclear cycle (from uranium extraction, enrichment, fuel production, power generation at Nuclear Power Plants (NPPs) and up to all stages of Spent Nuclear Fuel (SNF) & Radioactive Waste (RW) management) are generalized in materials of UNSCEAR, NEA OECD and CEC.

These data prove that under normal operation of nuclear facilities the exposure of the involved personnel throughout the nuclear power generation cycle, not to mention the population, varies within the dose range below the limits of detecting harmful health effects using epidemiological methods. Furthermore, additional population exposure doses due to the impact of nuclear industry facilities (even regarding the population living in the immediate vicinity of such facilities) are hundredths and thousandths of the background-origin exposure values. These doses are equally low if considering variations in the background exposure.

Exposure doses regarding the overwhelming majority of the nuclear industry personnel are at the level of the background-origin population exposure and do not exceed variations in the radiation background in different countries.

Actually over 300 thousand workers are employed in the RF nuclear industry of which 50 thousand people work in conditions of "potential additional exposure". However only < 5% of the personnel working in the high-exposure environment (2400 people) receive doses over 15 mSv/year. An excess over annual dose limits is identified only in individual cases. Cases of overexposure with clinical manifestations are extremely rare.

The present-day estimates of the health hazard to both the personnel and the population due to additional (extremely low) exposure resulting from the operation of nuclear industry enterprises are based on a hypothetical linear nonthreshold concept (in keeping with this concept, harmful effects on human beings are identified even at low radiation exposure doses). However epidemiological examinations of hundreds of thousands of people performed during more than a 50-year period could not confirm this concept. But it is namely this approach that was used in the most recent generalized investigations performed within the framework of CEC to assess the health damage throughout the nuclear industry cycle.

When analyzing this hypothetical damage, one should take into account that the largest contribution is due to additional exposure from globally dispersed long-lived radionuclides. Over 95% of additional collective exposure dose (and, correspondingly, of the hypothetical health risk) results from releases of ^{14}C, ^{85}Kr and ^{129}I into the environment (Figure 3) [2].

It is worthy of noting that the additional population exposure dose from global radionuclides released throughout the nuclear industry history is below 0, 0001 of the average population exposure dose due to the natural radiation background.

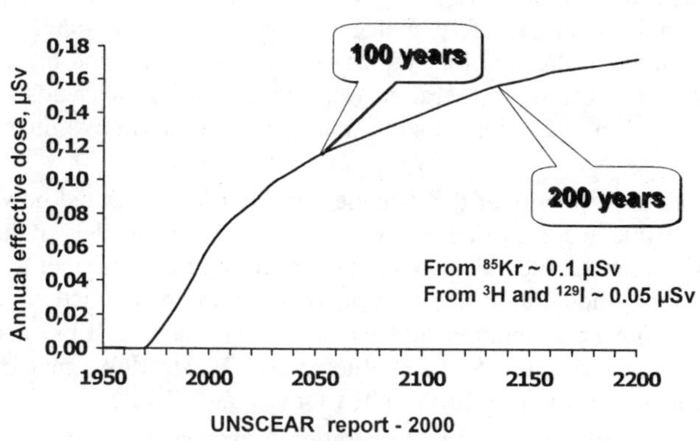

Figure 3. Average Annual Doses from Globally Dispersed ^{14}C of Nuclear Facilities [2]

2. Comparison between Radiation and Chemical Environmental Risks

Comparative analyzes of radiation & chemical environmental risks of the technogenic origin showed up a rather unfavorable situation regarding chemical pollution of the environment [4].

In Table 3 assessments of both the radiation risk and chemical air pollution risk are compared.

TABLE 3. Individual Annual Mortality Risks Regarding the RF Population [4]

Causes	Confirmed, million people	Risks	Fatal case/year
All causes	69 (men)	$1.5\cdot10^{-2}$ (average over 1996-1999)	1 060 000
Different casualties	69 (men)	$3.4\cdot10^{-3}$ (average over 1996-1999)	240 000
High-level environment pollution	70	$1\cdot10^{-3}$	70 000
Relocation zone around the Chernobyl NPP	0.1 (contaminated areas of the Ukraine, Russia and Byelorussia)	$8\cdot10^{-5*}$	8*
Population around MCC, SCC and PA "Mayak"	0.9	$6\cdot10^{-6}$ - $3\cdot10^{-7*}$	<3*
Population around NPP	0,5 – 1	$7\cdot10^{-7*}$	<0.7*

Comment: *hypothetical risks and fatal cases in the low-dose area in keeping with the nonthreshold concept.

According to the results of a large-scale investigation of the effects of atmospheric pollution on the population health published in 2000 [5], air pollution in West Europe (Austria, France and Switzerland) is responsible for 6% (40 000 cases) of total mortality per year. Note that about 50% of all fatal cases are attributable to traffic-related air pollution: more than 25 000 new cases of chronic bronchitis (adults); more than 290 000 episodes of bronchitis (children); more than 0,5 million asthma attacks; and more than 16 million person-days of restricted activities. Here it should be mentioned that average particulate matter concentrations in the air of the above countries are 5 to10 times below the Maximum Allowable Concentrations (MACs) in force in Russia.

Unfortunately, so far only isolated investigations of a similar-type were performed in the RF to detect chemical pollution effects on the population health. But even these results showed up serious problems. This thesis is illustrated by the risk assessment data due to chemical air pollution with respect to the population health presented in this chapter.

A comparison of both the methods and levels of practical protection of the population health & the environment against the radioactive contamination and chemical pollution has demonstrated their considerable distinctions and imbalance. This concerns all elements of the regulatory control: approaches to standardization, methods of determining admissible values of release/dumping, possibilities of organizing the environmental monitoring and the attitude toward the observance of the established regulations.

The distinctions begin at a level of approaches to standardization: when considering chemical pollution, the threshold concept is used; in the case of ionizing radiation one is guided by the nonthreshold concept. If the linear nonthreshold concept were not a scientific hypothesis, such a distinction could be considered as an unimportant one. When establishing MACs for different environmental pollutants in Russia, not only toxicological but also other characteristics are often employed as a limiting factor. It seems that this approach can result in lower MAC levels. However if comparing the risks of a theoretically equal adverse health effect (at the MAC level), one can see that the environment contamination at MAC levels by many carcinogenic chemical pollutants results in higher (by an order of magnitude and more) risks as compared to that of chronic population exposure at a level of 1 mSv; in some cases these risks reach an inadmissibly high level (see Table 4).

TABLE 4. Lifelong Carcinogenic Risks Due to the Effects of Chemical Air Pollution at MAC Levels [6]

Substance	Risk	Substance	Risk
Arsenic	$1.3 \cdot 10^{-2}$	Benzol	$2.9 \cdot 10^{-3}$
Cadmium	$5.5 \cdot 10^{-4}$	1,2 dichlorethane	$2.6 \cdot 10^{-2}$
Chromium (YI)	$2.2 \cdot 10^{-1}$	Nickel	$2.6 \cdot 10^{-4}$
Epichlorohydrin	$4.6 \cdot 10^{-3}$	Hexachloran	$1.5 \cdot 10^{-2}$
1-3 butadiene	$2.8 \cdot 10^{-1}$	Chloroform	$6.9 \cdot 10^{-4}$

The data presented are not a result of sampling most the unfavorable MAC-fixing cases; unfortunately, this is a type situation. The data of Table 5 demonstrates that carcinogenic risks due to chemical air pollution calculated at MAC levels currently in force in Russia reach extremely high values for most of the previously standardized substances [7].

TABLE 5. Carcinogenic Risks and MACs [7]

Risk	Surface waters		Atmospheric air		Work area	
	Absol.	%	Absol.	%	Absol.	%
>10-2	7	8,0	2	5,4	42	45,1
10-2-10-3	19	21,8	6	16,2	34	36,5
10-3-10-4	29	33,3	13	35,1	10	10,7
10-4-10-5	23	26,4	9	24,3	7	7,5
<10-5	9	10,5	7	19,0	0	0
Total	**87**	**100**	**37**	**100**	**93**	**100**

The actual system of the radiation environmental monitoring makes it possible to detect changes at a level of the natural background variations, being by 5–7 orders of magnitude below MACs levels. The system needs improvements only as an emergency response instrument. At the same time the monitoring over chemical air pollution gives no way of correct by assessing the concentrations of chemical pollutants in the environment and identifying the air pollution risk to the population health. Within the whole Roshydromet survey network about 70 chemical pollutants are controlled, but at most survey stations only 5–10 toxic substances are recorded. Often the concentration of basic pollutants in the air is not monitored, whereas, when detecting some pollutants, the sensitivity of applied methods is at their MAC levels. The results of detailed investigations demonstrate that the contribution of air pollutants controlled by the Roshydromet survey network into the integral risk to the population health makes up ≤ 30%.

The actual levels of both the radiation and chemical air monitoring are compared in Figure 4.

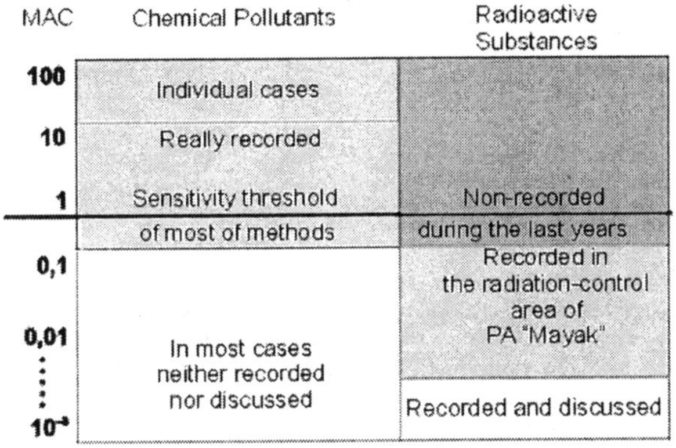

Figure 4. Monitoring over Air Pollution & Air Contamination in Russia

The RF law "On the Environment Protection" determines and distinguishes the notions of "the Environmental Emergency" and of "the Environmental Disaster" as follows [8]:

Under "the Environmental Emergency" one faces stable negative changes in the environment, which present hazard to the population health, whereas by "the Environmental Disaster" is meant "…deep and irreversible changes in the environmental status resulting in appreciable worsening of the population health and disturbances in the natural equilibrium...".

In keeping with the radiation safety criteria [9], territories with additional exposure dose varying from 5 mSv/year to 10 mSv/year are considered as the "the Environmental Emergency" area, whereas those of > 10 mSv/year additional exposure dose fall into the category of "the Environmental Disaster".

The radiation risks, as applied to territories of different environmental-emergency status, are compared in Table 6 [4].

TABLE 6. Comparison of the Radiation Risks in a case of Different Environmental Emergencies [4]

Indices	Parameters		
	Environmental disaster	**Environmental emergency**	**Satisfactory status**
Effective exposure dose, mSv/year	> 10	5-10	< 1
Risk	$> 5 \cdot 10^{-4}$	$(2,5\text{-}5) \cdot 10^{-4}$	$< 5 \cdot 10^{-5}$

It is intriguing that, if observing the criteria established by the Russian regulation (Table 6), the whole population of Finland receiving the natural-origin annual exposure dose of almost 8 mSv falls into the category of living within "the Environmental Emergency" conditions (Figure 5).

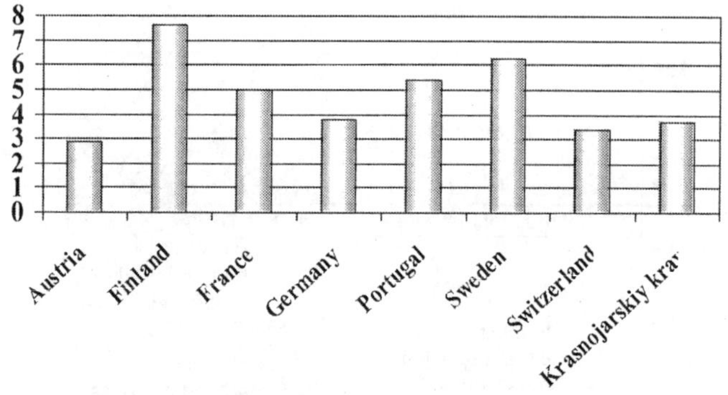

Figure 5. Natural-Origin Population Exposure Doses in Different European Countries (data of National Radiological Protection Board, U.K.) and Krasnojarskiy kray (data of Regional Radiological Center attached to State Sanitary Surveillance Center in Krasnojarskiy kray)

By way of conclusion an inference can be drawn that, to solve tasks of the radiation risk analysis, the science & methodic base of initial data on the radiation and radioecological control over the population exposure is much more developed than that of the chemical risk analysis.

3. References

1. *Radiation Safety in the Russian Federation in 1998. Analytical Information* (1999). RF Ministry of Health, Moscow (in Russian).
2. *Sources and Effects of Ionizing Radiation* (2000). UNSCEAR 2000 Report to the General Assembly, vol.1, New York.
3. Agapov, A., Novikov, G., Aroutiounian, R., Linge, I. (2001) Ecological Safety of Atomic Power Engineering and Industry. *Cosmos. Information. New Technologies*, **4**, 29-32.
4. Bolshov, L., Aroutiounian, R. (2002) The Chernobyl Catastrophe, Nuclear Technologies and Environmental Problems. *In Proceedings of Symposium: "15 Years after Chernobyl: Lessons, Estimates and Prospects"*, Komtekhprint, Moscow, pp. 12-23 (in Russian).
5. Künzli, N., Kaiser, R., Medina, S., Studnicka, M., Channel, O., Fillinger, P., Herry, M., Horak, F., et al. (2000) Public-Health Impact of Outdoor and Traffic-Related Air Pollution: a European Assessment, *The Lancet*, **356**, 795-801.
6. Novikov, S., Rumiantsev, G,. Zholdakova, Z., Shashina, E., Ponomareva, O. (1998) Problem of Assessing the Carcinogenic Risk Due to Chemical Pollution of the Environment. *Hygiene and Sanitary*, **1**, 29-35 (in Russian).
7. Novikov, S., Porfiriev, B., Ponomareva, O. (2000) *Consulting Center on Risk Assessment*, Report of Nuclear Safety Institute (IBRAE) of the Russian Academy of Sciences, Moscow (in Russian).
8. *The Criteria of Assessing the Environmental Situation to Identify Territories of the Environmental Emergency Status and of the Environmental Disaster Status* (1992) The RF Minprirody Publishers, Moscow (in Russian).
9. Danilov-Danilian, V. (1992) *Radiation Safety Criteria. Normative Document.* Goskomecologija (State Environmental Committee), Moscow (in Russian).

ANALYSIS OF THE RADIATION POTENTIAL OF THE DECOMMISSIONED NUCLEAR SUBMARINES AND REACTOR UNITS

V. BARINOV, S. BOGATOV, V. DANILYAN, R. KALININ and P. SHVEDOV
Nuclear Safety Institute (IBRAE) of the Russian Academy of Sciences
Moscow, Russia

1. Introduction

By the beginning of 2002, 191 Russian Nuclear Submarines (NSs) were decommissioned. Part of them are stored afloat with/without fuel; other NSs are partly utilized and stored afloat in three-compartment units. The number and the actual status of the decommissioned Russian NSs are demonstrated in Table 1.

TABLE 1. Number and Actual Status of the Decommissioned NSs by the beginning of 2002

NS number	In the Northwest region	In the Far East region	Total
Decommissioned	115	76	191
Under waterborne storage with fuel	51	39	90
Under waterborne storage without fuel	16	13	29
Cut out into 3-compartment units	48	24	72*

*) Five 3-compartment units are stored with fuel

The potential radiation hazard of the decommissioned NSs is due to the activity of radionuclides concentrated largely within both the Spent Nuclear Fuel (SNF) and the metal structures of the reactor compartment activated during operation. In this paper the activity of the principal radionuclides regarding 73 decommissioned Russian NSs equipped with water-moderated reactors, as of 2002, is calculated and the forecast for 2012 is given. The calculation was based on an analysis of the history of reactor operation at full power during operational NS period. For these purposes, in every case, a "power card" was drawn up to illustrate operation of the Steam Producing Facility (SPF) at full power every month/year with consideration for the periods of time for factory repair and core-reloading operations. In keeping with the performed calculations, the fission-fragment activity of SNF stored within reactors of the decommissioned NSs, as of 2002, varies within 10^{15} - 10^{16} Bq, whereas the induced activity is about 10^{14} Bq. During the coming 10 years a decrease in SNF fission-fragment activity by a factor of 1,4-1,5, and of the induced activity by a factor of 1,9-5,5 (depending on the moment of the ultimate shutdown of reactor facility) is expected.

A.A. Sarkisov and L.G. LeSage (eds.),
Remaining Issues in the Decommissioning of Nuclear Powered Vessels, 113–124.

2. Methods of Calculating the Induced Activity in the Reactor Compartment (RC) of Decommissioned Russian Navy NSs with Consideration for the Actual History of SPF Operation at Full Power

The induced activity of structural and shielding materials, of the equipment and coolant impurities of nuclear installations can be determined by radioniclides (products of neutron activation) using the following reactions: (n,γ), (n,p), (n,α), (n,d), (n,2n), (n,n'). Among them the processes of thermal and epithermal neutron radiation capture is most important.

The induced activity depends on: the flux density and the energy spectrum of neutrons, the activation cross-section value, the chemical element concentration in materials, the relative concentration of the target isotope in chemical elements, the exposure time and the period after shutdown. When dealing with transport reactor units, the following reactions can be considered as basic activation reactions resulting in the generation of long-lived radionuclides with half-lives over 1 year:

$$ {}^{54}_{26}Fe(n,\gamma){}^{55}_{26}Fe \xrightarrow[KX]{2{,}73\,year} {}^{55}_{25}\underline{Mn} \qquad (5{,}8\%) $$

$$ {}^{54}_{26}Fe(n,p){}^{54}_{25}Mn \xrightarrow[KX,\gamma]{312{,}12d} {}^{54}_{24}\underline{Cr} \qquad (5{,}8\%) $$

$$ {}^{59}_{27}Co(n,\gamma){}^{60}_{27}Co \xrightarrow[\beta^-,\gamma]{5{,}2709y} {}^{60}_{28}\underline{Ni} \qquad (100\%) $$

$$ {}^{58}_{28}Ni(n,\gamma){}^{59}_{28}Ni \xrightarrow[KX]{2{,}9e5y} {}^{59}_{27}\underline{Co} \qquad (67{,}8\%) $$

$$ {}^{60}_{28}Ni(n,p){}^{60}_{27}Co \xrightarrow[\beta^-,\gamma]{5{,}2709y} {}^{60}_{28}\underline{Ni} \qquad (26{,}4\%) $$

$$ {}^{63}_{29}Cu(n,\alpha){}^{60}_{27}Co \xrightarrow[\beta^-,\gamma]{5{,}2709y} {}^{60}_{28}\underline{Ni} \qquad (69{,}1\%) $$

$$ {}^{63}_{29}Cu(n,p){}^{63}_{28}Ni \xrightarrow[\beta^-]{101{,}1y} {}^{63}_{29}\underline{Cu} \qquad (69{,}1\%) $$

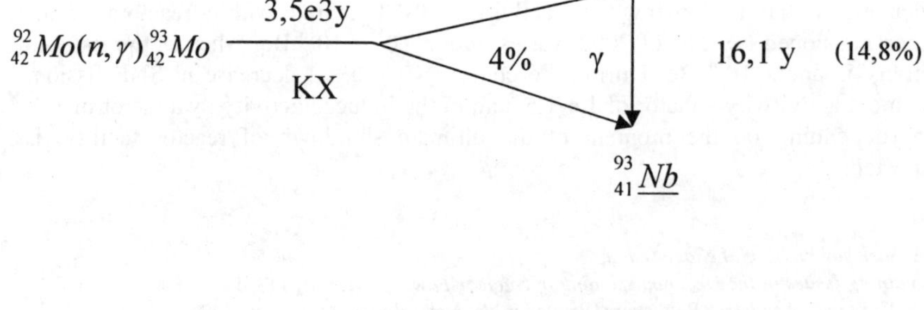

Concentrations of the target isotope in natural mixture are given in parentheses.

To calculate the induced activity, it is wise to introduce the notion of the activation integral, which characterizes the reaction rate. The activation integral q referred to one nucleus of the target isotope equals:

$$q = \int_0^{\infty} \sigma(E)\varphi(E)dE \quad (1)$$

where: σ(E) – cross-section for the activation reaction for E energy, φ(E) – energy spectrum of neutrons. When solving most of practical issues, secondary processes (such as, e.g., the target isotope burnup and the reaction product) can be neglected. In this case the induced volumetric activity A_V of the reaction product isotope with a disintegration constant λ in material possessing n_0 target isotope atoms in a unit of volume for exposure time T within a stable neutron flux after time t following the exposure termination is:

$$A_V = n_0 q(1 - e^{-\lambda T})e^{-\lambda t} \quad (2)$$

For a neutron flux unstable in time a number of time intervals l can be distinguished; thus, during j-interval flux density T_j can be considered as a stable value. In this case the integral induced activity is determined by superposition of induced activities of every j-interval. In this case:

$$A_V = n_0 \sum_{j=1}^{l} q_j(1 - e^{-\lambda T_j})e^{-\lambda t_j} \quad (3)$$

where: t_j – time after the termination of j-interval of exposure.

2.1. INDUCED ACTIVITY WITHIN THE RCs OF THE FIRST-GENERATION NUCLEAR SUBMARINES

Presented in Report [1] are the results of calculating the induced activity of construction and shielding materials of reactor compartment (two identical reactor units) for the first-generation nuclear submarines. The following nuclear submarine running model was used in the calculations:

- Operation time at full power ....40000 hours;
- Mean power level ....21 Mw;
- Operation mode: 60 days — operation at full power, 180 days – reactor shutdown;
- Cycle number 28;
- There were 2 long-term shutdowns of 3-years each to perform the core reloading, repair and modernization in the course of NS operation.

In Table 2 the results of calculating the integral induced activity of reactor compartment constructions are given. The largest contribution to the induced activity is due to reactor internals (as a result of a high concentration of nickel additions in

"X18H10T"-type of stainless steel used in reactor internals, as compared to high-carbon steel of the reactor vessel). Activities of the other reactor compartment elements (reactor vessel constructions, iron-water shielding tank, basements and the NS strong hull under the reactor compartment, as well as the steam generator baffle structures) are more than an order of magnitude less.

TABLE 2. Integral Induced Activity of Constructions in the NS Reactor Compartment, Bq [1].

Nuclide	Time following the reactor shutdown, years							
	0,5	5	10	30	70	100	500	1000
^{54}Mn	2,48+14*	6,49+12	1,13+11	1,04+4	-	-	-	-
^{55}Fe	7,1+15	2,26+15	6,32+14	3,85+12	1,44+8	6,89+4	-	-
^{60}Co	1,63+15	1,02+15	5,27+14	3,79+13	1,97+11	3,78+9	-	-
^{59}Ni	4,25+12	4,25+12	4,25+12	4,25+12	4,25+12	4,25+12	4,23+12	4,21+12
^{63}Ni	3,61+14	3,48+14	3,38+14	2,93+14	2,23+14	1,8+14	1,14+13	3,54+11
^{93}Mo	2,5+9	2,5+9	2,5+9	2,5+9	2,46+9	2,46+9	2,26+9	2,05+9
Сумма	9,89+15	3,64+15	1,5+15	3,41+14	2,27+14	1,86+14	1,57+13	4,6+12

*) Here and below the record of "2,48+14"-type should be understood as "$2{,}48 \cdot 10^{14}$"

Starting from the results of I. Engovatov, et al. [1] and knowing the power history, one can estimate "the activation integrals" Q' = n_0qV (where n_0 — target isotope nucleus number in a unit of volume, V – reactor compartment volume) for 21 MW power with consideration for radionuclides given in Table 2. Here, under the activation integral, an equilibrium activity (over infinite exposure time) of the overall reactor compartment is understood. Q value makes no physical sense since the effects of the target isotope and of the activation product burn-up over infinite period of time (neglected in equation (2)) become considerable. However this value makes it possible to satisfactorily estimate the reactor compartment activity, as applied to the actual history of the considered NS reactor operation at full power (see below).
Values of Q' recalculated with reference to one reactor for radionuclides listed in Table 2 are demonstrated in Table 3.

Given the NS operating time as an average annual indices during N years – average energy-production during i-year W_i [MW·h] and time at full power operation T_i [h], the activity of reactor compartment A for a radionuclide with λ disintegration constant for a given time t can be estimated as follows:

$$A(t) = Q\sum_{i=1}^{N}\frac{W_i}{T_i}(1-e^{-\lambda T_i})e^{-\lambda t_i} \tag{4}$$

where: t_i – period of time from the end of i-year of reactor operation up to the given time t.

TABLE 3. Reactor Compartment Activity Regarding NSs of the First Generation for 21 MW Power for Infinite Exposure Time (the last line contains Q' activation integral values per unit power)

Nuclide	^{60}Co	^{55}Fe	^{59}Ni	^{63}Ni	^{54}Mn	^{93}Mo
Q', Bq	4,7+15	2,0+16	5,0+16	6,0+15	1,0+14	1,4+12
Q, [Bq/MW]	2,2+14	9,5+14	2,4+15	2,9+14	4,8+12	6,7+10

The equation (4) contains a number of assumptions. In particular:

- radionuclide production during the operating period is assumed linearly proportional to power value for each radionuclide;
- as fuel burns up, changes in neutron energy spectrum do not taken into account;
- power history during one year is neglected;
- elemental composition of the steel is considered as constant for the whole compartment.

However for long-lived radionuclides whose production during the operational period varies in direct proportion to energy-production, actual consideration of the above-listed factors will result in a correction of 10-20% at the most. Thus, the precision of induced activity estimates within the reactor compartment using equation (4) can be considered as 20% worse than that of calculations performed by I. Engovatov, et al. [1].

To estimate the calculational accuracy (see [1]), special experimental investigations of the induced activity were performed. For this purpose samples of materials cut out of the vessel of '27 BM' bench installation (a surface prototype of the first-generation NSs, Obninsk town) were used. The bench installation was put into operation in 1956 and its energy-production over 30 years was 1898000 MW·h [1]. Samples of heat shield inside the reactor, the vessel interior lining and the vessel itself were studied. For samples of heat shield and the reactor interior lining (made of "X18H10T"-type steel of maximum specific activity) the calculated specific activity of samples for ^{60}Co and ^{63}Ni is about 2 times higher. The authors explain the discrepancy between the experimental and calculated results by distinctions between the actual elemental composition of steel and that used in the calculations. The authors do not mention whether or not they took into account the history of the unit operation at full power.

There is almost no way of taking into account the actual geometry of constructions (the authors [1] give steel specific activity calculations only with reference to two directions – "upward" and "stern" with the maximum specific activity difference a factor of 3). Therefore the calculated estimations of induced activity within the reactor compartment can be treated as true ones with accuracy of no less than an order of magnitude.

Note that to estimate the radiation potential of the reactor compartment as a whole, the consideration of only the induced activity of the reactor internals is sufficient within the accuracy of calculations. According to estimations [1]:

- the activity of the reactor vessel constructions makes up about 4%,
- the activity of iron-water shielding tank structures is equal to 1%,
- the activity of the strong vessel basement under the reactor baffles and that of constructions within steam generator baffles makes up a thousandth of a per cent compared to the induced activity inside the reactor vessel.

2.2. INDUCED ACTIVITY IN REACTOR COMPARTMENTS OF THE SECOND- AND THE THIRD-GENERATION NUCLEAR SUBMARINES

To estimate the induced activity within metallic constructions of the reactor compartment, the initial data of Reference [2] were used. Similar to the first-generation NSs, in this case over 95% of the induced activity is also concentrated within reactor internals. The results presented in Reference [2] related to the accumulation of basic radionuclides within reactor internals of the second-generation NSs at the time of reactor shutdown are demonstrated in Table 4.

TABLE 4. Activities of Basic Radionuclides within Reactor Internals of the Second-Generation NSs at the time of Reactor Shutdown

Radionuclide	**^{60}Co**	**^{54}Mn**	**^{55}Fe**	**^{59}Ni**	**^{63}Ni**
Activity, Bq	4,8+14	3,4+14	9,3+15	6,3+11	7,4+13

In the calculations the nonstop operation of two 90 MW BM-4-type reactors during 60000 hours at a power level of 20% from the rated value was assumed. The results of Q activation integral calculations using equation (4) are demonstrated in Table 5.

TABLE 5. Activation Integrals per Unit Power for Induced Activity of the Reactor Internals (Case of Second-generation NSs)

Radionuclide	**^{60}Co**	**^{55}Fe**	**^{59}Ni**	**^{63}Ni**	**^{54}Mn**	**^{93}Mo***
Q, Bq/MW	9,2+12	1,3+14	1,1+14	1,7+13	4,3+12	3,2+9

Reference [2] does not contain calculations for ^{93}Mo. Because stainless steel of the same "1X18H10T" -type determines the induced activity in the first- and the second-generation NSs, to estimate ^{93}Mo activity, the mean relation of activation integrals for cobalt & nickel radionuclides with respect to NSs of the first- and the second generation was used.

It is worthy of notice that in the case of the second generation NSs the activation integrals are 0,02...0,03 from those of the first generation NSs. This difference is due to some constructional distinctions of their reactors (BM-A and BM-4): in the reactor internals of the second-generation NSs the amount of "1X18H10T"-type stainless steel is considerably less.

To calculate the induced activity within RC of the third-generation NSs, activation integral data given in Table 5 were used.

Similar to previous calculations, the estimated accuracy of the results is an order of magnitude.

3. Estimates of Fission-Fragment Activity in the Fuel of NS Reactors

When performing evaluation calculations (wherein the linear dependence of fission isotopes with energy production of the core is assumed at the time of reactor shutdown), the data of S. Bogatov, et al. [3] can be used. The work comprises the results of calculating the fission-fragment activity of – the "Lenin" icebreaker sunken reactor compartment, - six NS reactors and – the steam producing facility of "Sevmorput" lighter-aboard ship. In these calculations the actual history of the reactor operation at full power was taken into account. Averaged evaluations of long-lived fission product activities normalized to unity of energy production are demonstrated in Table 6.

TABLE 6. Average Values of Normalized Activities of Long-lived Fission Products and their Standard Deviations at the time of Reactor Shutdown

Nuclide	**$T_{1/2}$**	**Activity, Bq/(GW*day)**
^{3}H	12,3 years	(7,5±1,6)+11
^{85}Kr	10,7 years	(1,8±0,6)+13
^{90}Sr	29,1 years	(1,1±0,1)+14
^{99}Tc	2,13+5 years	(1,70±0,04)+10
^{129}I	1,57+7 years	(2,8±0,1)+7
^{134}Cs	2,06 years	(2,7±2,3)+13
^{135}Cs	2,3+6 years	(8,5±2,1)+8
^{137}Cs	30,0 years	(1,2±0,1)+14
^{147}Pm	2,62 years	(3,6±1,0)+14
^{151}Sm	90 years	(2,6±1,1)+12
^{152}Eu	13,3 years	(2,3±1,7)+10
^{154}Eu	8,8 years	(4,4±2,5)+11
^{155}Eu	4,96 years	(3,0±1,5)+12

One can see that the hypothesis of linear dependence of the activity accumulation due to energy production with regard to the actual reactor fuel operating period gives a satisfactory result for most radionuclides (the accuracy of about 30 to 50%). The only exceptions are radionuclides of ^{134}Cs-type, since their activity depends nonlinearly on burnup. Considering that fuel operating periods and energy production of most of the NS reactors differ only slightly from those used by S. Bogatov, et al. [3], the accuracy of initial data presented in Table 6 is no less than factor 2.

In is worth noting that such a simplified approach cannot be used to estimate TransUranium Element (TUE) activity accumulations. The accumulation of TUE activities depends strongly on neutron spectrum & power density and varies in a non-linear fashion with fuel burnup. Simplified approaches, such as that used to estimate fission radionuclide accumulation (see Table 6) can lead to errors of up to 2 orders of magnitude and are inapplicable in principle. To obtain adequate evaluations, actual dependencies of the TUE accumulation on neutron spectrum in NS reactor, the degree of fuel burnup, the ^{238}U mass in every reactor and some other parameters should be used. In the given paper no estimate of TUE activity within NS reactors is performed.

4. Calculation Results

By way of example, in Table 7, power characteristic of operating reactor facilities for a first-generation NS during its running period is demonstrated. To perform calculations, similar-type tables were made for all NSs.

TABLE 7. Power Characteristic of Reactor Operation during NS Operating Period

Year	Starboard RU operation		Port side RU operation	
	Operation time, hour	Power generation, MW·h	Operation time, hour	Power generation, MW·h
1960	721	15136	721	15280
1961	2434	34402	2200	38325
1962	427	7515	402	5845
1963	826	14273	1466	39765
1964	1449	31298	1378	23481
1965	513	7170	790	7231
1966	Running repair of October 1965 through February 1967			
1967	3212	75224	3693	67528
1968	2286	58472	2358	63576
1969	1178	22007	1173	21030
1970	Mid-life repair of December 1969 through December 1972			
1971				
1972	349	7165	407	7957
1973	1968	40334	1928	31448
1974	1512	25663	1459	20596
1975	2547	43720	2546	54717
1976	1053	18607	500	17747
1977	1543	29275	1595	23821
1978	693	12107	691	11183
1979	128	1300	142	1856
1980	Running repair of November 1980-July 1981			
1981	1694	22100	1537	19365
1982	2194	28790	2182	24780
1983	2313	29860	2275	30020
1984	2151	26610	2095	26350
1985	0	0	0	0
1986	790	13577	135	2291

In Figures 1 and 2 are demonstrated, respectively, the integral activities of fission radionuclides in SNF and the induced activity in metal structures of the RCs in the First-generation NSs. In Figures 3 and 4 the same information is given regarding NSs of the second-generation, and in Figures 5 and 6 with respect to the third-generation NSs depending on the number of considered NSs.

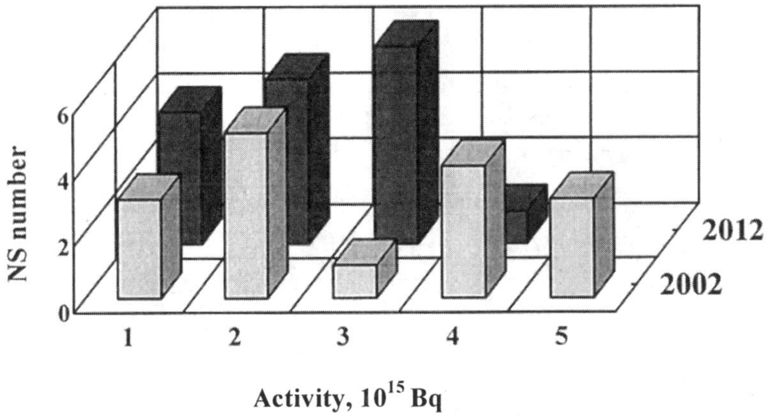

Figure 1. Activity of Fission Radionuclides in SNF of the Decommissioned First-Generation Nuclear Submarines

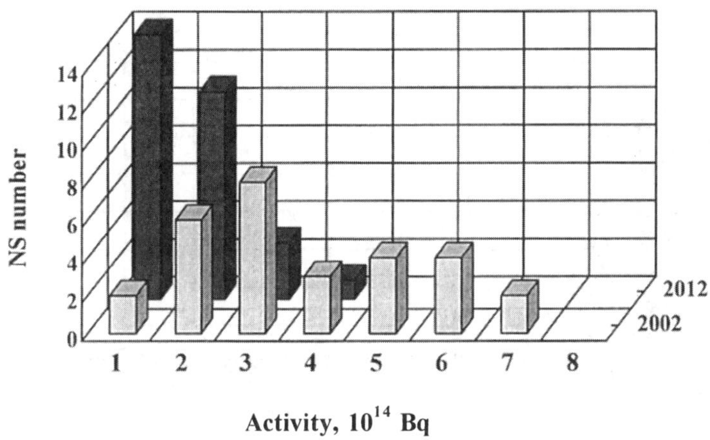

Figure 2. Induced Activity of RCs of the Decommissioned First-Generation NSs

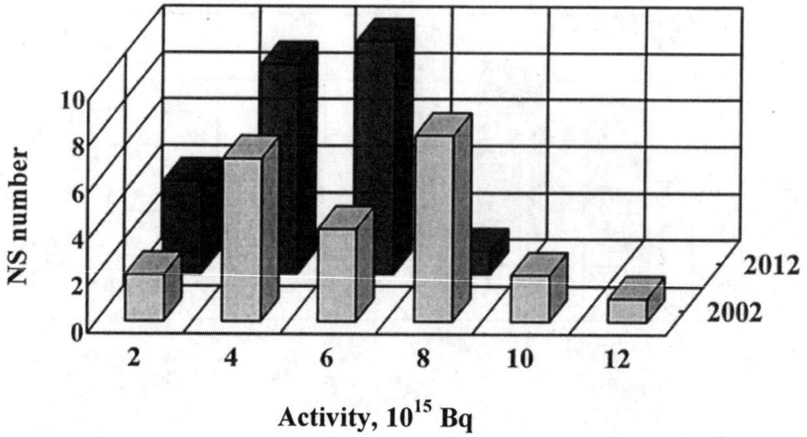

Figure 3. Activity of Fission Radionuclides in SNF of the Decommissioned Second-Generation Nuclear Submarines

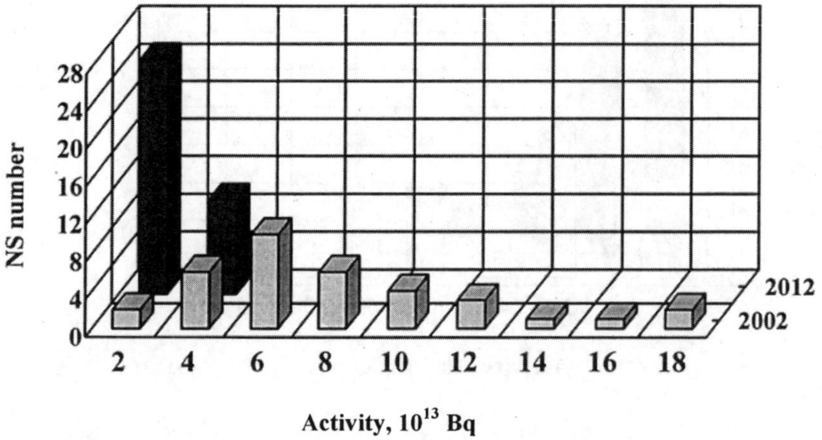

Figure 4. Induced Activity of RCs of the Decommissioned Second-Generation NSs

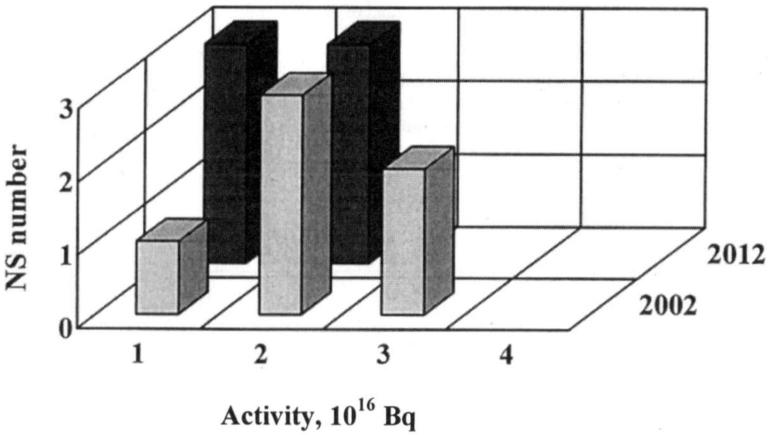

Figure 5. Activity of Fission Radionuclides in SNF of the Decommissioned Third-Generation Nuclear Submarines

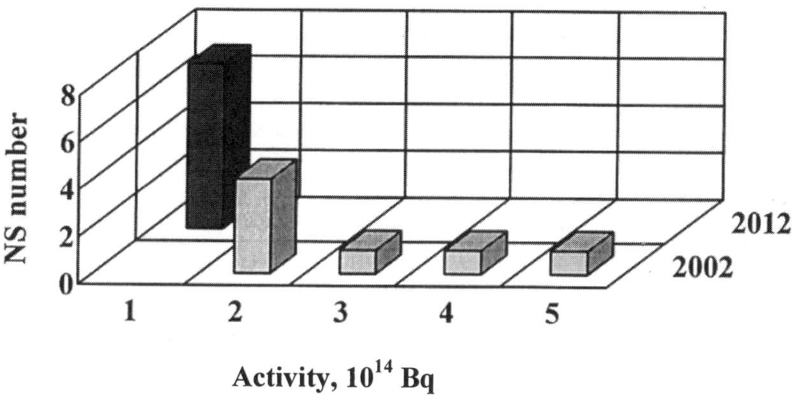

Figure 6. Induced Activity of RCs of the Decommissioned Third-Generation NSs

Calculated estimations of the mean radiation potential regarding different-generation decommissioned NSs are given in Table 8.

TABLE 8. Averaged Calculated Radiation Potential of the Decommissioned Nuclear Submarines

NS	Average SNF activity, Bq		SNF activity decay during the coming 10 years	Average induced activity in RCs, Bq		Induced activity decay during the coming 10 years
	2002	2012	2002/2012	2002	2012	2002/2012
First-generation NS	$2{,}3 \cdot 10^{15}$	$1{,}7 \cdot 10^{15}$	1,4	$4{,}3 \cdot 10^{14}$	$2{,}3 \cdot 10^{14}$	1,9
Second-generation NS	$5{,}4 \cdot 10^{15}$	$3{,}6 \cdot 10^{15}$	1,5	$7{,}3 \cdot 10^{13}$	$1{,}7 \cdot 10^{13}$	4,3
Third-generation NS	$1{,}6 \cdot 10^{16}$	$1{,}1 \cdot 10^{16}$	1,5	$2{,}2 \cdot 10^{14}$	$4{,}0 \cdot 10^{13}$	5,5

5. References

1. Engovatov, I., Mashkovich, V., Orlov, Y., Pologuikh, B., Khlopkin, N., Tsypin, S. (1999) *Radiation Safety, when Decommissioning Reactor Units of Civil and Military Purposes, MSTC Project №465-97*. PAMIS Publishers, Moscow.
2. Andreev, N., Vavilkin, V., Mitenkov, F., Sandler, N., Chibyshev, S. (1999) Analysis of the Process of Activity Release from Reactor Compartment of the Second-generation Nuclear Submarines under Long-term Waterborne Storage after De-fueling, in A. A Sarkissov (eds.), *Analysis of the Risks Related to the Decommissioning, Storage and Utilization of Nuclear Submarines*. KomTekh Publishers, Moscow.
3. Bogatov, S., Gavrilov, S., Danilian, V., Kisselev, V. (2001) *Estimate of Radionuclide Release with Consideration for Some Hypothetical Accidents at* Russian Navy Facilities, Preprint № IBRAE-2001-07. Nuclear Safety Institute (IBRAE) of the Russian Academy of Sciences, Moscow.

IMPROVING NUCLEAR AND RADIATION SAFETY DURING THE PROCESS OF RUSSIAN NUCLEAR SUBMARINE UTILIZATION

A.KIRIUSHIN, E. AKSENOV, V. VAVILKIN, N. SANDLER
Minatom Experimental & Design Office of Mechanical Engineering
Moscow, Russia

When organizing work on nuclear vessel utilization, the issues of ensuring both nuclear and radiation safety should be regarded as a fundamental condition at all stages of the utilization operations.

At the present time one can state that the whole utilization process is relatively stable is relatively stable: -the level of preparedness of both the industrial capacities and the involved personnel has improved and –the pace of work has appreciably accelerated. One has also managed to gain strict technological regulation of all performed work, establish a list of potential dangerous operations, develop technical requirements for to their realization; simultaneously, training of the involved personnel has been organized.

In addition to routine technological processes (for which regulation was a priori considered as necessary and indisputable), some key inventions were made the implementation of which allowed for raising both the nuclear and radiation safety to a higher qualitative level.

In the given paper an attempt is made to characterize both the role and place of the above actions in the technological sequence of utilization operations and also to demonstrate the amount and contents of R&D work to be performed at a stage prior to implementation.

It makes sense to consider the safety status, as well as the procedures & techniques ensuring safety of utilization operations with respect to every utilization stage, such as [1]:

- Nuclear Submarine (NS) decommissioning;
- storage with fuel of the NS, as a whole, and of the reactor facility, in particular;
- Spent Nuclear Fuel (SNF) unloading together with the attendant operations;
- NS cutting out, three-compartment unit forming (reactor compartment being placed in the middle) and Radioactive Waste (RW) storage;
- end compartment utilization.

Special attention should be given to the principal distinction of the Russian utilization procedures & techniques from the foreign ones: in the latter case SNF unloading and NS utilization are performed separately, NSs being transferred to plants after defueling only [2].

After the decommissioning, a sequence of operations is performed on Russian NSs aimed, first, at ensuring their nuclear safety. They include the following work:

A.A. Sarkisov and L.G. LeSage (eds.),
Remaining Issues in the Decommissioning of Nuclear Powered Vessels, 125–130.

- reactor facilities and servicing systems are brought to a condition complying with the requirements of both nuclear and radiation safety at minimum servicing;
- air is removed from the primary circuit and the facility is cooled down; pressure in the primary circuit is maintained at 10-15 MPa; the high-pressure gas system disconnected from the primary circuit; power supply cables for drivers of the control & protection system are cut off and their ends insulated; hand drivers for the compensating group of neutron absorbers are stopped and welded, pumps are cut off and their cables cut;
- a special water-chemical regime is introduced in order to minimize both the equipment corrosion and radionuclide release into the primary circuit medium;
- technical & organizational measures are undertaken to eliminate any unauthorized access to the NS Steam Producing Facility (SPF) compartments;
- temperature in the NS rooms is maintained at a level of ≥ 5° C.

An almost complete staff of highly skilled NS crews carries out the above operations.

During the following "waiting-for-utilization" stage of NS waterborne storage the amount of operational checks as well as the number of servicing personnel considerably reduces. Its professional skill also decreases. Appreciable worsening of the technical condition of NSs and their equipment due to both natural aging and the lack of necessary servicing further aggravates the situation. As mentioned above, at this stage NSs are stored afloat with fuel [3].

NS waterborne storage with SNF in the reactors is a forced measure resulting directly from mass decommissioning of Russian NSs during the last decades. The mean time of SNF storage within the decommissioned NSs is about 15 years, in some cases up to 35 years.

Thus, the cited data confirm the high-priority of the task of accelerating de-fueling operations at the decommissioned NSs from the standpoint of ensuring nuclear safety.

From a quantitative viewpoint, the hazard of NS storage with SNF can be characterized by radiation consequences of potential accidents and also by the value of radiation risk. The latter parameter is a complex characteristic of the radiation impact, which takes into account the consequences of both the radiation hazard and exposure. Moreover, an assessment of the radiation risk allows for distinguishing accidents which make the largest contribution to the potential hazard among all off-normal situations. The performed assessments have demonstrated that at a stage of NS storage with SNF the highest radiation risk characterizes the 'NS-sinking' accident in a case that its primary circuit is depressurized. This is due, first, to a high probability of the initial event occuring (accompanied, however, with relatively low radiation consequences). Any prolongation of NS waterborne storage with SNF raises both the probability of its sinking and, consequently, the radiation risk.

Only de-fueling guarantees non-initiation of nuclear accidents on the decommissioned NSs. However even at the actual NS de-fueling rate (up to 16-18 NS/year) these operations can only be completed by 2007 to 2010. Thus, the development & implementation of measures contributing to improvement of nuclear safety, as applied to waterborne NS storage with SNF, is necessary.

One of such measures consists of completely removing the water from the NS reactors. In case of water-moderated reactors this measure eliminates any probability for the reactor to reach the critical state under any actions with control rods up to their removal. It is important that such a measure eliminates the probability of initiating a nuclear accident not only when mishandling but also improves safety in cases of criminal negligence and/or terrorist attacks.

Prior to implementing the above "dry" SNF storage mode for Russian decommissioned NSs, careful calculated foundations were necessary. An analysis of chronological information related to NS decommissioning demonstrated that for most NSs of the first- and the second-generation SPF operation was completed five and more years ago. Therefore at the considered time the decay heat power in the core did not exceed 1 kW and, consequently, the temperature of Fuel Element (FE) surface varied between 150-180° C for admissible maximum of 350° C. Thus, according to the calculations performed, after one year following the shutdown of the Power Reactor Facilities (PRFs) of the first- and the second-generation NSs the temperature of the FE surface (for the case of "dry" core storage) decreased below 320° C.

In keeping with the joint resolution of the Russian Federation (RF) Minatom, Rossudostroenie and Ministry of Defense of June 20, 2001, removal of water from the reactor core was recommended for implementing at all Russian waiting-for-utilization NSs. Dewatering of the core & the primary circuit systems does not require special innovation and can be performed using the available technical resources; the procedure is rather simple and can be carried out without dismantling operations. At fully equipped PRFs the procedure is realized using regular systems; at partly dismantled PRFs the use of simple facilities and devices is suggested.

After dewatering some circuit elements (such as: reactor lower head area, steam generator shells, heat exchanger of circuits 1-3) still contain small amounts of coolant. However calculations of the primary circuit strength, when freezing (25 cycles), with consideration for the least resistant element, have revealed no destruction. By this its is meant that potential power supply failures accompanied with a decrease of the NS room temperature below 0° C will not result in the circuit depressurization.

Performed bench-tests of SNF "unloadability" after long-term dry storage have demonstrated that the efforts for SFA blasting & extracting changed only slightly. Note that, when testing, special corrosion products were introduced into the medium prior to dewatering. It was assumed that these particles would remain in gaps contributing to an increase of the blasting effort. However only a 12-16% increase in the effort was obtained and this stayed within the relevant technological requirements with a considerable margin. Repeatedly performed SNF unloading from dewatered reactors did not involve technical difficulties either.

SNF storage within dewatered reactors ensures complete nuclear safety of the reactor facilities; at the same time the hazard resulting from radionuclide release into the environment still persists. Taking into account equal the probability of occurring natural or technogenic accidents, one can state that the period of safe SNF storage in NS reactors depends on: -the SNF technical condition and -the pressurization of its "package" (by the latter the strong hull & the primary circuit structures of NSs are understood).

As long as NS the is stored afloat, its strong hull preventing the activity release into the environment can be considered as the first safety barrier. Its depressurizations (i.e. NS sinking or marine water ingress into SPF compartment) results in a contact between the environment and the primary circuit surfaces.

In such a case the second safety barrier – the primary circuit structures - becomes active. Corrosion processes on the contaminated equipment located within the compartment (reactor vessel & caisson of the metal-water protection tank) lead to some increase of radioactivity release into the environment. However, because of a relatively small surface and low corrosion rates such an increase is not large.

Depressurization of the primary circuit due to technical condition of PRF or to effects of corrosion during storage results in the a direct contact between fuel and the environment. Depending on fuel condition, either contact of the environment with FE claddings and other primary circuit surfaces (and, correspondingly, the transfer of active corrosion products contained on these surfaces) or a release of fission products directly into the environment (in the case of FE depressurization) is possible.

In such a situation the environment contamination level will depend on the degree of FE leaktightness determining the area of direct contact between the environment and fuel composition.

Thus, an estimate of potential duration of SNF safe storage within NS reactors and, indirectly, the priority for SNF unloading is determined by indices of the status (depressurization level) of the following three safety barriers [4]:

- NS strong hull;
- primary circuit;
- fuel element claddings.

Based on analysis of their status, the maximum duration of Russian decommissioned NS storage with fuel was assessed. Using these data, de-fueling of 17 decommissioned NSs in the course of three years was recommended. The recommendations were prepared in 1999; by the present time 12 NSs (out of the indicated 17 NSs) have been de-fueled. For other NSs the duration of waterborne storage complies with the time-schedule agreed to in the utilization program; by this we mean that it depends only on industrial capacities of the plants performing utilization operations.

Thus, implementing the technique of SNF unloading from dewatered reactors is an important innovation from the viewpoint of improving nuclear safety during nuclear vessel utilization operations.

NS de-fueling represents a rather complicated task. One should take into account: the real capacities of: floating servicing vessels, coastal SNF accepting complexes and storage facilities; the technical condition of reloading equipment and many other factors. At the preparatory stage an important amount of attendant operations should be performed. Many of them fall into the category of ‘potential by radiation-dangerous’ work. The documents regulating the SNF unloading process include a list of such actions together with technical requirements for their execution.

NS de-fueling is carried out after depressurization of both the NS strong hull & the reactor, i.e. when two of the above-considered safety barriers are inoperable. Therefore a set of technical & organizational measures has been proposed aimed at eliminating any probability of initiating accidents at the given stage. They include, in particular, SNF unloading from an already dewatered reactor guarantying nuclear safety of the PRF. For lack of moderator in the reactor, the probability of a reaching critical state is eliminated at any position of the control rods.

To substantiate SNF unloading from dewatered reactors, potential irradiation levels on the plate of fuel reloading installation were assessed. In keeping with these calculations, SNF unloading from a dewatered reactor of the second-generation NS can be performed 3 years following PRF shutdown, whereas for the third-generation NSs after 5 years due to reaching a rather satisfactory radiation situation: if observing the technological requirements, the maximum admissible levels established by Russian Radiation Safety Standards (RSS-99) in force are not exceeded.

Potentially, the subsequent waterborne storage of three-compartment units (including reactor compartments) presents a radiation hazard for the environment. If they should sink, radionuclide leak age from the unit into the environment is possible.

Analysis of both NS reactor facility embodiment and power compartment setting-up [5] allows for identifying principal surfaces for, which contact with the environment can result in radioactivity release. These are:

- a part of NS strong hull under the reactor;
- the outer surface of the reactor vessel and the surface of the caisson of metal-water protection tank facing the reactor;
- inner surfaces of both the reactor and of the reactor internals.

Loading of RWs (produced during NS utilization operations) into the Reactor Compartment (RC) contributes to an increase of its radioactivity (in keeping with our calculations, this "addition" makes up 1 – 3% of the integral RC activity).

Five years after NS PRF shutdown, the integral activity of constructional materials is determined by ^{55}Fe and estimated at $8{,}7 \cdot 10^4$ Ci (in the case of the second-generation NSs). After 25 years the activity will be determined by: ^{63}Ni ($1{,}6 \cdot 10^3$ Ci), ^{55}Fe ($4 \cdot 10^2$ Ci) and ^{60}Co ($4{,}8 \cdot 10^2$ Ci). During the subsequent storage only one ^{63}Ni becomes dominant.

The activity release rate depends on a number of parameters including:

- corrosion rate and corrosion product release;
- composition & dispersion of corrosion products;
- activity of the corrodent metal;
- area of surfaces contacting with marine water;
- efficiency of protective barriers separating the radiation source from the environment.

An increase in the area of the surfaces contacting with marine water is due to depressurization of safety barriers. The amount of radonuclides released into the environment depends on the type of safety barrier destructions. Because the most probable damage (determining depressurization) is a narrow gap or a local corrosion failure, the diffusion mechanism of radionuclide release is assumed. In keeping with this mechanism, only soluble products are released making up 10% at the most of the integral contamination.

There is a good reason to believe that any emergency sinking of three-compartment units (because of relative by low corrosion rates and the specific medium exchange type) will not lead to by appreciable worsening of the radiation situation even near the sunken facility. The results of monitoring the environment in the area of NS "Kursk" collapse as well as in the vicinity of NS "Komsomolets" confirm this conclusion.

However such an optimistic forecast does not mean that sinking of three-compartment units can be tolerated. One needs to ensure their reliable floodability and pressurization until (at least) coastal sites for reactor compartment long-term storage will be commissioned.

Issues of both transportation and storage of SNF unloaded from NSs, when performing utilization operations, should be considered as organizational measures rather than technical ones. In this case nuclear safety is ensured in full measure. If SNF unloading – SNF transporting time-schedule is well interrelated, no serious radioecological problem arises. Issues of SNF storage at coastal servicing enterprises require a special solution; however they should be considered within the framework of individual programs separately of NS utilization procedures & techniques.

Thus, from the aforesaid the following conclusions can be drawn:

1. Implementation of dewatered core storage for reactors of "waiting-for-utilization" decommissioned 'afloat' NSs ensures their nuclear safety in full measure. In addition, in such a situation both technical and maintenance requirements for the reactor facility storage are considerably simplified.
2. The actual technical condition of most decommissioned Russian NSs does not limit de-fueling due dates established in the utilization program. In keeping with the assessments performed, 5 NSs need to be de-fueled in the immediate future.
3. SNF unloading from NSs coming for utilization should be performed after primary circuit dewatering. This operation provides for additional safety when performing the most dangerous work related to nuclear safety under conditions of depressurization of both the NS strong hull and the reactor.

References

1. Mitenkov, F, Aksenov, E., et al. (1997) Priority Problems of Nuclear Submarine Decommissioning and Utilization, *Atomic Energy*, **82**, n°2, pp. 146-149 (in Russian).
2. Hutchison, J. (1999) *Nuclear Energy*. **38**, n° 5 pp. 315-319.
3. Aksenov, E., Vavilkin, V., et al. (1999) Improving Safety when Storing Vessel Power Reactor Facilities of the Decommissioned Nuclear Submarines Prior to Utilization, in A. Sarkisov and A. Tournyol du Clos (eds.), *Analysis of Risks Associated with Nuclear Submarine Decommissioning, Dismantling and Disposal*, Kluwer Academic Publishers, pp. 33-40.
4. Andreiev, N., Vavilkin, V.N., ibid, pp. 81 – 85.
5. Andreiev, N., Vavilkin, V.N., et al. (1999) *Proceedings of International Conference «Afterword to «The White Book» (A. Jablokov Commission)»* Nizhnij Novgorod, pp. 29-33 (in Russian).

ANALYSIS OF THE RADIATION RISK FOR THE POPULATION AT DIFFERENT STAGES OF NUCLEAR SUBMARINE UTILIZATION

ALEXANDER BLEKHER
Research Institute of Industrial and Marine Medicine
St. Petersburg, Russia

1. Introduction

The whole process of Russian Nuclear Submarine (NS) utilization represents a sequence of many operations of which the principal ones are:

- Decommissioning;
- NS waterborne storage with Spent Nuclear Fuel (SNF);
- SNF unloading;
- NS waterborne storage after SNF unloading;
- NS moving to special shipyards;
- Loading of additional Solid Radioactive Wastes (SRWs) into NS Reactor Compartments (RC);
- RC cutting out and preparing for long-term storage;
- Subsequent cutting of tail compartments into sections down to merchantable scrap.

Many utilization operations (e.g., SNF unloading, coolant removal, contaminated equipment dismantling, et al.) are potentially dangerous from both the nuclear and radiation standpoint. Some operations are performed at unsealed protective barriers, such as the reactor baffle and NS strong hull. As a result, when depressurizing the Nuclear Steam Producing Facility (NSPF), radionuclides are released directly to the atmosphere. In emergency cases serious consequences for both the population and the environment can occur, especially as the shipyards are located in the immediate vicinity to habitable areas. Thus, at "SMBE Zvezdochka" Federal State Unitary Enterprise in Severodvinck SNF is taken out of NS reactors at only about 600 m from Jagry-settlement.

During the utilization process many high-level Liquid Radioactive Wastes (LRW) and SRWs are being produced including the cut out RC. This results in a considerable radiation load for both the environment and the population of the concerned region.

The radioecological danger increases further due to waterborne storage of many NSs and cut out RCs (particularly of those containing SNF) especially as their storage in afloat conditions can last for tens of years.

A.A. Sarkisov and L.G. LeSage (eds.),
Remaining Issues in the Decommissioning of Nuclear Powered Vessels, 131–142.

The situation is aggravated because many decommissioned NSs have worn out basic equipment of their Power Reactor Facilities (PRF) and other structures.

However, despite the above facts, analyses of measurements taken in different environment media within the sites of leading atomic shipbuilding enterprises in the Russian North-West Region have demonstrated almost no contamination effect on the air, soil, plants, marine water, algae and bottom sediments due to the above radiation-dangerous operations. The activity in the environment media is close to levels created by both natural radionuclides and global fallout. E. g., the activity of such specific radionuclides as ^{60}Co, ^{90}Sr, ^{137}Cs, at al., in marine water within shipbuilding enterprise sites is appreciably below that of ^{40}K varying usually from 9 through 12 Bq/l. With the exception for some restricted areas immediately adjacent to NS utilization sites, gamma dose rates in the environment media in and out the enterprise site do not exceed the limits of natural radiation background variations.

Incidentally, various off-normal situations can occur at different NS utilization stages as well as during their repair. Such emergencies can be accompanied with uncontrolled radionuclide release into the environment media. Off-normal situations are possible during NS waterborne storage, SNF unloading and also when performing dismantling operations. They can result from internal reasons (technical device & equipment failures, flowsheet violations, personnel errors) as well as from external reasons (extreme weather conditions, vessel collisions, aircraft crashes, explosions (including diversions), et al). The situation is further aggravated because principal equipment at many decommissioned NSs is worn out.

In the given paper an estimate of the scale of the radiation consequences resulting from the principal types of potential accidents accompanied with radionuclide release into the atmosphere is presented. The list of accidents under consideration is given in Table 1.

TABLE 1. Principal Types of Considered Accidents

№	Accident type	Accident probability estimate
1	Sinking of NS with SNF	$2{\cdot}10^{-3}$ year^{-1}
2	Accidents related to SCR and the subsequent radionuclide release	$1{\cdot}10^{-5}$ case/operation
3	Primary circuit loss-of-coolant accident	$4{\cdot}10^{-4}$ case/repair
4	Fire within the power compartment of NS under utilization	$1{\cdot}10^{-2}$ case/operation
5	Unauthorized LRW dumping into water area	$(1\text{-}5){\cdot}10^{-2}$ year^{-1}
6	Fire in LRW storage facility	$(0,5\text{-}1){\cdot}10^{-4}$ year^{-1}
7	LRW storage facility collapse	$1{\cdot}10^{-6}$ year^{-1}
8	Sinking of the cut out reactor unit during haulage	$1{\cdot}10^{-2}$ case/operation

2. Bases of the Radiation Risk Assessment Methodology

By the radiation risk are meant the manifestations of stochastic effects due to the impact of ionizing radiation on human health. In keeping with the present-day concepts [1], they include both carcinogenic and genetic effects of the radiation exposure. To make decisions related to human safety on the basis of risk analysis, one has to establish a system of risk indices including safety standards, control levels, post-accident intervention levels, negligible risk levels, etc.

Up to now simplified risk analysis approaches have been applied in most cases. Generally, it consists in a comparison of the environmental contamination level or dose impact on human beings/biota with the corresponding maximum permissible values. Such a so-called "dose-approach" is commonly applied in the radiation protection wherein the effective dose is used as the base index.

However, when performing a detailed-level risk assessment, the parameter of effective dose cannot be used as a basic one. Due to averaging and time integration of some coefficients forming a part of the effective dose parameter, a phenomenon of the radiogenic cancer manifestations after a few-decade delay only is not taken into account. Such a delay is fundamentally important from economical, social and psychological point of view. Moreover, the efficient dose considers neither local nor age-specific peculiarities of a cohort of personnel/population for which the consequences of accidental radiation exposure are evaluated.

Thus, when estimating health damages to personnel/population in a case of radiation accident resulting in a relatively high exposure level, the dose approach is insufficient. To assess properly the damage and make well-grounded decisions on protective measures and rehabilitation procedures, both the risk assessment & analysis are necessary.

Depending on a specific task (optimization of the radiation protection under normal operating conditions, emergency situations, protective measure planning, insurance, payment of different indemnities, etc.), the following parameters are recommended as risk indices [2]:

- harm G to human health (lost years of healthy life) to assess damage to the population health and optimize protective measures;
- mortality/disease risk R (reduced to mortality risk) to assess risks when managing post-accident situations including the establishment of protective measure levels, pecuniary recompenses for life/health damage, etc.;
- damage q to human health expressed in lost years of life and referred to the year of a risk source impact (reduced damage) to perform the risk normalization;
- causation probability of initiating radiation-induced cancers to establish cause-and-effect relations between the exposure and cancerous disease/death (performed by medical commissions of experts);
- individual dose D_{rm} and collective equivalent dose S_{rm} for red marrow;

- individual E^p and collective $S^P{}_E$ incomplete effective doses (excepting the red marrow dose) to perform an a priori damage assessment;
- individual E and collective S_E effective doses to ensure the radiation protection in normal/close-to-normal conditions;
- death/morbidity events (deterministic exposure effects) to assess the damage (a priori as well as post factum).

Values E^p and $S^P{}_E$ are calculated similarly to the integral effective dose case but through excluding red marrow from the affected organs (tissues).

Most completely the risk (damage) for human health can be expressed via a generalized damage index, i.e. the shortening of expected healthy lifetime due to morbidity and (or) mortality resulting from the considered risk factor impact. As a rule, it is namely this risk index that is recommended as a basic value when calculating the economic damage due to taking measures on the population protection against the effects of one or another risk factor, upgrading the public health level, etc.

Death (morbidity) risk index should be used to perform a detailed risk assessment when managing post-accident situations including the establishment of protective measure levels, pecuniary compensations for life/health damage, etc.

To perform an a priori assessment of a radiation accident damage, some scientists [3] use an expected damage value $\mathfrak{R}$ calculated by multiplying the probability of an event into full damage due to the event realization. In this case the damage (in person-year) is expressed as follows [1]:

$$G = g_E\ E$$

where g_E – averaged damage coefficient for ionizing radiation, E – effective dose for one year or accident.

In keeping with [1], g_E = 1 person year/person Sv for the population, g_E = 0.8 person year/person Sv for professional exposure.

If applying the above reasoning to different practical situations including potential accidents at different NS utilization stages, the basic risk analysis results should include:

- results of analyzing conditions of the accident initiation and development;
- accidental risk assessment results.

3. Estimate of the Probability of Initiating Accidents at Different Nuclear Submarine Utilization Stages

Generally, when estimating the probability of initiating accidents, an approach based on statistical treatment of a large number of events (off-normal events, incidents or failures) is applied in all technogenic activity areas.

However NSs represent very sophisticated weapons systems, and the knowledge of only statistics related to an event with negative consequences is insufficient to perform probabilistic analysis. Thus, the only way to carry out a quantitative assessment of the probability/rate of initiating accidents is an expert estimate [4, 5]. Such-type assessments were considered in detail in a previous publication by the author [6] and are summarized below.

3.1. ACCIDENTS AT NS DECOMMISSIONED STAGE

The stage of NS decommissioning comprises a set of actions aimed at transferring the reactor facility and its supporting systems into a nuclear/radiation-safe status. Accidents that can potentially occur at this stage result from external impacts or intentional activities, however the probability of their initiation is considered as minimal. In addition, any radionuclide release into the environment at the decommissioning stage is impossible since the protective barrier system is still operable. Therefore accidents occurring at the NS decommissioning stage are not considered in the paper.

3.1.1. SINKING OF NS WITH SNF

The actual status of many NSs stored with SNF causes a rather high potential risk for the concerned utilization stage. Thus, the "afloat" period (including the operation time) of the first-generation NSs makes up 36,8 years on average, whereas the whole service life of the NSs under consideration is equal to 40 years in keeping with their design documentation.

In real practice NS sinking has occurred principally on open sea. At least two NSs sank during their waterborne storage period and shipyard repair operations. The conservative estimates of the probability of initiating such accident type make up about $(2\text{-}3)\cdot 10^{-3}$ case/year.

3.1.2. ACCIDENTS DUE TO SPONTANEOUS CHAIN REACTION (SCR) INITIATION FOLLOWED BY A RADIONUCLIDE RELEASE

When unloading SNF from NS reactors, 2 principal safety barriers (the strong hull and the pressurized primary circuit) become unsealed. In the case that SNF unloading is accompanied with violations in the flowsheet, the probability of SCR initiation increases. It should be stressed that SCR accident is most hazardous from the radioecological consequence viewpoint.

One possesses more or less certain information on 2 SCR initiation cases during nuclear fuel reloading operations at operable NSs, which consequences can be compared to those resulting from a potential SCR during SNF unloading from the decommissioned NSs. The relevant expert estimate gives $(1\text{-}3).10^{-3}$ cases per one SNF unloading operation out of a reactor still containing water.

If considering SNF unloading from already unwatered reactor, the above value should be multiplied by the probability of faulty filling of the reactor with water during SNF unloading period. Fortunately, such cases had never occurred in real practice, and the above factor can be assessed ideally as maximum 1 case per all subsequent SNF unloading operations. Such an estimate can be explained as follows: if a faulty reactor filling with water should occur during the unloading period, procedures and techniques of SNF unloading will be changed immediately up to carrying out the operation in question at a special slip or in dock. Thus, the most conservative estimate of the SCR initiation probability for future SNF unloading operations makes up $< 1 \cdot 10^{-5}$ case/operation at more probable value of $1 \cdot 10^{-6}$.

3.1.3. PRIMARY CIRCUIT LOSS-OF-COOLANT ACCIDENT DURING NS WATERBORNE STORAGE WITHIN SHIPYARD SITE

Accidents with large loss-of-coolant in the primary circuit and partial core collapse fall into the category of Maximum Design-Basis Accidents (MDBA) and are necessarily considered in NS design when creating safety systems. In real practice such-type accidents took place for the first-generation NSs during the initial period of their operation and were due mainly to numerous leaks in steam generators. Severe radiation consequences were recorded only within the through-pass corridors for the impossibility of realizing efficient evacuation of the NS crews. One full-scale MDBA (a break in the primary circuit) resulting in the primary circuit unwatering and core melting took place almost 40 years ago on open sea. Therefore the upper estimate of initiating MDBA in shipyard conditions makes up $< 4 \cdot 10^{-4}$ cases per one repair with the probable estimate of below $1 \cdot 10^{-4}$ case/repair.

3.1.4. FIRE IN POWER COMPARTMENT OF NS UNDER UTILIZATION

Different-type fires occur for NSs rather frequently because of their extremely high available power, enormous quantity of electric lines and considerable high-pressure air stocks. The collected experience demonstrates that, when utilizing NS, the fire initiation probability can be estimated at $1 \cdot 10^{-2}$ cases per operation.

3.1.5. CUT OUT REACTOR UNIT (RU) SINKING DURING HAULAGE

The probability of navigation accidents depends on the vessel tonnage, waterway traffic intensity and length, weather conditions and safety measures. Taking into account that the very first case of RU sinking during haulage will result in immediate tightening of safety measures, the probability of sinking is estimated at $< 1 \cdot 10^{-2}$ cases per operation.

3.1.6. UNAUTHORIZED DUMPING OF LIQUID RADIOACTIVE WASTES INTO SEA WATERS

During the utilization process many high-level LRW and SRW are being produced. Potential accidents resulting from Radioactive Waste (RW) management should be also considered within the framework of "NS utilization accident" category.

Thus, due to a navigation accident an important LRW dumping is possible when sinking a tanker or floating tank with LRW. The probability of such an event depends on technical status of LRW transporting vessels, waterway traffic, weather conditions, professional skill of the involved crews, conscientiousness in managerial control and surveillance over the operation. If observing all standard requirements, the probability of unauthorized dumping into the marine environment does not exceed $(1\text{-}5)\cdot 10^{-2}$ case · year^{-1}.

3.1.7. FIRE IN LRW REPOSITORY

In general case the probability of initiating a fire depends on the observance of fire safety requirements at every individual enterprise. Usually this index does not exceed $(0{,}5\text{-}1)\cdot 10^{-4}$ case·year^{-1}.

3.1.8 LRW REPOSITORY COLLAPSE

LRW repositories represent specially designed reinforced concrete storage facilities. Therefore their collapse can be due to extreme circumstances only, such as:

- fall of heavy aircraft/missile fragments followed by explosion of the remainder fuel;
- rare natural cataclysms (large force earthquake, lightning strike, tornado, landslide).

The probability of such an event exceeds $(10^{-6} - 10^{-7})$ year^{-1} in no case, thus, it can be taken equal to $1\cdot 10^{-6}$ case · year^{-1}with a safety margin.

4. Analysis of the Radiation Risk Resulting from Radionuclide Release into the Environment

Individual and collective population doses resulting from radionuclide release into the environment were calculated with reference to the following most severe accidents (for more details see [6]):

- a loss-of-coolant accident (MDBA) at nuclear vessel;
- Inflammation (local fire) within RC during utilization operations;

- fire (overall fire) during nuclear vessel utilization operations;
- SCR at a nuclear vessel.

Calculations were performed with respect to a conventional locality of 2·2-km area and 20,000 population, the distance from the source term to the nearest buildings being equal to 1 km. When analyzing the situation, most unfavorable radionuclide transfer conditions in both air and water were chosen. As a result physical-geographic peculiarities within NS utilization enterprise regions were smoothed, and the magnitude of negative effects depended mainly on relative positions of NS utilization enterprises and their nearest settlements.

The performed calculation of collective radiation risk indices is based on the results of the above-cited work. The calculation procedure is illustrated in Table 2. Figures presented in the last column characterize the magnitude of consequences for every considered accident using the IAEA Scale [7, 8].

TABLE 2. Scheme and Results of Calculating Both the Radiation Risk and Damage

Accident type	Time of accident onset	Collective dose, person*Sv	Collective risk per accident	Damage per accident, pers*year	Accepted probability of initiating accident, 1/year	Collective risk with consideration for the probability of initiating accident, 1/year	Damage with consideration for the accident probability, pers.*year/year	Magnitude (level) of negative consequences in keeping with the IAEA scale
Loss-of-coolant Accident	Day	$3{,}50 \cdot 10^{-02}$	$2{,}55 \cdot 10^{-03}$	$3{,}50 \cdot 10^{-02}$	$4 \cdot 10^{-04}$	$1{,}02 \cdot 10^{-06}$	$1{,}40 \cdot 10^{-05}$	Level 3 "Serious event"
	Night	$1{,}19 \cdot 10^{-01}$	$8{,}69 \cdot 10^{-03}$	$1{,}19 \cdot 10^{-01}$		$3{,}48 \cdot 10^{-06}$	$4{,}76 \cdot 10^{-05}$	
Local Fire	Day	$4{,}67 \cdot 10^{-02}$	$3{,}41 \cdot 10^{-03}$	$4{,}67 \cdot 10^{-02}$	$1 \cdot 10^{-02}$	$3{,}41 \cdot 10^{-05}$	$4{,}67 \cdot 10^{-04}$	Level 3 "Serious event"-
	Night	$2{,}34 \cdot 10^{-01}$	$1{,}71 \cdot 10^{-02}$	$2{,}34 \cdot 10^{-01}$		$1{,}71 \cdot 10^{-04}$	$2{,}34 \cdot 10^{-03}$	Level 4 "Accident within the facility"
Overall fire	Day	$2{,}25 \cdot 10^{+00}$	$1{,}64 \cdot 10^{-01}$	$2{,}25 \cdot 10^{+00}$	$1 \cdot 10^{-02}$	$1{,}64 \cdot 10^{-03}$	$2{,}25 \cdot 10^{-02}$	Level 4 "Accident within the facility"- Level 5 "Environmentally hazardous accident"
	Night	$1{,}05 \cdot 10^{+01}$	$7{,}67 \cdot 10^{-01}$	$1{,}05 \cdot 10^{+01}$		$7{,}67 \cdot 10^{-03}$	$1{,}05 \cdot 10^{-01}$	
SCR	Day	$3{,}07 \cdot 10^{+01}$	$2{,}24 \cdot 10^{+00}$	$3{,}07 \cdot 10^{+01}$	$1 \cdot 10^{-05}$	$2{,}24 \cdot 10^{-05}$	$3{,}07 \cdot 10^{-04}$	Level 5 "Environmentally hazardous accident"
	Night	$7{,}72 \cdot 10^{+01}$	$5{,}63 \cdot 10^{+00}$	$7{,}72 \cdot 10^{+01}$		$5{,}63 \cdot 10^{-05}$	$7{,}72 \cdot 10^{-04}$	

The population exposure risk coefficient was taken equal to $7{,}3\cdot10^{-2}$ 1/person·Sv [1, 9], the damage coefficient g_E to 1 person·year/person·Sv.

The ultimate risk index value was obtained with consideration for technical probability of initiating all of the concerned accidents.

An analysis of the calculation results allows for the following conclusion. Due to different air dispersion conditions between day and night, the radiation risk is higher if an accident occurs at night. Maximum individual & collective doses can be realized in the case of initiating SCR accident; when considering the case of the "overall fire at NS under utilization", the concerned doses are slightly below those of SCR accident. Correspondingly, the IAEA level of negative effects level ranks first in SCR accident case. However, when taking into account the low technical provability of SCR initiation & development, in the case of NS fire (both "overall fire" and "local fire" are considered) the radiation risk exceeds that of SCR accident.

With respect to the case of overall fire at NS under utilization the value of maximum collective radiation risk is $7{,}67\cdot10^{-03}$ 1/year.

The anticipated damage reaches also its maximum in the case of SCR accident and is $7{,}72.10^{+01}$ person·year; however with consideration for the accident probability the anticipated damage is maximal in the "overall fire" case ($1{,}05.10^{01}$ person·year/year).

5. Unresolved Issues of the Radiation Risk Assessment Methodology when Utilizing NS

The risk assessment task comprises many issues requiring an adequate solution. As applied to different NS utilization stages, they can be stated as follows: [10]

- Selection of a subject of the impact under consideration (an administrative unit/units wherein the risk is assessed) and performing investigations of its geographical, economic, social, etc. characteristics;
- Thorough estimate of the radiation and radioecological situation within the concerned territory; description of the available radiation & radioecological monitoring systems;
- Assessment of the integral technogenic risk within the concerned territory as well as of its structure and principal risk-forming factors;
- Studying the morbidity structure within the territory under consideration;

- Analyzing data on both the population density and living conditions which determine protectability in the case of a harmful effect occurrence;
- Obtaining data on the structure & number of critical population groups (for which individual, social and collective risk is to be assessed) as applied to the territory under investigation;
- Assessing the probability of initiating different-type NS utilization accidents due to any accident at other technogenic risk facilities within the studied territory;
- Application of the integrated risk method (i.e. the integral consideration of both the risk due to living within the contaminated territory (chronic exposure) and that of acute radiation exposure);
- Use of the method of assessing the efficiency of protective measures;
- Studying both social and psychological risk perception by different population groups within the concerned territory;
- Studying the risk limits and indicating these with respect to the concerned population;
- Applying economic equivalents of the environmental safety criteria when performing NS utilization programs and optimizing protective measures;
- Organizing a risk insurance system as applied to the whole NS utilization process.

6. References

1. *The 1990 ICRP Recommendations. Publication № 60.* (1994) (Translation from English) Energoatomizdat, Moscow, (in Russian).

2. Demin, V., Kut'kov, V., Golikov, V., et al. (1999) Risk Analysis when Taking Radiation & Social Population Protection Measures, *Atomnaya Energija (Atomic Energy)* **87**, **n°. 5**, 384-395 (in Russian).

3. Demin, V., Golikov, V., Ivanov, E., et al. (2001) Standardization & Comparison of Different Risk Factors with Reference to the Population Health, *Atomnaya Energija (Atomic Energy),* **90, n°. 5**, 385-398 (in Russian).

4. Pologikh, B., Elatomtsev, V. (1998) SCR Probability when SNF Unloading and Its Consequences. Afterword to the "White Book" (A. Jablokov's Commission). *Proceedings of International Conference, January 19-21, 1998*, Nizhniy Novgorod, pp. 151-156 (in Russian).

5. Ivanov, V., Kalinkin, A., Petrov, S. (1996) Statistical Analysis of Vessel Accident Data and Proposals on Classification of Their Radiation Consequences. *Proceeding of International Conference "300-th Anniversary of Russian Navy. Navy and Shipbuilding Nowadays". 26-29 February, 1996, St.-Petersburg, Russia.* St.-Petersburg, B2-1 — B2-8.

6. Blekher, A., Kuchin, N. (2001) Analysis of the Radiation Risk at Different Stages of NS Decommissioning and Multi-Purpose Utilization. *Proceedings of MORINTEKH-2001 Conference, St. Petersburg*, v. 2, pp. 251-254 (in Russian).

7. Bukritskiy, A., Fedulov, V. (1991) International Nuclear Event Scale as Applied to Nuclear Power Plants. *Atomnaya Energija (Atomic Energy)*, **v. 71, n°. 1** (in Russian).

8. International Nuclear Event Scale to Perform Rapid Data Communication in the Case of Initiating Any Important Event from the Safety Standpoint. (1990) *IAEA Journal*, **v. 32, N2**, Vienna, Austria (translated into Russian).

9. *Radiation Safety Standards (RSS-99) Sanitary Regulations 2.6.1. 758-99.* (1999) Ministry of Health of Russia, Moscow (in Russian).

10. Lisovskiy, I. (2001) Environmental and Technogenic Risk Assessment when Utilizing Nuclear Submarines. *Proceedings of MORINTEKH-2001 Conference, St. Petersburg*, v. 2, pp. 261-264 (in Russian).

RADIATION BURDEN IN REGIONS RESULTING FROM NUCLEAR SUBMARINE HANDLING AFTER DECOMMISSIONING AND THE EXISTING UNCERTAINTIES IN THEIR ESTIMATION

B. POLOGIKH
RSC "Kurchatov Institute"
1 Kurchatova sq., 123182 Moscow, Russia

1. General concepts

The utilization concept of nuclear-powered submarines (NPS) in Russia assumes that the process is finished by the complete dismantlement of all construction of the reactor compartment (RC) upon storing for approximately 70 years.

During the first one-two years following decommissioning, the vessel owner (the Navy of Russia) prepares it for the delivery to the enterprise, which is charged to carry out the utilization. The utilization process starts with the NPS anchoring in the sea bay pending defuelling of the reactors. Then, the spent nuclear fuel (SNF) is unloaded. The NPS is delivered to the workshop where the reactor compartment is cut out from the hull and turned into the sealed package with positive floatage. The package is launched afloat and temporary stored pending transportation to the coastal site of the long-term storage (LTSS). The sealed package with the RC should be stored at the LTSS for the complete holding period of approximately 70 years. Upon this period, the package should be taken out of the LTSS and delivered to the industrial site, where all constructions should be dismantled and sorted with respect to the residual activity level. Only small amounts of metal are assumed to have high activity level, which requires final burial.

Utilization is a process of handling radiation-hazardous objects. They can grow dangerous for the environment and population in case of an accident, when radionuclides accumulated in reactors and constructions of the reactor compartments are released in the environment. In this report, we estimate the expected levels of radiation burden for aquatic life and population owing to accidents with activity release in sea water and atmosphere at various stages of the protracted utilization process. We notice the conditional character of theses estimates resulting from limited scope of experimental information on radionuclide migration and suggest experiments aimed at substantial improvement of this information.

At the beginning of the utilization process, approximately two years upon the reactor shutdown, the reactor facility inside the RC contains the following amount of radioactive materials:

- (10-100) PBq of the SNF;
- (1-10) PBq in the steel internal constructions of the reactor;
- (0.1-1) TBq in sediments on the surfaces of these constructions;

A.A. Sarkisov and L.G. LeSage (eds.),
Remaining Issues in the Decommissioning of Nuclear Powered Vessels, 143–154.

- (0.1-1) GBq in the first-contour water;
- Approximately 1 GBq and 1 MBq in metal and water of the biological shield tank, respectively.

The obtained estimate of the radiation burden for the population (sanitary and hygienic aspect) uses Radiation Safety Regulations NRB-99 [1] and Main Sanitary Rules Providing Radiation Safety OSPORB-99 [2]. In addition, we attracted recommendations on the control concentrations of radionuclides in sea water (CC) suggested by the authors of [3] in two variants: for human protection (CC_1) and protection of aquatic life against overexposure (CC_2).

The authors relate the hazard for a human (CC_1) with the nuclide intake inside the human body with food. The values of CC_1 are calculated assuming annual consumption of the following contaminated food by an adult person: 40 kg of fish, 2 kg of shellfish, and 2 kg of clam. The internal exposure dose owing to radionuclides contained in this ration is limited by the annual dose of 0.1 mSv or 10% of the main dose limit [1].

The potential risk for the aquatic life is estimated using the values of CC_2. They restrict the chronic exposure of the aquatic life by the values of 0.1 and 0.5 Gy for the sea fauna and flora, respectively. This radiation burden are considered as ecologically insignificant, because they are close to one percent of the unit lethal dose that results in the loss of 50% of the exposed individuals.

Table 1 displays the values of CC_1 and CC_2 as well as total concentrations CC that allows for the simultaneous radiation effects for the population and aquatic life taken from [3]. The second column of the table contains the values of the Intervention Levels IL^{water}; as soon as these levels in the water of the opened pools are exceeded, certain measures for the population protection should be applied [1]. The numeric values of IL^{water} correspond to the annual individual exposure dose of 1 mSv; the values of CC_1 correspond to the annual dose of 0.1 mSv. Comparing the value of IL_i^{water} for the ith isotope with the value of $10{\cdot}CC_1$, we can see that the recommended values of $10{\cdot}CC_1$ are lower or close to IL_i^{water} with the only exception for ^{90}Sr. This seems to be quite coherent.

Table 1. Control water concentrations of radionuclides, Bq/kg

Nuclide	Population		Aquatic life	CC
	IL_i^{water} [1]	CC_1 [3]	CC_2 [3]	
^{60}Co	41	0.5	2	0.5
^{59}Ni	2200	34	850	34
^{63}Ni	930	14	1100	14
^{90}Sr	5	15	60	15
^{95}Zr	150	9	3	3
^{106}Ru	20	3.3	30	3.3
^{134}Cs	7.3	1.2	13	1.2
^{137}Cs	11	1.8	30	1.8
^{144}Ce	27	1.3	6	1.3
^{152}Eu	99	2.5	7	2.5
^{154}Eu	69	1.8	5	1.8

2. Accident with the sea water penetration into the shield tank

The NPS stored afloat can sink. There is a possibility that sea water in this case would fill the reactor compartment and penetrate inside the biological-shield tank. Owing to sea tides and wind effect on the sea level in the storage bay, water exchange between the shield tank and the external water area can appear; this will result in radionuclide release in the environment.

The probability of an accident increases several years after the reactor shutdown. Up to this moment, activity of water and metal constructions in the shield tank will be determined by the long-living products of neutron activation: ^{59}Ni ($7.5 \cdot 10^4$ years), ^{60}Co (5.27 years), and ^{63}Ni (96 years). The maximum activity contribution comes from ^{60}Co (by the order of magnitude higher than ^{63}Ni and by several hundred times higher than ^{59}Ni). For this isotope, the established value of IL^{water} and the recommended value of CC are significantly lower than those for $^{59,63}Ni$ (see Table 1). Therefore, we assume that all water activity in the tank (1 MBq) and activity of the tank metal constructions (1 Gbq, see above) are related only with ^{60}Co.

Assuming water mass in the tank equal to 10 t, we estimate the average possible concentration of ^{60}Co in the tank water 100 Bq/kg. This value is four times lower than the recommended value of IL_{Co}^{water}. Therefore, according to the definition 3.12.1 OSPORB-99 [2], tank water is not a liquid radioactive waste (LRW). A single release of this water outside the NPS with dilution by the factor 50 of (mixing with 500 m^3 of sea water) would result in the decrease of ^{60}Co concentration in sea water to the value of CC_2. Mixing of the released water with 2000 m^3 of sea water lowers the ^{60}Co concentration to the value of CC_1. The considered volume 2000 m^3 amounts to only 10% of the total volume of the sunken NPS. The dilution of the released water by the factors of 50 and 200 would normally take a only short time to occur.

^{60}Co content in metal and surface sediments exceeds water content by the factor of 10^3. However, cobalt radionuclides are locked in the crystal lattices of metal and surface oxides. The relative velocity of cobalt release into sea water is restricted by the values on the order of 1 $year^{-1}$ and $10^{-3} year^{-1}$ for the desorption from sediments and the process of general steel corrosion in sea water at the temperature 1-10 °C, respectively.

Sediments hardly contain more than 1% of the total ^{60}Co activity. Thus, we can estimate possible cobalt release into sea during the first year following the accident to reach approximately 10 MBq; further annual release of ^{60}Co from main metal amounts approximately to 1 MBq. It is quite possible that the protracted release of 10 MBq of ^{60}Co will be accompanied by its dilution by sea water of the total water at least $2 \cdot 10^4$ m^3; therefore, ^{60}Co concentration in the close neighborhood of the NPS would not exceed neither IL_{Co}^{water}, nor CC_1 and CC_2.

We arrive at the conclusion that an assumed accident with sea water penetration in the biological shield tank of the NPC reactor facility approximately two years after the reactor shutdown does not result in any significant radiation burden for the region population and aquatic life.

3. Accident with water discharge from the reactor in the sea

For several years following the NPS decommissioning, the first contours of the non-defuelled NPS reactors are filled with water coolant. During this period, an accident with the NPS sinking might result in sea water penetration inside the internal space of the reactor vessel, in addition to the considered above water penetration inside the reactor compartment and the biological shield tank. In result of such an accident, a system of two volumes V_1 and V_2 appears, where V1 is the volume initially filled with the primary coolant and V_2 is the volume of the neighboring sea area of the sunken NPS. Two volumes are linked by the water-exchange path and have an outlet to the water area of the sea bay through the volume V_2.

The volume V_2 can be represented as a water space between the coast and the sunken NPS, which is similar to a pier parallel to the coastline.

For the established water-exchange regime assuming absence of absorption-desorption effects at the internal surfaces of the system, we can calculate the concentration of the ith radionuclide in the volume V_2 as

$$C_{2,i} = \frac{\omega_1 A_{0,i}}{(\omega_2 - \omega_1)V_2}[\exp(-\omega_1 t) - \exp(-\omega_2 t)] \cdot \exp(-\lambda_i t), \text{ Bq/l} \qquad (1)$$

where $A_{0,i}$ is the initial amount of ith radionuclide in the volume V_1, Bq;
ω_1 and ω_2 are the relative velocities of water and radionuclide leakage from the volumes V_1 and V_2, respectively, day^{-1};
λ_i is the nuclide decay constant, day^{-1}.

Radionuclides in question have large lifetimes. Therefore, it is convenient to consider average concentrations of radionuclides in the volume V_2, which for the time intervals $T >> \omega_1^{-1}, \omega_2^{-1}$ can be calculated as

$$\overline{C_{2,i}} = A_{0,i} \cdot (\omega_2 V_2 T)^{-1} \cdot \{\omega_1, \omega_2[(\lambda_i + \omega_1) \cdot (\lambda_i + \omega_2)]\}^{-1}, \text{ Bq/m}^3 \qquad (2)$$

An aquatic life individual in volume V_2, which accumulates radionuclides in the body with the coefficient CA, is exposed during the time T to the radiation burden D_i (owing to the absorption of the ionizing radiation energy E_i, MeV/decay)

$$D_i = 1.6 \cdot 10^{-13} \psi_i E_i \overline{C_{2,i}} (CA) T, \text{ Gy} \qquad (3)$$

where ψ_i is the fraction of energy E_i absorbed in the «source» (aquatic life body);
$1.6 \cdot 10^{-13}$ is the unit conversion coefficient for energy units, J/MeV

The preliminary analysis showed that the typical content of radionuclides in the first-contour water two years after the reactor shutdown corresponds to the data in Table 2. This set of nuclides determines internal exposure of aquatic life; the main contribution comes from beta-particles ($\psi_\beta \approx 1$, while $\psi_\gamma << 1$). To estimate typical radiation burden, we chose fish, shellfish, and.

Table 2 displays the calculated values of $\overline{C_{2,i}}$ and $D_i^{(\beta)}$ for T = 1 year in the volume $V_2 = 10^5$ m^3 assuming $\psi = 1$, $\omega_1 = 0.04$ day^{-1}, and $\omega_2 = 0.2$ day^{-1}. The values of CA are taken from [4]. The values of $D_i^{(\beta)}$ for $^{90}Sr + {}^{90}Y$ are calculated using CA for yttrium that exceeds CA for strontium by the factor $10\text{-}10^3$.

Table 2. Radionuclide concentration and aquatic life exposure doses in case of water leakage from the reactor in the sea

Radionuclide	$T_{1/2}$, years	E_β, Mev/decay	A_0, GBq	$\overline{C_2}$, mBq/l (during the 1st year)	$D_i^{(\beta)}$, μGy per year		
					fish	shellfish	algae
^{60}Co	5.27	0.096	0.20	27.4	13.2	65.9	132.3
$^{90}Sr+^{90}Y$	29.1	1.124	0.03	4.1	0.50	23.2	23.2
^{95}Zr	0.18	0.117	0.02	2.1	0.02	6.1	3.7
$^{106}Ru+^{106}Rh$	1.01	1.419	0.02	2.6	0.04	37.6	37.6
^{134}Cs	2.06	0.157	0.04	5.5	0.44	0.1	0.2
^{137}Cs	30	0.180	0.15	20.6	1.90	0.6	0.9
$^{144}Ce+^{144}Pr$	0.78	1.282	0.10	12.8	4.10	411.1	411.1
^{152}Eu	13.3	0.083	0.01	1.4	0.18	4.2	1.8
^{154}Eu	8.8	0.247	0.01	1.4	0.52	12.1	5.2
Total			0.58	77.9	20.90	560.9	616.0

It follows from Table 2 that the exposure doses $D^{(\beta)}$ for the aquatic life in the sea water area $V_2 = 10^5$ m^3 are determined mainly by $^{144}Ce+^{144}Pr$, ^{60}Co, $^{106}Ru+^{106}Rh$, and $^{90}Sr+^{90}Y$. The total radiation burden for the first year following the accident does not exceed 0.56 and 0.62 mGy for sea fauna and flora, respectively. These values are substantially lower than the conditionally permissible annual burden 100 and 500 mGy for fauna and flora, which are determined by the recommended values of CC [3]. The concentrations $\overline{C_{2,i}}$ are lower than the values of CC_1 and CC_2 by a factor of several dozens (for ^{60}Co), hundreds ($^{144}Ce+^{144}Pr$), and thousands ($^{106}Ru+^{106}Rh$) (see Table 1).

For the constant value of ω_1, the average concentrations $\overline{C_{2,i}}$ and radiation burdens $D_i^{(\beta)}$ are virtually inversely proportional to the value of ω_2. The obtained results correspond to the water exchange rate in the volume V_2 once per five days ($\omega_2 = 0.2$ day^{-1}). Assuming even slower exchange rate in V_2, once per 30 days, we obtain the values of $D_i^{(\beta)}$ approximately six times greater than those in Table 2. Nevertheless, these increased doses are significantly lower than the conditionally admissible levels.

In general, we can conclude that water discharge from the non-damaged reactor in the sea would not produce unacceptable (exceeding the admissible levels) radiation burdens for the population and aquatic life of the region where the utilized NPS are located.

3.1. DESORPTION AND DISCHARGE OF RADIONUCLIDES FROM THE SURFACES OF THE INTERNAL CONSTRUCTIONS OF THE REACTOR INTO SEA WATER

Possible consequences of the considered accident with the NPS sinking and sea water penetration inside the reactor are not restricted by the discharge of radionuclides from the water of the first contour. Sea water comes in contact with the surface oxide films on constructions of the first contour and destroys them; as a result, radionuclides from the surface sediments are desorbed in the water. Later, water contacts the main metal of the constructions, induces its corrosion and further release of radionuclides accumulated in metal constructions owing to neutron activation.

Considering the same system of the linked volumes V_1 and V_2, we introduce the radionuclide source $A_i^{(n)}(t) = \alpha_1 \cdot A_{0,i}^{(n)} \exp[-(\alpha_1 + \lambda_i)t]$ that describes their content in the surface sediments with the total initial activity $A_{0,i}^{(n)}$ decreasing with time owing to decay (λ_i) and desorption with the relative velocity α_1. Again, we neglect the effects of the repeated absorption-desorption in the volumes V_1 and V_2, and arrive at the following time dependences for nuclide concentrations produced by the source $A_i^{(n)}(t)$ in the considered volumes:

$$C_{1,i}^{(n)} = \frac{\alpha_1 A_{0,i}^{(n)}}{(\omega_1 - \alpha_1)V_1}[\exp(-\alpha_1 t) - \exp(-\omega_1 t)]\exp(-\lambda_i t), \quad (4)$$

$$C_{2,i}^{(n)} = \alpha_1 A_{0,i}^{(n)} V_2^{-1} \omega_1 f(t), \quad (5)$$

where

$$f(t) = \left[\frac{\exp(-\alpha_1 t)}{(\omega_1 - \alpha_1)(\omega_2 - \alpha_1)} + \frac{\exp(-\omega_1 t)}{(\alpha_1 - \omega_1)(\omega_2 - \omega_1)} + \frac{\exp(-\omega_2 t)}{(\alpha_1 - \omega_2)(\omega_1 - \omega_2)}\right] \cdot \exp(\lambda_i t) \quad (6)$$

We are looking for the concentrations averaged over the time interval T = 1 year, when $\omega_1, \omega_2 << T$, $\lambda_i << \omega_1, \omega_2$; with high accuracy, the average concentration of the ith radionuclide in the volume V_2 can be estimated as

$$\overline{C_{2,i}^{(n)}} = \frac{\alpha_1 A_{0,i}^{(n)}}{\omega_2 V_2 T}\left\{\frac{1 - \exp[-(\alpha_1 + \lambda_i)T]}{(\alpha_1 + \lambda_i)} - \frac{\omega_1 + \omega_2}{\omega_1 \omega_2}\right\}. \quad (7)$$

It follows from (7) that $\overline{C_{2,i}^{(n)}}$ weakly depends on α_1. Therefore, for various values of α_1 chosen for estimated, we shall obtain close average concentrations $\overline{C_{2,i}^{(n)}}$.

Table 3 presents the calculation results for $\overline{C_{2,i}^{(n)}}$ with $\alpha_1 = 0.01$ day^{-1}, T = 1 year, and the same values of other parameters that were used in the previous section ($\omega_1 = 0.04$ day^{-1}; $\omega_2 = 0.2$ day^{-1}, and $V_2 = 10^5$ m^3). The relative radionuclide content is assumed to be the same that in the reactor water (see Table 2), but the nuclide list is shortened: we show only the nuclides that give the total contribution to the radiation burden at least 80% of the total sum.

Table 3. Values of $\overline{C_{2,i}^{(n)}}$ for T = 1 year related to the nuclide desorption from the reactor construction surfaces

Nuclide	$A_0^{(n)}$, TBq	IL^{water}, Bq/kg	$\overline{C_{2,i}^{(n)}}$, Bq/kg	$\overline{C_{2,i}^{(n)}}/CC_1$	$\overline{C_{2,i}^{(n)}}/CC_2$
^{60}Co	0.20	41	18.9	37.8	9.5
^{90}Sr	0.03	5	2.9	0.2	0.05
^{106}Ru	0.02	20	1.5	0.5	0.05
^{144}Ce	0.10	27	6.9	5.3	1.15
Total	0.35		30.2	43.8	10.75

It follows from Table 3 that for all nuclides $\overline{C_{2,i}^{(n)}} < IL_i^{water}$; for the total sum, $\sum\left(\frac{\overline{C_2}}{IL^{water}}\right)_i = 0.14$. These values show that the level of the volume V_2 contamination by radionuclides is comparatively low. At least, water in the volume is very far from being attributed to LRW (see [2]). However, the ratio $\overline{C_2}/CC_1$ for ^{60}Co and ^{144}Ce reaches approximately the values 40 and 5, respectively; therefore, it is worth to ban fishing (fish, shellfish, etc.) in the volume V_2 for feeding purposes. The possible radiation burden on the body of aquatic life individuals reaches 10% of the lethal level ($\overline{C_2}/CC_2 \approx 10$).

4. Radionuclide discharge in water from the steel construction of the reactor

Destruction of the surface sediments on the internal reactor constructions in the sea water let it contact directly with austenitic steel, which contains activation radionuclides. Among these nuclides, ^{60}Co is potentially the most hazardous one. With respect to the external media (V_2 volume), cobalt discharge is not ceased with the complete destruction of the steel surface sediments. Instead, it continues owing to the new source—stainless steel exposed to corrosion in the sea water. The average concentration of radionuclides (in particular, ^{60}Co) produced by the new source in the volume V_2 can be calculated according to the expression (7) with the source $\alpha_1 A_{0,Co}^{(n)}$ substituted by $\alpha_2 A_{0,Co}^{(st)}$. Here, $\alpha_2 A_{0,Co}^{(st)}$ is the total activity of ^{60}Co in the steel volume of all internal reactor constructions, which hardly amounts to more than 5 PBq (see above); α_2 is the relative velocity of ^{60}Co discharge in the sea water that filler the reactor, 10^{-4} $year^{-1}$. The velocity of the cobalt discharge in the sea water is determined by the velocity of the general corrosion of the austenitic stainless steel in the rest sea water at the temperature approximately 5 °C; the latter amounts to 1 μm/year [5].

With these values of α_2 and $A_{0,Co}^{(st)}$ and the same values of ω_1, ω_2, and V_2, we obtain the cobalt concentration averaged over the period T = 1 year $\overline{C_{2,Co}^{(st)}}$ = 58.7 Bq/kg. This concentration exceeds $\overline{C_{2,Co}^{(n)}}$ (see above) by the factor 3.1, but it still does not exceed $10 \cdot IL^{water}$ = 410 Bq/kg. The contamination by radioactive cobalt is not excessive, but such level requires banning sea fishery in the bay with the sunken object. The radiation burden for aquatic life in the volume $V_2 = 10^5$ m^3 can amount to 30% of the lethal value.

The estimates of the expected radiation consequences of the nuclide desorption from the surface sediments on the reactor constructions and of their discharge owing to corrosion of the main construction metal use the assumed values of α_1 and α_2, and disregards possible fractionation of the nuclide content related to differences in their chemical properties. Therefore, the estimate methodology is conservative in nature.

It is important to recognize the level of this conservatism. It is necessary to carry out experiments on formation of radionuclide content and concentration in sea water flooded into special tanks with individual constructions extracted from the utilized reactor or in the water flooded directly in the vessel of the reactor after its final shutdown.

5. Accident with the NPS sinking and sea water penetration in the reactor with the SNF

Flooding of the reactor containing the SNF by sea water would result in depressurization of fuel claddings and discharge of the core radionuclides in the water. The scheme of radionuclide migration similar to those considered in two previous sections is now determined by the velocity of radionuclide discharge from the spent fuel assemblies $\alpha_f A_{0,i}^{(f)}$, where $A_{0,i}^{(f)}$ is the initial nuclide activity in all fuel assemblies of the reactor; α_f is the relative velocity of the nuclide release from the fuel assembly. According to the experiment [6], $\alpha_f \approx 0.01$ year^{-1}.

The average concentration $\overline{C_{2,i}^{(f)}}$ of the ith nuclide in the volume V_2 of sea water can be calculated according to the expression (7) with the substitutions $\alpha_l \to \alpha_f$ and $A_{0,i}^{(n)} \to A_{0,i}^{(f)} A_{0,i}^{(n)}$. Table 4 displays the obtained results for T = 1 year, $\omega_l = 0.04$ day^{-1}, and $\omega_2 = 0.2$ day^{-1} assuming that accident occurred two years after the reactor shutdown following its operation for 50,000 hours at the average heat power level of 20 MW. The table contains the isotopes that form the main contribution to the radiation burden.

Table 4. Values of $\overline{C_{2,i}^{(f)}}$ for T = 1 year after the accident with the SNF corrosion in the sea water

Nuclide	$A_{0,i}^{(n)}$, PBq	$\overline{C_2^{(f)}}$, kBq/kg	IL^{water}, Bq/kg	$\overline{C_2^{(f)}}/CC_1$	$\overline{C_2^{(f)}}/CC_2$
$^{90}Sr+^{90}Y$	4.4	5.53	5	368	92
$^{106}Ru+^{106}Rh$	0.6	0.52	20	158	17
^{134}Cs	3.0	3.10	7.3	2580	239
^{137}Cs	4.5	5.63	11	3130	188
$^{144}Ce+^{144}Pr$	6.2	4.9	27	3770	818
Total	18.7	19.68		10006	1354

It follows from Table 4 that caesium and cerium concentrations exceed the corresponding control levels CC_1 and CC_2 by the factors of several thousands and several hundreds, respectively. Immediately after the accident, sea fishery in the bay with the sunken NPS should be banned. We can expect loss of aquatic life of various species in the volume $V_2 = 10^5$ m^3. It should not violate the radioecological balance in the bay as a whole, for the total water volume in the bay amounts to 10^9 m^3 exceeding the volume V_2 by four orders of magnitude.

The accident requires immediate intervention, which assumes rapid salvaging of the sunken NPS on the water surface. Maximum concentrations $C_{2,i}^{(f)}$ are reached 3-5 months after the moment when the nuclide migration regime with α_f = const is established. Figure illustrates the time dependence of the concentrations with $\alpha_f = 0.01$ year^{-1}. The value of α_f requires additional experimental proof.

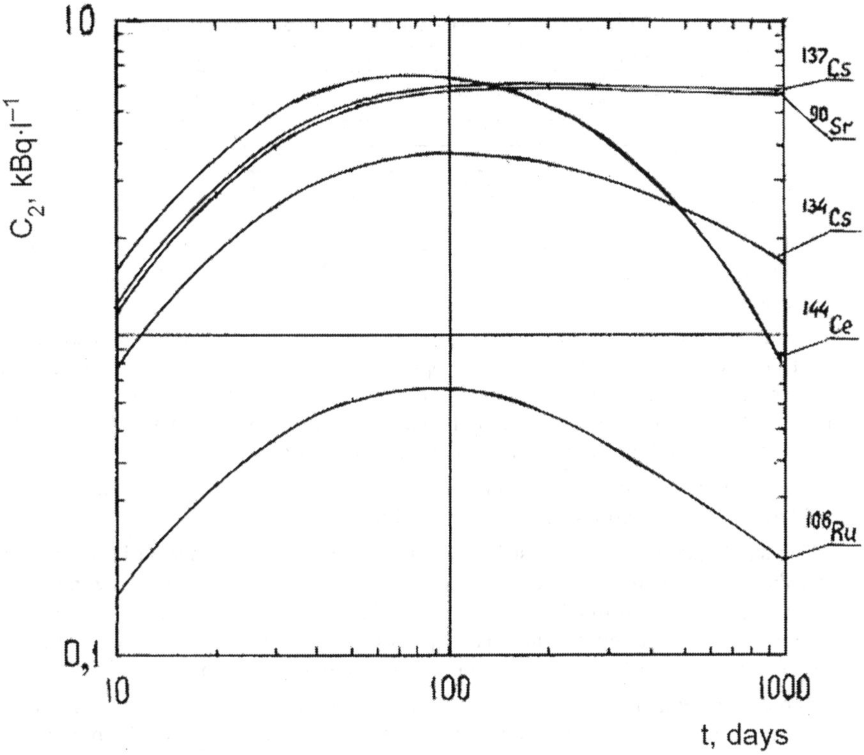

Figure 1. Time dependence of radionuclide concentrations following the accident with the SNF contact with sea water

6. Nuclear accident during defuelling

Operations of the reactor defuelling can trigger, though with very small probability, the self-sustaining chain reaction (SCR) of fission. The SCR is hazardous owing to the possibility of the enhanced release of radionuclides in the environment in case when both the reactor and the reactor compartment are opened at the moment of an accident. The SCR flash would destroy part of the spent fuel assemblies (SFA) accompanied by the release of radionuclides in the atmosphere.

In principle, the SCR flash is possible only if there is some water in the active zone. Water produces steam, which carries radioactive materials. The analysis of the data obtained from actual accidents and special experiments [7] shows that the fraction of the radionuclides released to the atmosphere by the SCR flash with respect to the total amount of radionuclides in the active zone might amount to 10% of noble gaze isotopes, 1% of iodine, bromium, ruthenium, tellurium, and caesium isotopes, and 0.2% of other isotopes. Table 5 displays the activity of the released long-living nuclides in case of an accident occurred three years after the shutdown of the reactor, which operated for 50,000 hours at the average heat power level of 20 MW (see above).

Table 5. Release of nuclides in the atmosphere following the SCR flash.

Beta-nuclides	TBq	Alpha-nuclides	GBq
^{90}Sr	8.3	^{238}Pu	82
^{106}Ru	3.2	^{239}Pu	13
$^{134,137}Cs$	22.2	^{240}Pu	8
^{144}Ce	4.6	^{241}Am	22
^{147}Pm	12.6		
Total	50.9		125

The scale of the nuclide release is lower than the release level followed the Chernobyl accident in April-May, 1986 by the factors 2,000 for beta-sources and 1,000 for alpha-sources.

The emergency release of the radioactive materials will result in the appearance of a tail of the contaminated air in the wind direction. The resulting hazards are 1) external exposure from the tail and its trace formed by the fallout from the tail and 2) internal exposure of humans and animals owing to radionuclides inhaled with the contaminated air or consumed with contaminated water and food.

The external exposure doses from the passing tail with radionuclides amounts to 10-20 mSv. The inhalation hazard is slightly higher.

The stable state of the atmosphere during the post-accident period is less favorable. For the built-up area (McErloy model), the air tail released at the height 30 m from the ground conditionally reaches the surface level at the distance 400 m from the source point. The human internal exposure dose owing to nuclide inhalation in this point amounts to 40 mSv of beta radiation and 80 mSv of alpha radiation.

Table 6 presents the exposure doses for the individuals resulting from the radioactive fallout. Radiation burden would require temporary evacuation of the population from the territories within the trace 3-4 km distant from the activity release point.

Table 6. Expected exposure doses within the radioactive trace, mSv

Dose for the period	Distance from the source, km						
	0.4	1	2	3	5	10	20
1 month	16.7	8.2	3.5	2.0	1.0	0.34	0.12
1 year	200	98.3	42.0	24.0	12.0	4.1	1.40

7. Storing of the RC package in the LTSS

The RC packages stored at the LTSS are radiation-hazardous objects owing to presence of activation radionuclides (in particular, ^{60}Co) and uranium fission products in the sediments on the surfaces of the reactor internal constructions (see Table 3).

The LTSS is a ground-state construction. We do not expect that this construction will be sunken in sea water for long time period and part of the RC packages will loose impermeability so that sea water would penetrate inside the reactor. However, if this occurs, the level of the possible local contamination of sea water by radionuclides would not exceed the above estimates for the volume $V_2 = 10^5$ m^3 in cases of sinking of a defuelled NPS or RC. The corresponding radiation burden is not excessive with respect to the violation of the natural ecological balance.

Natural disasters with temporary flooding of the LTSS site are more probable. Short-term water contact with radionuclide sources will result in partial radionuclide release from the surface sediments on steel constructions.

We cannot exclude emergency penetration of rain (snow) water into individual depressurized RC packages. Such an accident might result in water penetration inside the reactor with overflooding its internal volume. Finally, a radioactive stream would appear. The LTSS site as an enterprise aimed at the RW management should be equipped with special runoff ditches entering the system of collecting and recycling of the LRW. The LTSS substantiation solutions are the integral part of the storage project.

8. Discussion. Problems

The performed estimates disregard the moderated character of sea water entry in the reactor following flooding of the reactor compartment and ignores the process of gradual replacement of the initial (redistillated) water inside the reactor by sea water when the active zone is not defuelled. The estimates also ignore partial sedimentation of radionuclides from sea water in course of the migration process from the source to the sea water area (V_2 volume) and in the sea. As a result, we overestimate the radionuclide concentrations in the sea water area of the emergency object $C_{2,i}$.

In spite of the conservative nature of the estimates, the obtained results make it possible to state that the emergency discharge of radionuclides in the sea in the region of the NPS utilization would not result in excessive radiation burden for the environment (including the population) provided the NPS reactors are defuelled. When fuel is present in the active zones, an accident might result in locally increased radiation burden. Avoiding such situations requires immediate intervention measures mentioned in the report. It is worth noting here that the hazard related to the SNF is the consequence of depressurization both in the RS and the first contour of the reactor facility in course of the NPS sinking on shallow water (close to the berth). Such event combination is highly unlikely.

The performed calculations use the values of the relative velocities of desorption α_1 and corrosion release of radionuclides α_2, which require experimental confirmation. It is also necessary to confirm the presence (or absence) of the fractionating effect in the nuclide yield depending on their chemical attribution. The experiments will be representative provided they will use real (spent) reactor constructions. We are looking for investigations with a reactor of the utilized NPS intentionally flooded by sea water.

Sometimes, the reactor vessel can be used as a storage of individual removable parts. In particular, we mean reactivity control rods, which contain radioactive isotopes $^{152,154,155}Eu$. The most hazardous are the ^{152}Eu and ^{154}Eu isotopes (see Table 1). They can compete with ^{60}Co in contributing to the total radiation burden. Therefore, experiments on long-term stay of neutron-absorbing rods with europium in sea water are required.

An accident with sinking of the non-defuelled NPS requires immediate intervention with the NPS salvaging during 1-2 months following the accident. During this time interval, sea water contact with the cladding surface result in corrosion destruction of claddings. The information was obtained in a single experiment [6]. It should be improved by additional investigations of the SFA behavior in the vessel filled by sea water at the temperature close to 5 °C.

The estimates of potential radiation hazards for the population and the environment of the NPS utilization region used the control concentrations suggested in [3].

They seem to be more or less substantiated. However, careful expertizing with further improvement (if required) is desired. The improved values might be accepted as a regulating document.

References

1. *Radiation Safety Norms, NRB-99.* Minzdrav of Russia (1999), Moscow (in Russian).
2. *Main Sanitary Rules Providing Radiation Safety OSPORB-99.* Minzdrav of Russia (2000), Moscow (in Russian).
3. Sazykina, T.G and Kryshev, I.I. (1999) Estimate of Control Concentration of Radionuclides in Sea Water Accounting for Hygienic and Radioecological Criteria, *Atomnaya Energiya* **87**, 302-304 (in Russian).
4. *Sediment K_d's and Concentration Factors for Radionuclides in the Marine Environment.* IAEA Technical Rep. Ser. No 247 (1985), Vienna.
5. Gerasimov, V.V. (1999) *Prognozing of Metal Corrosion,* Metallurgiya, Moscow (in Russian).
6. Dozhdikov, S.I., Zhuravkov, A.M., and Zolotkov, A.A. Corrosion (1991) Resistance of Spent Nuclear Fuel in Sea Water, in *Proc. Int. Scientific Seminar of the USSR Nuclear Society, September 24-28, 1990, Nuclear Power in the Sea. Ecology and Safety,* Kurchatov Institute, Moscow (in Russian).
7. Nucleonics (1965) **23**, 55-59.

AUTOMATED RADIOLOGICAL MONITORING AT A RUSSIAN MINISTRY OF DEFENCE NAVAL SITE

P. D. MOSKOWITZ
Brookhaven National Laboratory
Upton, NY

J. POMERVILLE
U.S. Department of Defense
Washington, DC

S. GAVRILOV, V. KISSELEV, V. DANILYAN
Institute for Nuclear Safety
Moscow, Russia

A. BELIKOV, A. EGORKIN
Russian Navy
Moscow, Russia

Y. SOKOLOVSKI
ICC Nuklid
Saint Petersburg, Russia

M. ENDREGARD
Norwegian Defence Research Establishment
Oslo, Norway

M. KROSSHAVN, C. V. SUNDLING, H. YOKSTAD
Institute for Energy Technology
Oslo, Norway

ABSTRACT

The Arctic Military Environmental Cooperation (AMEC) Program is a cooperative effort between the military establishments of the Kingdom of Norway, the Russian Federation, and the United States. This paper discusses joint activities conducted over the past year among Norwegian, Russian, and U.S. technical experts on a project to develop, demonstrate and implement automated radiological monitoring at Russian Navy facilities engaged in the dismantlement of nuclear-powered strategic ballistic missile launching submarines.

A.A. Sarkisov and L.G. LeSage (eds.),
Remaining Issues in the Decommissioning of Nuclear Powered Vessels, 155–161.

Radiological monitoring is needed at these facilities to help protect workers engaged in the dismantlement program and the public living within the footprint of routine and accidental radiation exposure areas. By providing remote stand-alone monitoring, the Russian Navy will achieve added protection due to the defense-in-depth strategy afforded by local (at the site), regional (Kola) and national-level (Moscow) oversight. The system being implemented at the Polyarninsky Russian Naval Shipyard was developed from a working model tested at the Russian Institute for Nuclear Safety, Moscow, Russia. It includes Russian manufactured terrestrial and underwater gamma detectors, smart controllers for graded sampling, radio-modems for offsite transmission of the data, and a data fusion/display system. The data fusion/display system is derived from the Norwegian Picasso AMEC Environmental Monitoring software package. This computer package allows monitoring personnel to review the real-time and historical status of monitoring at specific sites and objects and to establish new monitoring protocols as required, for example, in an off-normal accident situation. Plans are being developed to implement the use of this system at most RF Naval sites handling spent nuclear fuel.

1. INTRODUCTION

This paper discusses joint activities conducted over the past year among Norwegian, Russian, and U.S. technical experts on a project to develop, demonstrate and implement remote radiological monitoring at Russian Navy facilities engaged in the dismantlement of nuclear-powered strategic ballistic missile launching submarines. Radiological monitoring is needed at these facilities to help protect workers engaged in the dismantlement program and the public living within the footprint of routine and accidental radiation exposure areas. By providing remote stand-alone monitoring, the Russian Navy will achieve added protection due to the defense-in-depth strategy afforded by local (at the site), regional (Kola) and national-level (Moscow) oversight.

AMEC Project Area 1.5 enhances Russia's ability to monitor radiation levels during submarine dismantlement through analysis of risk factors, equipment procurement, and improved personnel training. Several successful dosimetry-training programs have been conducted for Russian military officers at DOD and DOE facilities. Technologies and approaches for managing radiation hazards have also been shared. In addition, radiation-monitoring equipment has been transferred. Finally, a graphical interface software system was created and installed in 2000 to improve radioecological monitoring at sites handling and storing spent nuclear fuel from dismantled ballistic missile submarines.

2. PICASSO AMEC ENVIRONMENTAL MONITORING SYSTEM

The PICASSO System was developed by the Institute for Energy Technology, OECD Halden Reactor Project, Halden, Norway. Picasso is tailor-made for developing dynamic graphical user interfaces. Data can be received from data acquisition systems, simulators, databases and/or other input sources, and the Graphical User Interface (GUI) is continuously updated to visualize the current state. PICASSO has been used for simulator systems and process surveillance and control systems in various application areas such as nuclear power plants, oil production platforms, telecommunication networks, electric power distribution, ship bridge systems, paper mills and environmental monitoring.

The PICASSO AMEC Environmental Monitoring System (Figure 1) includes:

- Multiple terrestrial and underwater smart gamma sensors to take regular radiation readings;
- "Smart" sensors that automatically increase the rate of sampling in the case of unusual readings
- Radio-modems for off-site transmission of data
- A computerized data fusion/display system to compile data and allow for real-time monitoring as well as historical comparisons.

Figure 2. presents a sample display screen of the PICASSO AMEC Environmental Monitoring System.

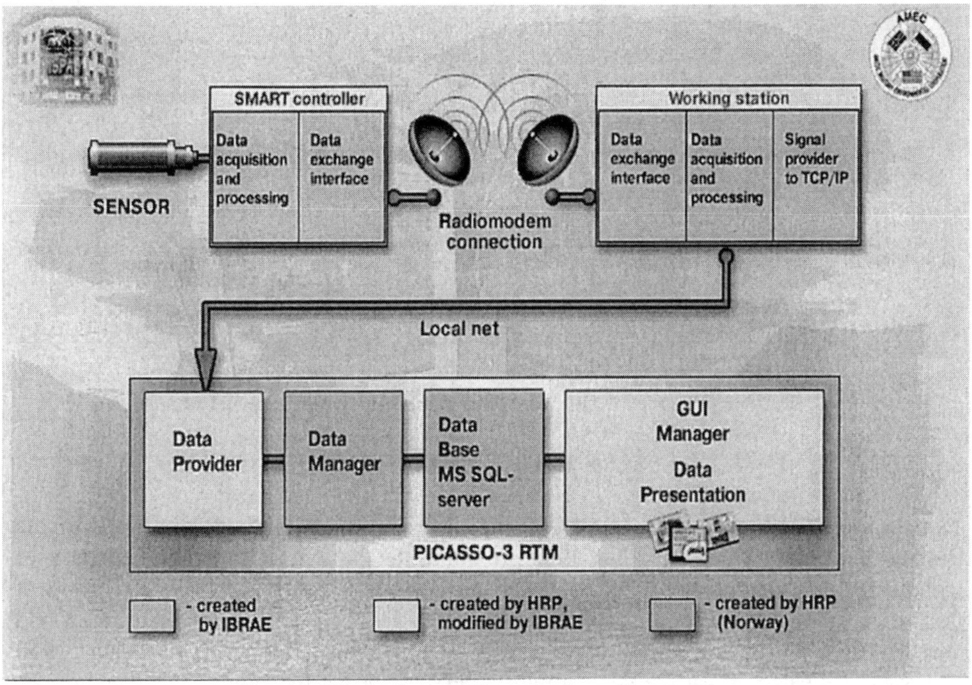

Figure 1. Block scheme of the system's working model.

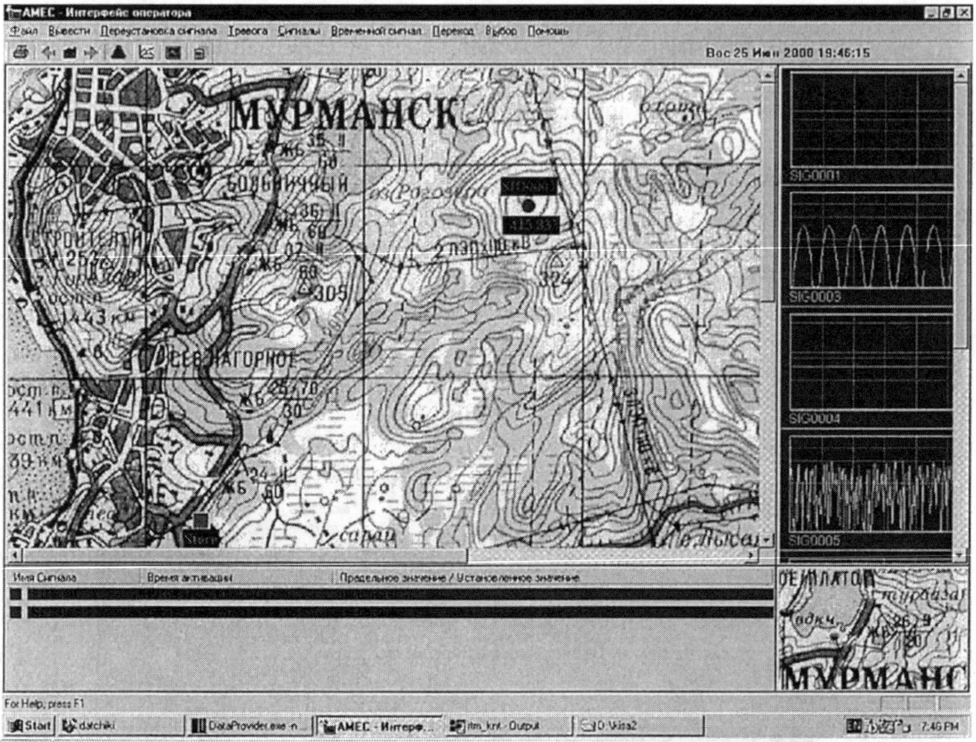

Figure 2. PICASSO AMEC Environmental Monitoring System Display Screen.

3. POLYARNINSKY RUSSIAN FEDERATION NAVAL SHIPYARD

The Russian Federation has almost 25 shipyards (in the Northwest proximate to the Barents Sea/Arctic Ocean and Far East proximate to the Sea of Japan and the Sea of Okhotsk/Pacific Ocean) servicing its naval nuclear operations. The Naval Shipyard at Polyarninsky was established in 1935. This facility serviced naval vessels during World War II. After World War II, the Navy expanded the facility. Beginning in August 1950 and was dedicated exclusively to military vessels, primarily submarines (Figure 3). Information presented below on this shipyard was derived from several open literature sources (1-4).

As the first nuclear powered submarines were delivered to the Northern Fleet at the end of the 1950s, the yard was modified for the docking and repair of these vessels. Tenders, service ships and dry docks were acquired, including the floating dock PD-63. At the present time, there are two covered floating docks at the yard constituting a total quay length of 700 m. The yard has a surface area of 41 330 m² (446 000 sq. ft.). There are approx. 3000 employees at the yard.

Repair and maintenance operations have been carried out on first, second and third generation nuclear submarines since1962. Polyarninsky has participated in minimal dismantlement operations. Although refueling equipment is available at Polyarninsky, the future of refueling activities at Polyarninsky is also uncertain.

Polyarninsky operates a naval waste storage facility, and is the homeport for two service vessels engaged in submarine maintenance/defueling operations: the *Pinega* class *Amur* and the *Vala* class TNT-12. Solid radioactive waste is placed into containers and stored in an area specifically dedicated to this purpose. Two hundred containers and some large pieces of contaminated material have been placed outside the actual storage site that is full. There are plans to expand the storage facility or build an additional one, but so far there is no money allocated for this.

Liquid radioactive waste is stored in two floating tanks at the quay. The capacity of this storage is approximately 150 m^3 (5300 cu. ft.). There are plans for establishing a storage facility for spent nuclear fuel in an existing tunnel near the shipyard, but no firm decision has been made. While the yard does possess equipment for the removal of fuel from both operational and inactive submarines, it is currently not in use.

Figure 3. The naval yard Polyarninsky lies close to the city of Polyarnyy. The shipyard has a storage facility for solid radioactive waste and two floating tanks containing liquid radioactive waste.

4. PICASSO AMEC ENVIRONMENTAL MONTORING SYSTEM AT THE POLYARNINSKY NAVAL SHIPYARD

The main goal of control for radiation pollution of the environment at naval sites, such as Polyarninsky, servicing and decommissioning nuclear powered submarines is to effectively manage radioactive materials during normal operation and accidental situations

Decommissioning of nuclear powered submarines is a long-term and laborious process. It includes several major steps.

- Storage of floating nuclear submarines, brought out from active service.
- Off-loading of reactor cores under shipyards or base conditions.
- Off-loading, storage and transportation of irradiated nuclear fuel and radioactive waste.
- Dismantlement of the reactor compartment.
- Cleaning contaminated equipment.
- Long-term storage of three compartments blocks with reactors on water (with or without cores).

During each stage, radionuclides can be released to the environment and must therefore be managed. All of these enumerated stages of nuclear submarine decommissioning are very characteristic of the Polyarninsky Shipyard where the radiation monitoring system, PICASSO-AMEC will be implemented. Facilities to be monitored include the existing spent fuel, solid and liquid radioactive storage facilities as well as a new low-level waste storage site to be constructed as part of the AMEC Project 1.4 program (1) and a new mobile processing facility to be constructed as part of the AMEC Project 1.3 Program (2). These projects are described in detail in other reports presented at this conference.

At this site, the current radiological control system is limited and does not correspond to contemporary requirements for environmental safety. More specifically, the current radiation monitoring system is primarily based on field measurements taken with hand-held equipment. This equipment does not support comprehensive measurements of temporally and spatially variant radiation levels that are characteristically observed at this kind of a facility. Measurements of dynamic changes are particularly important because of the nature of the activities conducted at the site and the potential need for predictive forecasts in an accident situation.

Analysis of impact of nuclear and radiation facilities to the environment shows that it is necessary to create complex and differential systems for permanent control taking into account the specific characteristics of operating conditions of the special units of shipyards. Automated systems, like PICASSO-AMEC, are very important as initial indicators of accidental situations and as providers of data for analytical, forecast and decision-making centers.

Radioecological monitoring must operate for ecological sites and must be based on landscape zoning for technical sites and surrounding areas (water areas) taking into account prevailing ways that the contamination may spread. This approach was not used before.

The most convenient form of data presentation is mapping system displaying information on radiation background, sites with higher content of artificial radionuclides, exposure rates, radioactive contamination and radiation threat level. The forecast functions are very important for both onsite and offsite locations.

In this context, the PICASSO-AMEC system being implemented at the Polyarninsky shipyard will help update and modernize radiation control practices at this site and will help provide better protection to the workers at the site and to civilian populations living near the site. Radiological monitoring is needed at this facility to help protect workers engaged in the dismantlement program and the public living within the footprint of routine and accidental radiation exposure areas.

5. CONCLUSION

Use of the PICASSO AMEC system will provide multiple layers of monitoring (both human and remote, locally and from offsite) that will increase the RF Navy's ability to detect elevated radiation levels and decrease the response time if a problem should arise. Further, by providing remote stand-alone monitoring, the Russian Navy will achieve added protection due to the defense-in-depth strategy afforded by local (at the site), regional (Kola) and national-level (Moscow) oversight of its radiological monitoring program. Plans are being developed to implement the use of this system at most RF Naval sites handling spent nuclear fuel.

6. REFERENCES

1. Bellona Report nr.2:96. Written by: Thomas Nilsen, Igor Kudrik and Alexander Nikitin, http://spb.org.ru/bellona/ehome/russia/nfl/short/htm
2. Russia: Naval Nuclear Reactors: Nuclear Fleet:Northern Fleet, http://cns.miis.edu/cns/projects/nisnp/naval/nucflt/norflt/ff_noflt.htm
3. Submarine Launched Ballistic Missile Facilities, http://sun00781.dn.net/nuke/guide/russia/facility/slbm/index.html
4. The Russian Northern Fleet, http://www.bellona.no/imaker?id=12675&sub=1
5. B.T. Spargo, V. Engoy, V. Balkunov, P. Krumrine, S. Backe and S. Gorin, "Solid Radioactive Waste Treatment Initiatives for Nuclear Submarine Decommissioning Under the AMEC Program," Waste Management 01, Tuscon, Arizona, USA (2001).
6. Griffith, T. Engoy, A. Diashev, P. Schwab, and A. Nazarian, "Storage of Solid Radioactive Waste from the Northern Fleet of the Russian Navy under the AMEC Program," Waste Management 01, Tuscon, Arizona, USA (2001).

ANALYSIS OF THE RADIOECOLOGICAL SITUATION WITHIN BAYS OF NUCLEAR SUBMARINE AND FLOATING REACTOR COMPARTMENT STORAGE

V. DANILYAN
IBRAE RAS
52 Bolshaya Tulskaya, Moscow 115191, Russia

V. VYSOTSKY, A. BORISSOV, AND D. SALKO
Russian Navy

Decommissioning of a large number of vessels with nuclear-powered engines (NPE) resulted in appearance of a new class of nuclear- and radiation-hazardous objects—decommissioned and utilized nuclear submarines (NPS). They are disposed at the interim storage sites (ISS); each site should provide nuclear, radiation, and radioecological safety for the stored objects and the sites as a whole, as well as their long-term safe existence.

According to existing technology NPS utilization is performed in several stages: defuelling; dismantling of the nose and stern compartments; formation of the three-compartment autonomous floatable module containing the empty reactor vessels and equipment contaminated by radioactive materials in its central part (certain blocks are used as interim storages of the SRW).

Nuclear safety for the decommissioned NPS (unauthorized increasing the power of the NPE) is provided through unplugging of the electric drives for the control and shield systems of the reactor, and through mechanically securing the drives for manual control, lifting of the compensating grids, draining and unplugging of the circulation pumps of the primary system [1]; in the utilized NPS, nuclear safety is provided through defuelling of the reactor active zones (AZ). The satisfactory radiation situation is provided through the sorbent removing from the activity filters; drainage and cleaning of the systems and steam generators, pressure reduction in the system of air and gaze supply to the atmospheric level, pressurization of the pipe tips and equipment of the primary system. Radioecological safety is provided through welding of scuppers and other technological holes in the light and solid hull of the utilized NPS; deactivation of the external surfaces of three-compartment modules and 30-year warranty for their impermeability.

General radiation and radioecological safety of the ISS is provided through their distant location from the dwelling sites, zonal distribution, optimal disposition of nuclear- and radiation-hazardous objects aimed at their mutual screening, and providing fire safety and floodability.

The nuclear potential of the NPS is determined by the residual amount of fissible materials; the radiation potential is determined by the amount of radioactive materials; the radioecological potential is determined by the power rate of the activity sources. Currently, out of 188 decommissioned NPS, 63 are utilized, 90 are defuelled, and 125 are pending utilization.

A.A. Sarkisov and L.G. LeSage (eds.),
Remaining Issues in the Decommissioning of Nuclear Powered Vessels, 163–179.

All decommissioned NPE are disposed at 12 ISS: 8 at Arctic Fleet (AF) sites and 4 at Pacific Fleet (PF) sites. Each ISS stores from 1 to 23 vessels; the NPE of the stored vessels contain (1-20)·10^{15} Bq of radioactive materials (Tables 1, 2).

TABLE 1. Number of decommissioned and utilized NPS in the ISS

ISS name	Arctic Fleet		ISS name	Pacific Fleet	
	NPS	Block		NPS	Block
Ura bay	13	-	Pavlovskogo bay	23	-
B. Lopatka bay	6	-	Razboinik bay	-	19
Olen'ys bay	1	-	Postovaya bay	3	-
Ara bay	15	-	Krasheninnikova bay	16	-
Berpich'ya bay	2	-			
Gremikha nay	18	1			
Belomorskaya Navy base	2	2			
Saida bay		37			
Total	57	40	Total	42	19

Comment: The ship-repairing plants of the Arctic Fleet store 21 NPS and 3 blocks, the same for the Pacific Fleet—5 NPS and 10 blocks.

TABLE 2. Nuclear and radiation potential of the decommissioned NPS, 10^{15} Bq

NPS type		Activity of the reactor active zones	Induced* activity of construction materials
Generation	NATO classification		
First	November	2,8 ± 0,8	0,32 ± 0,7
	Hotel	1,1 ± 0,3	0,25 ± 0,3
	Echo-II	2,5 ± 0,7	0,55 ±1,2
Second	Charlie – II	-	0,045 ± 0,006
	Victor – I	6,2 ± 1,7	0,055 ± 0,021
	Victor – II	2,9 ± 0,3	0,058± 0,016
	Victor – III	6,6 ± 2,8	0,102 ± 0,028
	Papa	3,2 ± 0,6	0,019 ± 0,003
	Julliet	4,0 ± 0,7	0,00004 ± 0,00001
	Yankee	3,2 ± 1,0	0,058 ± 0,022
	Delta – I	6,2 ± 1,0	0,120 ± 0,040
	Delta – II	-	0,150 ± 0,020
	Delta – III	-	0,210 ± 0,020
Third	Typhoon	22,0 ± 0,2	0,33 ± 0,06
	Shark	8,3 ± 0,3	0,12 ± 0,02
	Oskar	12,0 ± 0,1	0,15 ± 0,02

Comment: *Activity of three-compartment blocks after the BPS utilization. Activity of nuclides absorbed at the internal surfaces of the 1st contour is lower by 2–4 orders of magnitude

Decommissioned NPS introduce two kinds of radiation fields that permanently affect the environment. The fields of the first kind (nearby) are related to the gamma radiation of the nuclides inside the NPS (block) and those firmly fixed on the external surface of the light and solid hulls; the second kind of fields (distant) are produced through the diffusion of the man-caused radionuclides in the marine environment under effects of various natural factors.

The impact range of the nearby field of the above-water part is determined by the range of gamma-rays in air, which, depending on their energy can amount to dozens and hundreds meters. In the under-water part, the impact range is restricted by 2-3 m from the vessel surface owing to virtually complete gamma quanta absorption by sea water. Impact on fauna and flora is determined by the scattered external radiation.

The distant field is formed under conditions of the external environment and has a complicated structure. Radionuclide behavior depends on chemical content and aggregate state of the radionuclides; it is determined by gravitational, hydrophysical, chemical, and biochemical processes, but in most cases—by the aggregate exposure of all factors. Depending on the intensity of the contamination sources, orographic peculiarities of the storage site, and the state of the marine environment, man-caused radionuclides might be found at the distances of several dozens, hundreds, or even thousands of meters from the NPS. They affect the environment and the human body through external and internal radioactive exposure.

The storage sites are equivalent to each other with respect to radioecological hazard. Some of them contain clean three-compartment blocks, others store blocks together with NPS, and some ISS store blocks, NPS, and vessels with the damaged cores. Most of the ISS are located at sites formerly used as Navy NPS bases, where residual radioactive contamination of the environment is present. The NPSs can be considered radioecologically clean only upon utilization and transformation into three-compartment blocks free of removable surface contamination. In other cases, virtually all decommissioned NPS should be considered as sources of radioactive contamination of various intensities.

Water penetration into the NPS (block) is highly undesirable, because surface contamination is present inside each object upon decommissioning or utilization. The average density of β-particle flux in the reactor compartment can amount to 100-200 decays/cm^2·min (α-contamination is virtually absent), and for individual NPS, it can amount to $5 \cdot 10^4$ decays/cm^2·min.

β-activity is absent from the passageways and places for the staff. The exceptions are some bottlenecks (pipelines, hold), where the highest levels of gamma radiation are observed. The contamination is always local, the typical spot size is 0.1-1 m^2. The maximum contamination occurs in the equipment enclosures (from several hundred to 5,000-7,000 decays/cm^2·min) and in the pipelines and hold of the reactor compartment (from several hundred to 2,000 decays/cm^2·min). For other compartments, the contamination distribution is as follows: the 3rd floor of the reactor compartment—25–120 decays/cm^2·min (maximum 700 decays/cm^2·min); the 2nd floor—25–600 decays/cm^2·min; the neighboring compartments—20–90 decays/cm^2·min (individual points); light and solid hulls—20–30 decays/cm^2·min (individual points).

There is also always a small amount of water (condensate) in the hold, which contains radioactive materials with specific β-activity of 0,5-50 Bq/l and α-activity of 0.005-0.5 Bq/l. In the three-compartment modules, the ER in the compartments neighboring to the reactor is within 5-90 μR/h, β-activity does not exceed 20 decays/cm^2·min. However, upon loading SRW, the ER might increase to 100–150 μR/h, and surface contamination—to 3,000 decays/cm^2·min.

In addition to the surface contamination, the equipment of the reactor compartment has internal contamination. For the 1st-generation NPSs equipped with PWR reactors, the surface contamination is determined by the corrosion products of ^{60}Co and ^{54}Mn (^{60}Co gives the main contribution). For the 2nd-generation NPS, the radioactive contamination is mainly formed by the fission nuclides ^{137}Cs, ^{134}Cs, ^{144}Ce, ^{106}Ru, ^{90}Sr, and ^{90}Y (^{60}Co share is within 40%).

For all objects, radioactive contamination is usually restricted to the space of the reactor compartment. The specific design and operational features of the NPE determine its intensity; therefore, the ER value varies in a rather broad interval. The radiation intensity is maximum in the equipment enclosures (4-5 mR/h); at the 3rd floor of the reactor compartment it reaches 0.1-0.8 mR/h; at the 2nd floor—0.4-2.0 mR/h; in the hold—0.6-70 mR/h; (predominantly, 30-50 mR/h). In the neighboring compartments, the ER usually does not exceed 0.1 mR/h (Tables 3, 4 and Fig. 1).

TABLE 3. Exposure rate of gamma radiation in the reactor and neighboring compartments of the decommissioned and utilized NPS (10-year hold), µR/h

Compartments of the NPS/module	NPS generation		Module generation	
	First	Second	First	Second
Neighboring nose	8 – 14	12 – 18	5 – 10	10 – 20
Reactor (2nd floor)	70 – 600	100 – 800	50 – 12000	80 – 14000
Neighboring stem	7 – 90	10 – 120	8 – 25	12 – 50

Comment: The dominating level of gamma radiation corresponds to the lower limit of the ER variation; approximately 20–30% of all objects feature the upper limit of the radiation level

TABLE 4. Maximum exposure rate of the gamma radiation on the surfaces of the above-water and under-water parts of the decommissioned and utilized NPS (atmospheric media), µR/h

Ser. #	NPS		Ser. #	Three-compartment module, part		
	Above-water part			Above-water		Under-water
	10 cm	100 cm		10 cm	100 cm	10 cm
142	23	20	140	20	20	8,000**
152	27	25	141	110	68	25,000**
172	7	5	143	20	16	20,000**
180	8	4	178	1,500	560	15,000**
290	430*	410*	908	70	40	5,000**
442	22	20	151	65	34	11,000***
702	35	30	153	10	10	20,000***
-	-	-	154	230	130	75,000***
-	-	-	155	55	30	38,000***
-	-	-	156	15	15	24,000***
-	-	-	416	20	20	23,700***

*The NPS ser.# 290 was broadside to broadside of K–413 at the moment of the NPE accident in the Chazhma bay on August 10, 1985.
**The first-generation NPS.
***The second-generation NPS.

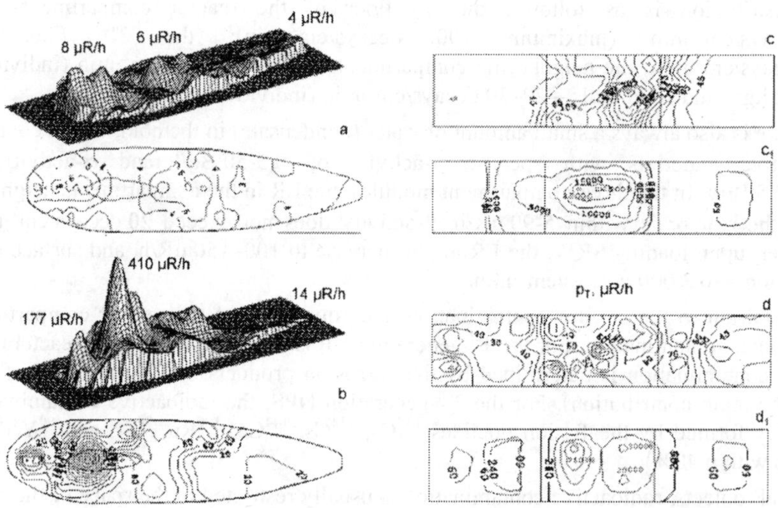

Figure 1. State of gamma field at the decommissioned (a—typical NPS, b—NPS exposed to radioactive discharge) and utilized (c, d) NPS (c, d—above-water part, c_1, d_1—underwater part; c—block filled with SRW)

The radiation levels from the NPS of the first and the second generations differ by 20–30% owing to differences in the reactor types and equipment layout. The maximum values of ER in air media are observed in the lower parts of modules under the reactor bottom. However, when the submarines are afloat, the gamma radiation is absorbed by sea water and the ER value in any direction at the distance 2 m from the hull does not exceed the background reaching 1,5±0,5 μR/h.

There is always small removable radioactive surface contamination on the solid hull of the NPS in the region of the reactor compartment. In the underwater part, the contamination is caused by the activation of the construction materials of the biological shield tank (BST) by thermal neutrons. At present, the accumulated radionuclides from corrosion can be discharged in sea water in case of depressurization of the BST (under unfavorable conditions, up to 10 GBq/year). In the above-water part of the reactor compartment, there is residual contamination caused by the reloading of the reactor AZ; the contamination is localized in corrosion materials (Table 5).

TABLE 5. Specific activity of the man-caused radionuclides in the corrosion elements of the solid hull of the decommissioned NPS, Bq/kg

Ser. #	Sampling location	Radionuclide, permissible level			
		^{60}Co	MPC	^{137}Cs	MPC
142	Reactor compartment	520 ± 80	1.4	18 ± 6	< 0.1
152	Reactor compartment	18 ± 6	< 0.1	110 ± 35	< 0.1
172	Reactor compartment	300 ± 60	0.8	74 ± 30	< 0.1
180	Reactor compartment	814 ± 130	2.2	76 ± 30	0.1
442	Reactor compartment	37 ± 12	0,1	290 ± 60	< 0.1
702	Reactor compartment	7 ± 4	< 0.1	22 ± 7	< 0.1
290	Nose	1330 ± 260*	3.6	37 ± 12*	< 0.1
290	Reactor compartment	7580 ± 600*	20.5	370 ± 70*	0.1
290	Stern	530 ± 80*	1.4	110 ± 35*	< 0.1

Comment: MPC is the maximum permissible concentration. According to RKVS-90 (Navy) [2], the permissible contamination of beaches and soil of the flood zones by ^{60}Co is 370 Bq/kg; the same figure for ^{137}Cs is 2,590 Bq/kg.

*The NPS ser.# 290 was broadside to broadside of K–413 at the moment of the NPE accident in the Chazhma bay on August 10, 1985.

The data in Table 5 shows that corrosion products near the reactor compartment of the decommissioned NPS contain ^{60}Co and ^{137}Cs, but only the ^{60}Co concentration reaches 2 MPC according to the norm adopted by the Russian Navy, ^{137}Cs concentration does not exceed 0.01 MPC. These concentrations should be considered as maximum possible values when estimating activity release into the environment with sedimentation of the corrosion products to the bottom. Actually, the activity release would be substantially lower, because it is reduced by the light hull of the NPS and high sorbing ability of the iron oxides (desorption factor of ^{60}Co and ^{137}Cs to the sea water is below 10^{-5}). The situation in the pipelines that have unavoidable contact with the environment is similar.

The calculations show that the NPSs prepared for interim afloat storage can annually release, owing to various reasons, up to 20 kBq of radioactive materials (RM) into the water area; the ISS storing 15–30 vessels can release annually up to 200-600 kBq. However, this amount of RM is not hazardous for the environment, because it would be dissolved in the water area and be below the level of permissible concentration.

As a result, the permanent factor that determines dose burden for the ISS staff is gamma radiation from the above-water part of the decommissioned and utilized NPS, The data in Table 4 show that for 80% of objects, the ER does not exceed the permissible level for the population of 30 μR/h (at 1 m height). Only for individual three-compartment blocks, it can be several times higher. This was observed when equipment used as a radiation shield was removed, or when the block was used to store SRW; when the NPS had a radiation accident or when it had a damaged core.

Despite the existing variations in the radiation situation at each individual object, their optimal arrangement in the water area allowing for mutual screening reduces the effect to a minimum and keeps the ER at the neighboring territory and water area at background level for many years (see Fig. 2, 3 and Table 6).

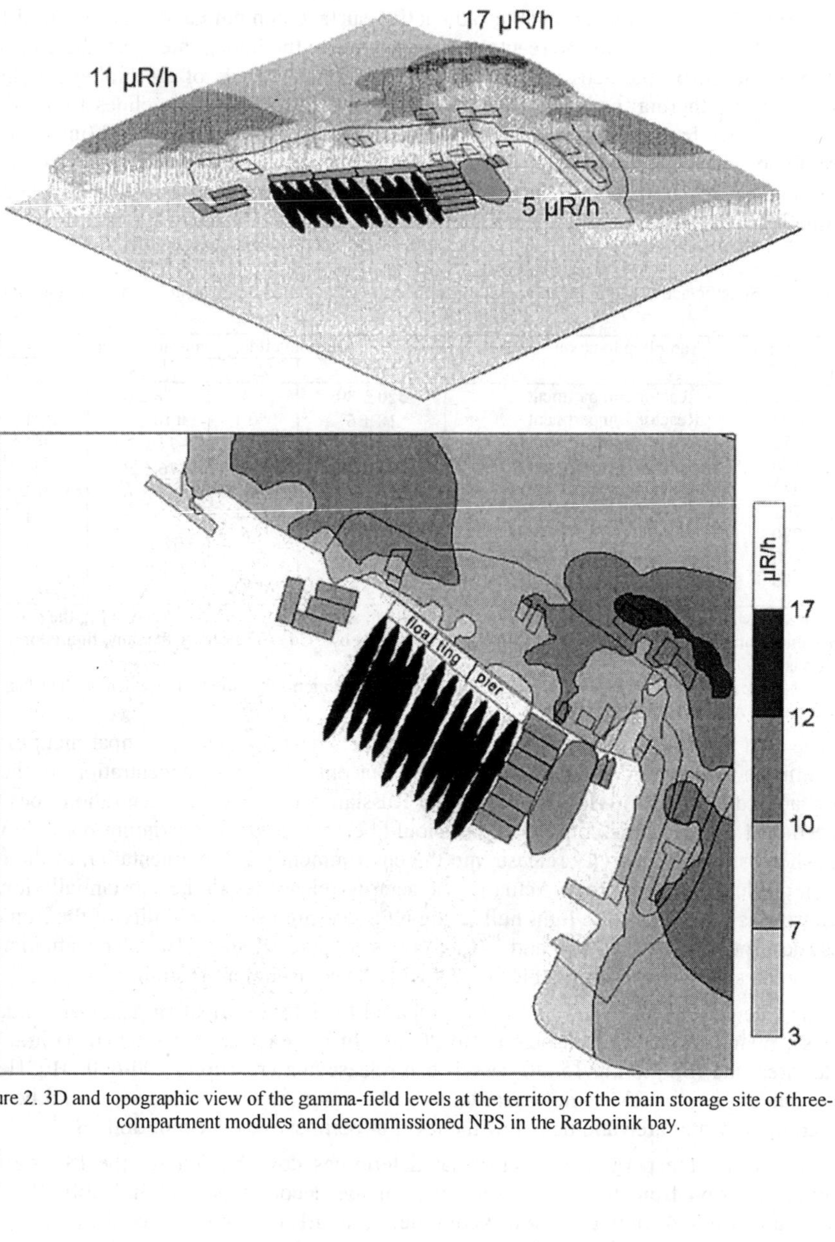

Figure 2. 3D and topographic view of the gamma-field levels at the territory of the main storage site of three-compartment modules and decommissioned NPS in the Razboinik bay.

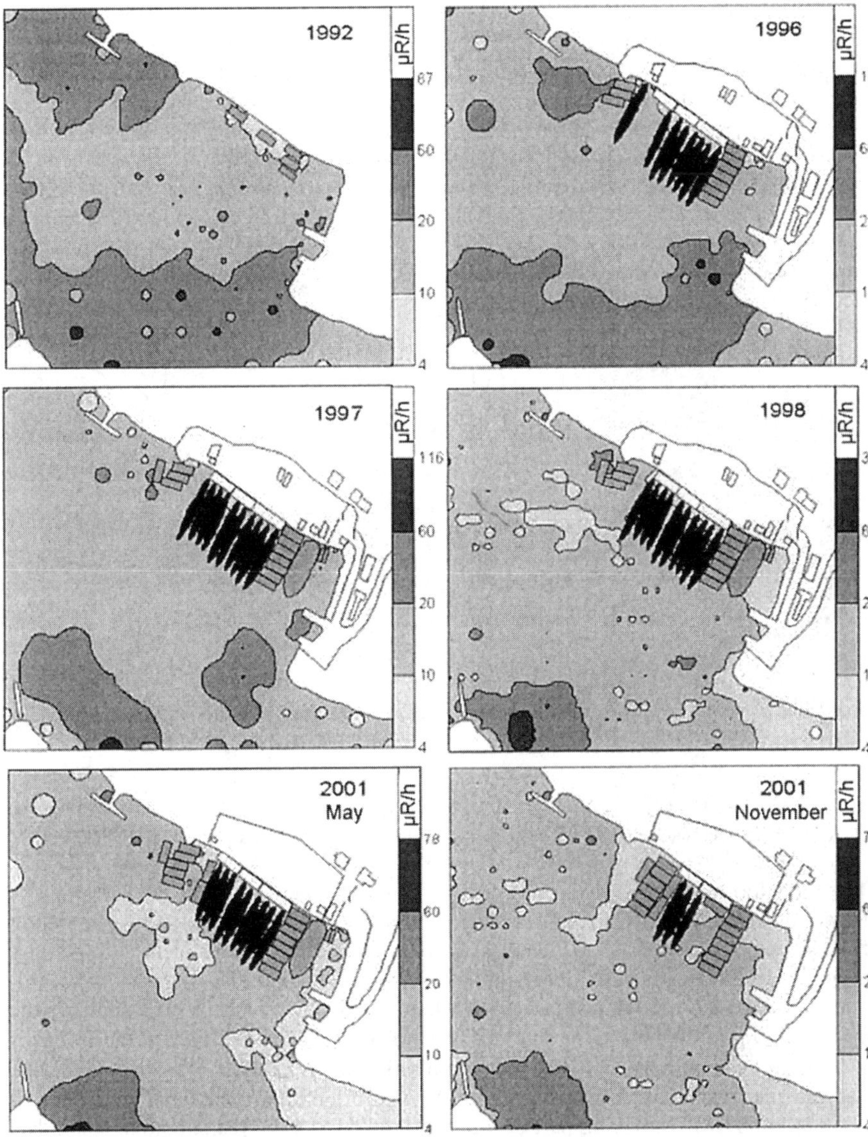

Figure 3. Gamma-field dynamics in the bottom sediments at the main storage site of three-compartment modules and decommissioned NPS in the Razboinik bay.

TABLE 6. Dynamics of the gamma ER variation in the ISS at the Razboinik bay (Pacific Fleet), μR/h

Observation region	1990*	1991	1993	1995	1997	1999	2001
Territory	8-14	7-13	9-15	8-16	6-15	9-14	8-15
Sea water**	1-2	1-2	1-2	1-2	1-2	1-2	1-2
Bottom sediment	11-14	10-13	9-14	10-15	11-13	9-14	10-15

*ER before disposition of the NPS and three-compartment modules.
Measurements were carried out at the depth 5 m from the surface.

The data in Figs. 2 and 3 show that, despite the difference in the annual arrivals of three-compartment modules and NPS to the main storage site at the Razboinik bay, the level of the gamma field in the bottom sediments did not grow worse during 10 years; alternately, as a result of the natural rehabilitation processes, the radiation situation in the entire water area of Chazhma bay has improved (the ISS is located in Chazhma bay, where the accident with the radioactive release into the environment occurred in 1985 at NPS K-431). Currently, the background level is maintained in the water area of Krasheninnikova bay (Kamchatka region), Postovaya bay (Khabarovsk region), and Severnaya bay (Primorsky region, all NPSs were removed from the site in 1988, Fig. 4).

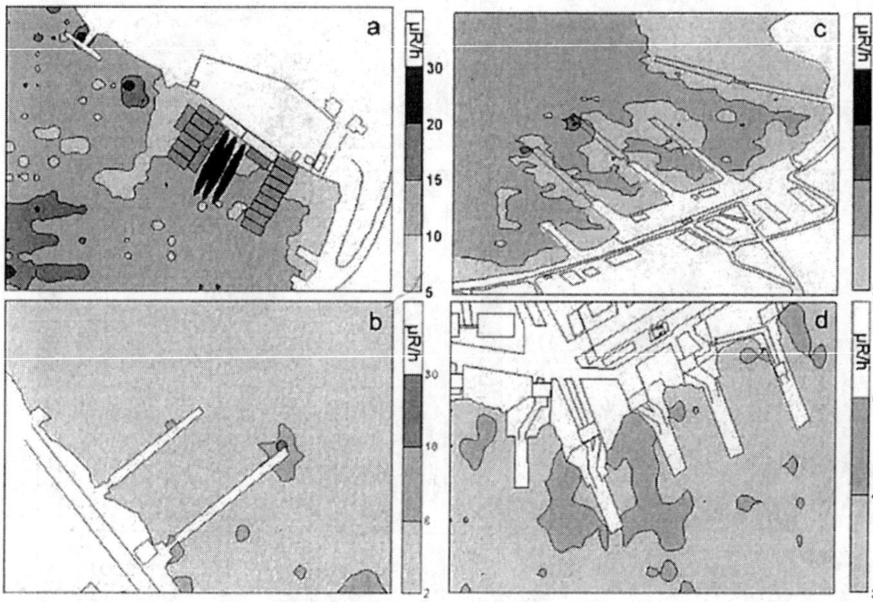

Figure 4. Gamma-field state in the bottom sediments of the main NPS storage sites of the Pacific Fleet in the Razboinik bay (a), Severnaya bay (b), Postovaya bay (c), and Krasheninnikova bay (d).

The radiation and radioecological situation is more complicated in the locations of NPS with damaged cores. Currently, the maximum surface contamination is observed at NPS K-431 of the Pacific Fleet (Pavlovskogo bay), which had the accident with activity release into the environment in Chazhma bay in 1985. The exposure in the most hazardous points inside the reactor compartment is kept at the level of 2-150 R/h (outside the compartment, up to 50 mR/h); the flux densities of α- and β-particles are at the levels of 10^1-10^2 decays/cm^2·min and 10^3-10^5 decays/cm^2·min, respectively [4, 5]. As a result, the ER at the pier varies within the interval 0.5-4 mR/h and decreases at distances of 150-200 m to 60 μR/h, covering part of the ISS territory. For 17 years, bottom sediments have contained stable spots of radioactive contamination with maximums under the NPS. Their external edges at the level 60 μR/h spread for distances of 100-300 m from the objects (Fig. 5). Affected by the bottom streams, the area of radioactive contamination changes its form, but the edges have not grown for the past decade and are limited by the internal water area of the ISS.

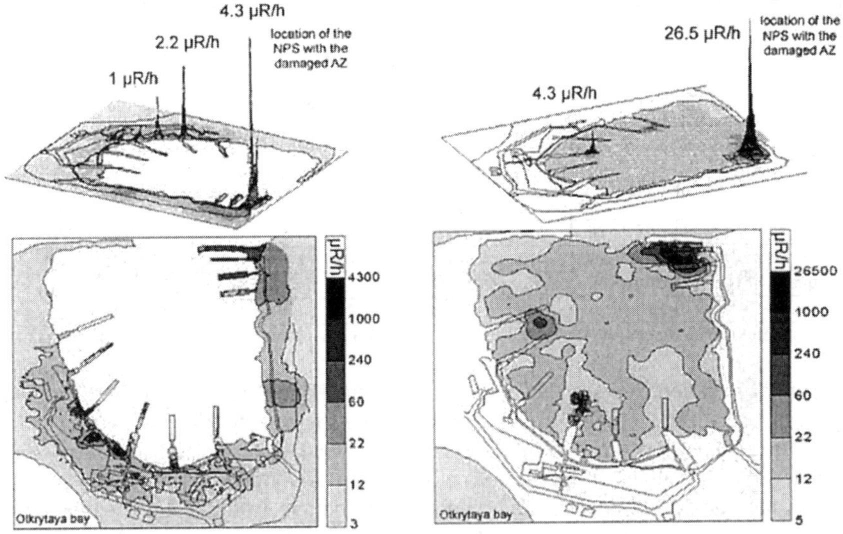

Figure 5. 3D and topographic view of gamma field at the territory and bottom sediments of the Pavlovskogo bay

Long-term investigations at the main ISS of the Arctic and Pacific Fleets confirm the previous assumption that the conservation of the primary (background) status during long-term storage of the NPS is a difficult problem. Actually, the environmental objects at each ISS contain certain amounts of fission or induced radionuclides (see Table 7 [3]).

TABLE 7. Specific activity A_{sp} of radionuclides in the bottom sediments of the Nikol'skoe estuary in the responsibility zone of the FSUE "Zvezdochka" and the background values (Arctic Fleet), Bq/kg

Sampling point	Nuclide	1989	1990	1991	1992	1993	1994	1995	1996	1997	1998	1999	2000	2001
NPS anchoring	A_{sp}	610	444	600	450	558	-	589	430	422	-	520	440	480
	^{137}Cs	32	15	30	15	5	-	8	9	10	-	11	5	6
	^{60}Co	21	bt	12	bt	bt	-	1	1	bt	-	bt	1	bt
NPS anchoring	A_{sp}	760	540	800	558	622	555	574	478	574	670	530	580	560
	^{137}Cs	56	33	60	42	22	89	21	25	34	37	16	38	22
	^{60}Co	91	14	23	12	bt	6	2	1	3	3	bt	4	2
NPS anchoring	A_{sp}	-	-♣	-	-	-	629	535	640	626	700	520	600	580
	^{137}Cs	-	-	-	-	-	36	18	29	22	46	11	19	14
	^{60}Co	-	-	-	-	-	bt	bt	bt	bt	6	bt	bt	bt
NPS anchoring	A_{sp}	635	7806	630	590	422	370	650	625	580	615	540	690	630
	^{137}Cs	53	81	115	35	17	41	24	33	31	28	21	32	24
	^{60}Co	bt	bt	330	14	bt	bt	10	3	bt	1	bt	3	2
Dvinsky bay (background)	A_{sp}	330	240	320	230	302	346	304	367	343	270	375	250	350
	^{137}Cs	7	bt	bt	bt	bt	2	bt	bt	2	2	1	2	2
	^{60}Co	bt	bt	bt	bt	bt	bt	bt	bt	bt	bt	bt	bt	bt
White sea (background)*	A_{sp}	-	-	-	-	-	-	-	-	-	-	-	-	-
	^{137}Cs	10	-	-	-	-	12	11	9	-	11	8	9	9
	^{60}Co	bt	bt	bt	bt	bt	bt	bt	bt	bt	bt	bt	bt	bt

Comment: bt — below the measurement threshold; * — data from [22] averaged over 10 points. Measurement accuracy 30%.

The data in Table 7 show that the specific activity of ^{60}Co and ^{137}Cs in the bottom sediments at the locations for NPS anchoring and preparation for utilization exceeded the background level by the factor 2–100. The man-caused radionuclides in the environment were observed at SRE "Nerpa", but their concentration was substantially lower (Table 8 [7]).

TABLE 8. Content of the man-caused radionuclides in the environmental objects at the ISS of the SRE "Nerpa" (Arctic Fleet, 2000), Bq/kg(m^3).

Environmental object	Sampling point											
	point # 1				point # 2				point # 3			
	A_{sp}	^{90}Sr	^{137}Cs	^{60}Co	A_{sp}	^{90}Sr	^{137}Cs	^{60}Co	A_{sp}	^{90}Sr	^{137}Cs	^{60}Co
Soil	265	28	13	< 5	480	<64	18	< 7	610	58	39	< 6
Sea water	-	< 40	< 90	< 90	-	< 40	< 70	< 90	-	< 30	< 80	< 80
Aquatic life	740	22	6	< 3	648	24	5	< 3	634	25	6	< 3
Ground	372	7	< 9	< 6	380	7	< 9	< 7	383	6	< 9	< 8

Comment. Measurement accuracy is 30%. Points # 1 and # 3 are outside the ISS, point # 2 is in the ISS water area.

Similar situation exists at the ISS for the Pacific Fleet (Table 9). The NPSs with damaged cores (Pacific Fleet, Pavlovskogo bay) are the sources of ^{60}Co (one vessel) and ^{137}Cs and ^{90}Sr (another vessel). Contamination of sea water is determined by the desorption of radioactive materials from the surfaces of the light and solid hulls, and by the sedimentation of the corrosion materials and paints on the bottom (radionuclide concentration in rust reaches 10^8 Bq/kg, in paints — $6 \cdot 10^5$ Bq/kg [4]). The positioning of both vessels at the same pier resulted in the mixed radionuclide content found in the environmental objects (see Tables 10 and 11 [8]).

TABLE 9. Contents of natural and man-caused radionuclides in the environmental objects at the main ISS of the Pacific Fleet, Bq/kg(l)

Environmental objects	Razboinik bay			Pavlovskogo bay								Krasheninnikova bay		
	NPS, 3-compartment modules			Damaged NPS					Non-damaged NPS			Non-damaged NPS		
	^{40}K	^{137}Cs	^{60}Co	A_{sp}	^{137}Cs	^{90}Sr	^{60}Co	^{144}Ce	A_{sp}	^{137}Cs	^{90}Sr	^{40}K	^{137}Cs	^{60}Co
Sea water														
surface	9,5	0,005	bt	0,16	0,015	0,010	0,012	0,019	0,150	0,007	0,005	8,7	0,003	-
bottom	9,6	0,005	bt	1,58	0,290	0,048	0,190	0,070	0,160	0,008	0,006	-	-	-
Algae	147	bt	260*	-	122	-	130	-	233	1,1	0,9	92	bt	bt
Fish	137	bt	bt	74	bt	-	bt	-	68	0,05	0,2	59	bt	bt
Shrimps	63	bt	bt	78	bt	-	bt	-	37	0,35	0,2	-	-	-
Scallops	81	bt	bt	74	bt	-	bt	-	41	bt	bt	-	-	-
Starfishes	63	bt	bt	-	-	-	-	-	26	bt	bt	44	bt	bt
Bottom sediments	407	12	75	$5,4 \cdot 10^4$	$2,0 \cdot 10^4$	$3,8 \cdot 10^3$	$3,2 \cdot 10^4$	$5,4 \cdot 10^2$	$7,4 \cdot 10^2$	18,5	9,2	126	4,4	bt
MPC	-	0,005	0,2		7,7	0,8	86,5	0,02	-	0,01	0,002	-	0,002	<0,001

Comment: MPC — maximum permitted concentration estimated according to [2]; bt — below the measurement threshold 1.1 Bq/kg; * — in result of the NPE accident at K-413 in 1985.

TABLE 10. Specific activity of sea water and man-caused radionuclides in the bottom sediments at various distances from the pier for NPSs with damaged NPEs

Sampling point	Distance, m			
	1	30	100	700
Sea water, Bq/m^3				
Water surface (background = 152 ± 75)	159±80	143±70	151±75	144±60*
Bottom layer (background = 200 ± 120)	1513±350	318±150	225±110	220±110*
Bottom sediments, Bq/kg				
^{60}Co	32000 ± 8000	1200 ± 400	20 ± 10	20 ± 10
^{90}Sr	3800 ± 1200	200 ± 70	10 ± 5	10 ± 5
^{137}Cs	20200 ± 7000	800 ± 300	20 ± 10	20 ± 10

TABLE 11. Specific activity of the man-caused radionuclides near the NPSs with the damaged NPEs, Bq/m^3

Radionuclide	Pier with the damaged NPS			Background
	Water surface, 1 m	Bottom layer, 1 m	Water surface, 700 m	Water surface, 2500 m
^{60}Co	11 ± 4	189 ± 60	< 10	< 7
^{90}Sr	11 ± 4	48 ± 12	< 10	3 ± 1
^{137}Cs	15 ± 5	289 ± 80	< 10	5 ± 2

The data in Tables 10, 11 show that the specific activity of the man-caused nuclides in sea water at the surface near the NPSs exceeds the background by a factor of 1.5–3, while in the bottom layer — by a factor of several dozens. However, at distances over 100 m, the concentrations of the man-caused radionuclides are comparable to other areas of the bay. The ground concentration of radionuclides near the NPS exceeds background by the factor 1,000–2,000, but at a distance of 100–300 m, it does not exceed the general contamination level of the bottom sediments at the ISS formed in the former NPS base, where a radiation accident with the NPS occurred.

Radioactive contamination of sea water and bottom sediments in the region of the anchored NPSs with damaged cores (up to 300 m in radius) resulted in changes of the radionuclide content of fauna and flora. The specific activity of the green algae increased on average by a factor of 3-5; its ^{137}Cs, ^{60}Co, and ^{90}Sr content amounted to 30-60, 6-12, and 10-25 Bq/kg, respectively, which exceeds the background values by a factor of 5-20. At the same time, an excessive contamination of kelp and starfishes was not observed. The ^{137}Cs and ^{60}Co concentrations in the shellfish viscera increased by a factor of 2-4 and 1.5-2, respectively, but the concentration of ^{90}Sr remained at the same level (the concentration in the clamshell increased by a factor of three). Such an increase in the radionuclide concentration in the consumable marine products does not exceed 0.1-0.3 MPC [9].

In general, the results of the long-term investigations make it possible to arrive at the conclusion that the impact of the decommissioned and utilized NPSs on the radioecological state of the ISS and neighboring settlements is insignificant except for the objects with the damaged NPS (Table 12).

The estimate of the radioecological hazard outside the ISS performed in terms of the step dependence of the form: $\leq 10^{-6}$ — insignificant hazard; 10^{-6}–10^{-5} — low hazard; 10^{-5}–10^{-3} — moderate hazard; 10^{-3}–10^{-2} — high hazard; $\geq 10^{-2}$ — excessive hazard; (limits $\geq 10^{-2}$ and $\leq 10^{-6}$ correspond to the probabilities of natural death and death as a result of natural disasters, respectively) makes it possible to conclude that the risk level for the population is insignificant (radiation risk $< 1 \cdot 10^{-6}$).

The risk levels in the radiation safety regime zones depend on the radioecological situation. In 75% of ISS, there is no significant man-caused radioactive contamination (insignificant hazard); in 20% of ISS, the contamination is comparable to the background level (low hazard); in 5% of ISS, the contamination exceeds the background (moderate hazard). At the water areas, the risk exists only in the zones of radioactive contamination of the bottom sediments (in a one-meter bottom layer); the hazard varies from moderate to high (Table 13).

TABLE 12. Radioecological state at the interim storage sites for the nuclear-powered vessels of the Pacific Fleet

Bay, parameters, ER	Water area				Territory				Dwelling zone
	RSRZ			SPZ	RSRZ			C33	Monitoring zone
	Background + 0.1 (B + 0.1)	B + 0.6	B + 2.4	B + 0.1	B + 0.1	B + 0.6	B + 2.4	B + 0.1	B + 0.1
Pavlovskogo bay									
Contaminated area,%	5	2	0.4	no	1	< 0.1	< 0.01	no	no
Soil, MPC:									
^{137}Cs	0.01-0.1	0.1-1.0	1-10	< 0.005	0.2-0.8	1-10	10-500	0.005	< 0.005
^{90}Sr	< 0.003	0.01-0.1	0.1-2.0	< 0.003	0.05-0.3	0.1-03	1-10	< 0.003	< 0.003
^{60}Co	0.01-0.1	1-10	10-80	< 0.001	0.01-0.1	0.3-1.0	1-10	< 0.001	< 0.001
Fauna, flora	< 0.01	0.01-0.1	0.1-0.3	< 0.001	0.01-0.1	0.1-1.0	1-3	0.01	< 0.001
Abrek bay*									
Contaminated area,%	no	no	no	no	no	no	no	no	no
Soil, MPC:									
^{137}Cs	< 0.005			< 0.005	< 0.005			< 0.005	< 0.005
^{90}Sr	< 0.003			< 0.003	< 0.003			< 0.003	< 0.003
^{60}Co	< 0.001			< 0.001	< 0.001			< 0.001	< 0.001
Fauna, flora	< 0.001			< 0.001	< 0.001			< 0.001	< 0.001
Razboinik bay									
Contaminated area,%	no	no	no	no	no	no	no	no	no
Soil, MPC:									
^{137}Cs	< 0.01			< 0.01	< 0.005			< 0.005	< 0.005
^{90}Sr	< 0.005			< 0.005	< 0.003			< 0.003	< 0.003
^{60}Co	0.01-0.1			0.1-1.0	< 0.001			< 0.001	< 0.001
Fauna, flora	< 0.001			0.05-0.1	< 0.001			< 0.001	< 0.001

TABLE 12 (Continued).

Bay, parameters, ER	Water area				Territory				Dwelling zone
	RSRZ			SPZ	RSRZ			SPZ	Monitoring zone
	Background + 0.1 (B + 0.1)	B + 0.6	B + 2.4	B + 0.1	B + 0.1	B + 0.6	B + 2.4	B + 0.1	B + 0.1
Severnaya bay									
Contaminated area,%	< 0.01	no	no	no	no	no	no	no	no
Soil, MPC:									
^{137}Cs	< 0.01			< 0,005	< 0,005			< 0,005	< 0,005
^{90}Sr	< 0.005			< 0,003	< 0,003			< 0,003	< 0,003
^{60}Co	0.1-0.3			< 0,001	< 0,001			< 0,001	< 0,001
Fauna, flora	< 0.001			< 0,001	< 0,001			< 0,001	< 0,001
Postovaya bay									
Contaminated area,%	< 1	< 0.1	< 0.01	no	no	no	no	no	no
Soil, MPC:									
^{137}Cs	< 0.01	< 0.01	< 0.01	< 0,005	< 0,005			< 0,005	< 0,005
^{90}Sr	< 0.005	< 0.005	< 0.005	< 0,003	< 0,003			< 0,003	< 0,003
^{60}Co	0.01-0.1	0.1-1.0	1-30	< 0,001	< 0,001			< 0,001	< 0,001
Fauna, flora	< 0.01	< 0.01	0.1-0.3	< 0,001	< 0,001			< 0,001	< 0,001
Krasheninnikova bay									
Contaminated area,%	no	no	no	no	< 0,01	no	no	no	no
Soil, MPC:									
^{137}Cs	< 0.005			< 0,005	0,01-0,1			< 0,005	< 0,005
^{90}Sr	< 0.003			< 0,003	< 0,05			< 0,003	< 0,003
^{60}Co	< 0.001			< 0,001	< 0,001			< 0,001	< 0,001
Fauna, flora	< 0.001			< 0,001	< 0,001			< 0,001	< 0,001

Comment: RSRZ — radiation safety regime zone; SPZ — sanitary protection zone. Surface nuclear-powered vessels are stored for 15 years.

TABLE 13. Additional man-caused radiation risk at the ISS and neighboring territories as a result of NPS storage, 10^{-6}

Objects, radiation zones	Territory			Water area		
	RSRZ	SPZ	MZ	RSRZ	SPZ	MZ
MPS anchoring sites	1-20	1-5	< 1	1-10*	< 1	< 1
Utilization shops	1-10**	< 1	< 1	5-15***	1-5	< 1
Settlements	-	-	< 1	-	-	< 1

Comments: 1. RSRZ — radiation safety regime zone; SPZ — sanitary protection zone; MZ — monitoring zone. * — radiation risk in the one-meter bottom layer of sea water in the Pavlovskogo bay (Pacific Fleet) at the anchoring point of the NPS with the damaged NPE reaches $100 \cdot 10^{-6}$; ** — radiation risk at the Chazhma SRE reaches $100 \cdot 10^{-6}$; *** — radiation risk in one-meter bottom layer of sea water reaches $1000 \cdot 10^{-6}$.

2. According to the data of the UN SCEAR, the world average effective equivalent dose caused by the gamma radiation from natural radionuclides is 0.35 mSv, which corresponds to the radiation risk $3.5 \cdot 10^{-6}$. The total effective dose allowing for the internal exposure and cosmic radiation is 2.4 mSv (risk $24 \cdot 10^{-6}$, interval $(5–100) \cdot 10^{-6}$ — average hazard). Radiation hazard of the natural background results from the ambiguous application of the linear nonthreshold model (LNT) for the dependence of the radiobiological effect on the exposure dose in the low-dose region. This assumption contradicts the fact of the long-term safe existence of human beings in their natural environment [8].

Application of the generally adopted risk criteria to the problem of radioecological safety for storage of the NPS decommissioned from the Navy made it possible to estimate its role and place in the hierarchy of all hazards for the population (Table 14).

TABLE 14. Risk hierarchy in everyday human activities and in the NPS storing activities.

Everyday activities	NPS activity	Risk, 10^{-6}	Hazard
Natural mortality of the population	-	10000	High
human death caused by smoking, consuming alcoholic beverages, using cars and civil planes	NPS with the damaged NPE (personnel)	100-1000	High, moderate
Using electricity, X-rays, surgery, swimming, coal mining, fishing, and other industrial activities	NPS with the damaged NPE (personnel)	10-100	Low
Using railway transport, bicycles, risk in building industry, fires, in hunting, in result of civilian traumas, in manufacturing and nuclear industries	NPS utilization (personnel, population of the cohort B)	1-10	Low
Human death in result of natural disasters	Residing near the NPS storage sites	≤ 1	Insignificant

Currently, the NPEs of the decommissioned NPSs contain large amount of nuclear fuel, and three-compartment blocks contain radioactive materials. However, the ER at the territories of the ISS does not exceed the level permitted for the population with the only exception being for NPSs with the damaged cores (Primorsky region, Pavlovskogo bay). During the entire storage period, the concentration of the man-caused radionuclides in the environment is within the background level in the neighboring settlements; while in various environmental objects it is within the interval 0,01–0,001 MPC.

Normalization of the radiation situation in the storage sites for NPSs with damaged cores is provided by their spatial separation from other objects and regulation of the staff presence at the site. However, starting from the accident and up to the present moment, the NPSs with the damaged cores that have radioactive contamination of the light and solid hulls are the permanent source of radioactive contamination of the marine environment. According to their radioecological impact, they are only local contamination sources, because the man-caused radionuclides are not released outside the ISS territory in concentrations exceeding the MPC. The scales of such events usually do not exceed 100–300 m from the object pending utilization. The maximum radioactive contamination, which exceeds the permissible level by one-two orders of magnitude, is observed for the bottom sediments, to a lesser extent, for sea water (several MPC or lower), and in insignificant amounts, for flora and fauna (below MPC).

Currently, all anchoring sites and the ISS for decommissioned NPSs are separated into prohibited zones (located 3–20 km far from the dwelling sites) where recreation activity and fishing are prohibited. The activity analysis from environment sampling shows that man-caused radioactive contamination outside the ISS zones is absent. Thus, excessive radiation impact for the population in the regions where the NPSs are utilized is not observed during the period under consideration.

In general, during the last decade of intensive NPS decommissioning, utilization, and storing, the radioecological situation in the coastal and sea areas of the Arctic and Pacific regions can be characterized as normal. The intensive activity aimed at the minimization and elimination of an unfavorable radiation impact on the environment and the population provides for the preservation of the natural balance in the environment. Afloat NPS storage does not produce any radioecological hazard for the population of Russia and neighboring countries.

At the same time, it is worth noting that protraction of the period for NPSs utilization to over 20 years would inevitably increase the radiation hazard, first of all, for the personnel and the population dwelling in the closest vicinity to the NPS location. The reasons are malfunctions in the NPS life-support systems that appear in course of the long-term afloat storage and can result in the increased release of man-caused radionuclides into the environment.

Despite the satisfactory radioecological situation at the storage sites of decommissioned and utilized NPS, any variant of afloat NPS storage is unacceptable with respect to hygienic issues, because it leads to a potential hazard of the radioactive contamination of the marine environment.

References

1. *Organization and Maintenance of the Water-Chemical Regime of the 1st Contour of the NPE with Non-Defuelled Active Zones at the Vessels Decommissioned from the Navy for Utilization (I-150-71-94).* Navy, Moscow (in Russian).
2. Sarkisov, A.A. et al. (2001) *Analysis of State, Environmental Regulations and Monitoring, Formation of the Control System for Ecological Safety in the Regions Involved in the NPS Utilization Processes After Decommissioning and SNF or RW Management.* IBRAE RAS, Moscow (in Russian).
3. *Manual for the Control Over the Radioactive Contamination of the Environment and the Internal Exposure of the Personnel at the Vessels With Nuclear-Powered Engines.* Voenizdat, MOD RF, Navy (1991), Moscow (in Russian).
4. Vysotsky, V.L. and Danilyan, V.A. (1999) Impact of Radiation Factors on the Choice of the Utlization Methods of the Vessels with the Damaged NPE, in L. LeSage and A. Sarkisov (eds.), *Problems of Decommissioning and Utilization of Nuclear Submarines,* IBRAE, Moscow, pp. 409–418 (in Russian).

5. Mazokin, V.A., Netecha, M.E., Stanislavsky, G.A. et al. (2002) Emergency NPS. Insulating and Providing Safety During Long-Term Storage, in *Proc. Int. COnf. "Ecological Problems of Utilization of Nuclear Submarines and Development ofNUclear Power Industry in the Region"*, Vladivostok, p. 35.

6. Danilyan, V.A., Vysotsky, V.L., Nikitin, V.S. et al. (2001) *Impacts of Nuclear Submarine Dismantlement and Salvaging on the Ecological State of the Site of the Russian State Center of Nuclear Shipbuilding.* Preprint IBRAE-2001-07, Moscow (in Russian).

7. Lyamin, P.Ya., Alexeeva, V.R., Rimdenok, R.E. et al. (2001) *Status of Nuclear and Radiation Safety at the SRE "Nerpa".* FSUE "Nerpa", Snezhnogorsk (in Russian).

8. Danilyan, V.A., Vysotsky, V.L., Maximov, A.A., and Sivintsev, Yu.V. (2000) Impact of the Nuclear Sunmarine Utilization on the Radioecological State in the Far-East Region, *Atomnaya Energiya* **2**, 451–474 (in Russian).

9. *Regulations for Radiation Safety (NRB-99). SP 2.6.1. 758 — 99*, (1999), Minzdrav of Russia, Moscow (in Russian).

THE FORECAST OF THE POSSIBLE RADIOECOLOGICAL CONSEQUENCES AND RISK OF DAMAGE TO THE HEALTH OF CRITICAL GROUPS OF THE POPULATION FROM SUNKEN NUCLEAR SUBMARINES AND OTHER RADIATION-DANGEROUS OBJECTS ON THE BOTTOM OF THE OCEAN

IGOR LISOVSKY
St.-Petersburg State Technical University
St.-Petersburg, Russia

Now at the bottom of Ocean are 6 nuclear submarines, lost as a result of failures - 4 Russian and 2 American.

For the period of operation of the nuclear underwater fleet (i.e. from middle 50 years) the Navy USA has lost 2 nuclear submarines: “Tresher” in April, 1969 and “Scorpion” in May, 1968. Both nuclear submarines sunk in areas of the intensive sea communications and active fishery.

In 1970 the soviet nuclear submarine “K-8” sunk in the gulf of Biskay, in 1986 the nuclear submarine “K- 219” sunk in the Atlantic ocean, in the area of Bermuda islands. In 1981 the nuclear submarine “K-27” was dumped near New Land island (gulf of Stepovogo). In 1989 the nuclear submarine “Komsomolets” sunk in the Norway sea.

In 2000 the nuclear submarine “Kursk” sunk in the Barents sea. Before salvaging in October 2001 the nuclear submarine was at the depth of 108 m near the coast of the Kola peninsula.

For an estimation of the ecological and radiation damage of the sunk nuclear submarines the appropriate methodology is used. For the first time such estimates were executed for the sunk nuclear submarine “Komsomolets” **[1, 2]**.

The working criterion - control concentration of the radionuclides in sea water (KK) - is necessary for an operative radiological estimation of the importance of results of determination of technogenic radionuclides concentration in sea water which is considered allowable concerning influence on man. The approaches to the decision of such task were developed and are partially realized, before scientific and normative base, in work **[3]** and are reconsidered in view of last achievement for sea radioecology and modern normative base in the field of radiation safety in works **[4, 5]**.

Estimation of the possible radioecological damage, which was developed for the sunk nuclear submarines “Komsomolets” showed that these consequences are non-significant **[2, 3]**.

In June 2001 in the Russian centre of science “Kurchatov institute” the estimations of the greatest possible emergency discharge of the radionuclides from the salvaging of the sunk nuclear submarine “Kursk” were carried out. Results of accounts in comparison to control concentration **[3-5]** are submitted in the table 1.

A.A. Sarkisov and L.G. LeSage (eds.),
Remaining Issues in the Decommissioning of Nuclear Powered Vessels, 181–187.

In the table the limits of annual receipt of the radionuclides and recommended levels of intervention in the Russian norms of the radiation protection are given also.

Comparison of the settlement data and sizes of control concentration **[4, 5]** allows one to note, that for Sr-90, Zr-95 + Nb-95, Ru-106 + Rh-106, Sb-125 and Cs-134 the settlement data in all cases were below offered sizes of control concentration. The settlement concentration of Cs-137 exceeds KK by approximately three times, as offered by the authors **[3-5]** for hygienic criteria.

TABLE 1. The comparative data about settlement and control concentration of the radionuclides in sea water for the salvaging of the sunk nuclear submarine. "Kursk"

Radionuclide	*K, Bq/l	**KK, Bq/l [3]	**KK, Bq/l [4]		**KK, Bq/l [5]			***, Bq/year	****, Bq/kg
			KK_f	KK_r	KK_g	KK_e	KK_u		
^{90}Sr	6,6	1,48	8,2	35,1	15	60	60	1,3 10^4	5,0
$^{95}Zr + {}^{95}Nb$	0,1	14,8 and 37	25,1 + 25,9		130 and 140	3 and 8	3 and 8	1,8 10^5 + 3,1 10^5	1,5 10^2 + 2,4 10^2
$^{106}Ru + {}^{106}Rh$	0,9	14,8 – ^{106}Ru	31,9		178 – ^{106}Ru	30 – ^{106}Ru	3,3 – ^{106}Ru	2,0 10^4 – ^{106}Ru	2,0 10^1 – ^{106}Ru
^{125}Sb	0,4	25,9	1					1,6 10^5	1,3 10^2
^{134}Cs	0,7	11,1	1,1	1,1	1,3	13	1,2	5,3 10^4	7,3
^{137}Cs	4,8	22,2	1,5	1,5	1,9	30	1,8	7,7 10^4	1,1 10^1
$^{144}Ce + {}^{144}Pr$	5,9	14,8 – ^{144}Ce	1,7 – ^{144}Ce		9,6 – ^{144}Ce	6 – ^{144}Ce	1,3 – ^{144}Ce	2,6 10^4 + 2,1 10^5	2,7 10^1 + 1,4 10^2
^{147}Pm	6,8		3,5		19,2	13		5,3 10^5 ·	5,3 10^2
Pu-α	0,2	7,4 – ^{239}Pu	110 – ^{239}Pu	110 – ^{239}Pu	110 – ^{239}Pu	0,25 – ^{239}Pu	0,03 – ^{239}Pu	2,4 10^3 – ^{239}Pu	5,6 10^1 – ^{239}Pu

* Settlement concentration
** Control concentration
*** Limit of annual receipt according to the Russian norms of the radiation protection
**** Level of intervention according to the Russian norms of the radiation protection
KK_f – Control concentration for fishering areas
KK_r Control concentration in view of a diet of the inhabitants of the Kola peninsula
KK_g – Hygienic control concentration
KK_e – Radioecological control concentration
KK_u – Universal control concentration

The settlement concentration Ce-144 exceeds KK for areas of a fishing **[4]** and universal KK for areas mixed (but not fish) craft **[5]**. The settlement concentration of Pm-147 exceeds KK for areas of a fishing only **[4]**. And, at last, settlement concentration of Pu-α exceeds only one size KK - universal KK Pu-239, offered for areas of the mixed craft **[5]**. By comparison with settlement concentration with KK, offered in work **[3]**, it is necessary to understand that D. Gusev (1975) proceeded from an other meaning of an annual limit of a dose and, most important, it was based in substantiations KK on limits of doses for critical bodies, instead of on an effective dose as in works **[4, 5]**. Hence, taking into account concurrence of results of the analysis (excess KK) of two independent sources **[4, 5]** it is expedient to pay attention only to the settlement concentration of Cs-137 and Ce-144.

It is necessary to note, that the salvaging of the nuclear submarine "Kursk" has passed in complete conformity with the technical project, without occurrence of emergencies, therefore, values given in the table of size and have remained academic, reflecting expected, possible level of the radiation damage.

The research near the sunk nuclear submarines "Tresher" (1963), "Scorpion" (1968) and "Komsomolets" (1989) have not revealed a significant concentration of the radionuclides in the sea environment. So, by results of inspections near the nuclear submarine "Tresher", executed by the Navy USA in 1965, 1977, 1985 and 1986, established, that in bottom sediment near to these objects (at a distance no more than 30 - 50 m) the low concentration of corrosion products of Co-60 and Ni-63 (maximal concentration Co-60 has made $1,4*10^{-2}$ Bq/g) are observed only. The amount of uranium, plutonium and fission products both in adjournment, and in samples of sea biota do not differ from that caused by global radioactive losses **[6]**. The same conclusion is made on results of inspections in 1968, 1977 and 1986 for the area near the submarine "Scorpion", where the maximal concentration Co-60 in the bottom sediment has made $4,3*10^{-2}$ Bq/g [7].

As a result of repeated expeditions in 1989 -1995 to a place of wreck of the nuclear submarine "Komsomolets" is established, that the content of fission and corrosion products, and also uranium and plutonium in sea water, tests of the bottom sediment and samples of biota in immediate proximity to the submarine do not differ from that caused by global radioactive losses. The special measurements have estimated the output of the Cs-137 in sea water to have a maximum $4*10^{10}$ Bq/year **[8]**. The given numerical meanings of concentration and output of the radionuclides testify to practically complete safety of the sunken nuclear submarine for the sea environment.

Conclusions about absence of real radiological danger of sunken nuclear submarines **[9, 10]** coincides with results of research of Japanese scientists **[11]**.

Special interests call attention to the nuclear weapons at the bottom of ocean. In main it is the nuclear weapon, sunken together with submarines. From the point of view of the authors **[12]** the nuclear weapons on the sunken submarines functionally are no longer weapons (table 2).

TABLE 2. Terms and degree of destruction of the nuclear weapon in the sea water

Submarine	Response time on the bottom, years (before 1997)	Part of the corroded plutonium, %
«K-278»	8,5	> 93
«K-219»	11	> 96
«K-8»	27	= 100
«K-129»	29	= 100

Thus, the quantitative estimations, executed to the present time, of the consequences of sunken nuclear submarines testify to their radiation safety for short-term and their limited radioecological influence. In 1987 and 1997 2 radionuclide power installations with $5{,}5*10^{16}$ Bq stroncium-90 were lost lossed in the sea near Sakhalin island.

Gosatomnadzor estimates showed that after 120 years (maximum possible time for keeping the safety of the radionuclide power installation) residual activity will be more than $3*10^{15}$ Bq. According to the same evaluations, output of the radionuclides to the Okhotsk sea water can begin much earlier - in 50-80 years and to make from $1{,}2*10^{16}$ to $0{,}8*10^{16}$ Bq.

For realization of the state control behind the safety of potentially dangerous underwater objects in Russian Federation the **state databank** about such objects and the **geoinformation system** for cartographical maintenance of its activity was created **[13-15]**. The special normative base was created too, whose basic components are:

- The federal law "About safety of underwater potentially dangerous objects of Russian Federation";
- The rule "About the State register of underwater potentially dangerous objects of Russian Federation";
- The rule "About the declaration of safety of underwater potentially dangerous object of Russian Federation";
- The rule "About examination of safety of underwater potentially dangerous object of Russian Federation".

The basic purpose of these documents is the prevention of extreme situations in water areas, the source of which can be potentially dangerous underwater objects.

As a possible approach to a system of criteria of danger for potentially dangerous underwater objects, the following classification in Table 3 is offered.

TABLE 3. Classification of underwater potentially dangerous objects by a degree of danger

Possible consequences of failure	**Class of danger and necessary measures**		
	C (The qualitative estimation of danger, incidental monitoring, intervention is inexpedient)	**B (Quantitative estimation of danger, periodic monitoring, intervention in dependence of the plans of development of territory)**	**A (State expert appraisal of safety, constant monitoring, intervention is necessary)**
Ecological consequences	Satisfactory ecological situation	Extreme ecological situation	Ecological disaster
Effects of damage to health	Stochastic effects	Somatic illness	Fatal cases
Infringement of conditions of life	The conditions of habitability are infringed for no more than 1 administrative formation during no more than 1 year	The conditions of habitability are infringed for more than 1 administrative formation or the period of infringements exceeds 1 year	The conditions of habitability of the population of 1 or more subjects federation are infringed or the evaquation is required
Material damage	Complete indemnification during a year	Complete indemnification during life of one generation	Complete indemnification during life of several generations
Territorial scale of potential effects	The consequences on leave for territory of the subject of federation	The consequences are distributed to territory 2 and more subjects of federation	The consequences have transboundary character
Social perception of risk on the given kind of danger	The separate publications in in mass media	Series of the articles and discussion in mass media, statement of public organizations	The danger is considered priority, the bodies of authority accept the special decisions, subjects of federation and international organizations agree on individual share in financing protective measures

It is possible to accept, that potentially dangerous underwater objects of a given class of danger (A, B, C) at present as one classification. The appropriated class of danger is not necessary to examine as having a constant size. This parameter can change in any situation depending on the occurrence of new information about the object (as in the case with the sunken nuclear submarine "Komsomolets" or with containers with radioactive waste) or with the occurrence of new social, economic and political factors (as in the case with the scuttled radionuclide sources). This is one of the basic problems of the planned annual updating of the State Register of potentially dangerous underwater objects.

At the present time the managers of EMERCOM of Russia consider possible and expedient the creation of international databanks of potentially dangerous underwater objects at the bottom of the sea and preparation of the coordinated normative documents on an estimation and declaration of safety of these objects with participation of all interested countries. Such database could, in particular serve for a substantiation of acceptance of the decisions on development of the investment projects in the marine water areas and would promote prevention of possible failures and accidents, connected to dangerous objects at the bottom of the Ocean.

References

1. Lisovsky I. In: *Radionuclides in the Russian Arctic*. Report of Russian experts for the AMAP, ed. by Balonov M. and Tsaturov Yu., 1997
2. *RADIOACTIVE CONTAMINATION IN THE RUSSIAN ARCTIC*. AMAP - Data Centre Report 1997, pp.38-43, 44-45, 149-153
3. Gusev D.I. Hygienic criterions for the estimation of the radioactive pollution of the coastal sea water. - In: *Impact of Nuclear Releases into the Aquatic Environment Symp*. IAEA. Vienna, 1975, pp.363-374
4. Ganul M.N., Kuchin N.L., Sergeev I.V. Kusnetsov Ju.V., Legin V.K. About control concentration of the technogenic radionuclides in sea water. *Ecological chemistry*, 1999, 8(3): pp. 197-203.
5. Sazikina T.G., Krishev I.I. Estimation of control concentration radionuclides in sea water with the account hygienic and radioecological criterions. *Atomic energy*, 1999, v. 87, iss. 4, pp. 302-307
6. Sheldon R., Michne J. - *Deep Sea Radiological Environmental Monitoring conducted at the of sife of the Nuclear - Powered Submarine «Thresher» Sinking*. Knolls Atomic Power Lab., Rept. KAPL - 4748, 1993.
7. Shellon R., Michne T. - *Deep Sea Radiological Environmental Monitoring conducted at the site of the Nuclear-Powered Submarine «Scorpion» sinking*. Knolls Atomic Power Lab., Rept. KAPL - 4749, 1993.
8. Gladkov G.A., Sivintsev Ju.V. Radiation situation in the region of the wreck nuclear submarine «Komsomolets». *Atomic energy*, v.77, iss.5, 1994, pp.379-386.
9. Lisovsky I, Dovgusha V. The estimation of the radioecological danger for the operation of the salvaging of the sunken nuclear submarine «Kursk». *4th International Conference on the Marine Intellectual Technologies*. St.-Petersburg, 20-22 September 2001. Proceeding of the conference, v. 2, pp. 255-260
10. Lisovsky I, Dovgusha V. Radiological danger of the sunken nuclear submarines. *Life and Safety*, 2001, 3-4, pp. 377-385

11. Kobayashi T., Togawa O., Odano N. and Ishida T. Estimates of Collective Doses from a Hypothetical Accident of a Nuclear Submarine. - *J. of Nuclear Science and Technology*, 2001, v. 38, No. 8, p. 658-662.

12. Tomashevsky L.G., Pertsev S.F., Filippovsky V.I., Lobyntsev V.V. About nuclear arsenals "for terrorists" in waters of World ocean. *Monitoring.Human and Indusrial Safety.* 1997, N 2, pp.14-17 (in russian)

13. Lisovsky I. State Control of Radio-Ecological Danger of Sunken Nuclear Objects in Russia. *Nuclear Risks, Environmental, and Development Co-operation in the North of Europe.* Proceedings from the conference in Apatity, 1999, ISBN 91-7991-789-6, pp. 109-115

14. Lisovsky I, Vialishev A. Underwater potentially dangerous objects of Russian Federation - the State control of Safety. *Proceeding of the Sixth International Conference on Ship`s and marine Structures in cold Regions (ICETECH`2000)*, 12-14 September 2000 St.-Petersburg, pp 575-584.

15. Lisovsky I, Vialishev A. The state control of the security of the potentially dangerous underwater objects. *Proceeding of the Second International Conference «Navy and shipbuilding nowadays» (NSN`2001).* St.-Petersburg, Russia, pp 205-209

RADIATION IMPACT ON THE ENVIRONMENT WHEN HANDLING FUEL OF DISPOSITIONED NUCLEAR SUBMARINES AT GMP "ZVEZDOCHKA"

Y. SIMANOVSKIY, G. ZEMBILGOTOV, V. SAFUTIN,
N.TIKHONOV, A. TOKARENKO
LI "Vnipiet"

M. IGNATIEV
TsNII TS SPV

G. NIKISHIN
TsNII KM "Prometey"

1. Current radiation situation at enterprise "Zvezdochka"

One of the main ways aimed to reduce radiation hazard in the RF North-West region is an urgent disposition of ships with Nuclear Propulsion Units (NPUs) decommissioned from the Navy. At the same time, disposition is feasible only after Spent Nuclear Fuel (SNF) has been unloaded from the reactors. Accordingly, decree № 518 (1998) of the RF Government put a task to intensify SNF unloading from Nuclear Submarines (NSs) to be dispositioned. For that purpose a land-based complex of premises was designed at the GMP "Zvezdochka" area to unload SNF from NSs to place it into shielding shipping casks TK-18 for shipping them by rail to PO "Mayak" for reprocessing. Currently the complex is near completion.

Simultaneously with the designing of the complex an environmental impact assessment in the "Zvezdochka" area required by the SNF operations was conducted as per Minatom's task.

The complex for SNF unloading from NSs is located in the area of existing special operations in which radiation and nuclear hazardous works are performed during NS repair, including works on refueling of reactors, decontamination and repair of contaminated equipment, collection / storage and treatment of LRW, SRW etc. The special area is located 1 km away from the main site.

Radiation monitoring for the personnel, and production rooms is provided together with radiation monitoring and nuclear safety services.

The complex comprises the following buildings and premises (see Figure 1):

1. A building for putting SNF into casks located on a steel-concrete basement on a special embankment.
2. A TK-18 cask interim storage area with access rail tracks. The area consists of 2 similar sections each of which accommodates 30 casks. The area is served with a traveling gantry crane to load and unload casks into and out of the building, into

A.A. Sarkisov and L.G. LeSage (eds.),
Remaining Issues in the Decommissioning of Nuclear Powered Vessels, 189–197.

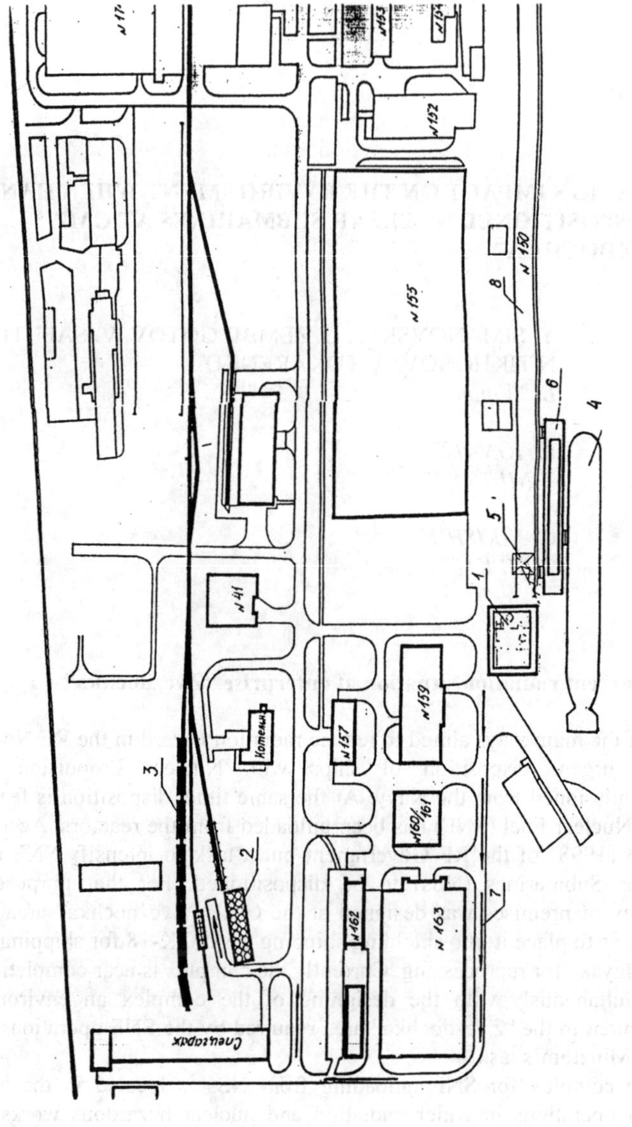

Figure 1. GMP "Zvezdochka" general layout fragment

container-wagons TK-VG-18 and on to the trailer.

3. Access rail tracks for assembling rolling stocks.
4. NSs from which SNF is unloaded.
5. Gantry crane.
6. Floating station for the radiation monitoring.
7. Transportation route for loaded shipping casks TK-18 to the interim storage site.
8. An additional area of special embankment with a 80 t gantry crane.

The main sources of radioactive airborne particle release into the air are NSs under repair, strictly speaking, ventilation systems fitted in reactor compartments (RC) to facilitate repair works. Specific activity of the RC air prior to purification is $1 \cdot 10^{-9}$–$1 \cdot 10^{-12}$ Ci/l but prior to the release into the atmosphere the air passes through filters FRI to be purified down to a specific activity of $1 \cdot 10^{-11}$–$1 \cdot 10^{14}$ Ci/l. A total volume of air released from all the RC ventilation systems is $1 \cdot 10^{8}$ m^3/year. Total annual activity value released to the atmosphere is $1 \div 1{,}5 \cdot 10^{-2}$ Ci as per regular measurements. The radionuclide composition of these releases (in %) is the following: ^{137}Cs – 69, ^{90}Sr – 14, ^{60}Co – 4; other radionuclides: ^{54}Mn, ^{55}Fe, ^{58}Co, ^{106}Ru, ^{125}Sb, ^{134}Cs, ^{144}Ce, ^{152}Eu 1-2 for each. Airborne particles also have α-emitters, including (in %) $^{239\text{-}240}$Pu – 0,001; ^{241}Am + ^{238}Pu – 0,004; ^{242}Cm – 0,005.

In addition to that compressed air and nitrogen are released during NS repair and disposition, which can contain radioactive impurities. Gases are released through a cleanup system; therefore, their volumetric activity is approximately $1 \cdot 10^{-14}$ Ci/l for airborne particles, and for ^{85}Kr it can reach $1 \cdot 10^{-9}$ Ci/l. Up to 1000 m^3 of compressed gases can be released from one NS in which case $2 \cdot 10^{-8}$ Ci of radioactive airborne particles and up to $2 \cdot 10^{-3}$ Ci of ^{85}Kr will annually get into the air.

An average annual activity of the air within the NS site varies within the range of $(4 \div 7{,}5) \cdot 10^{-15}$ Ci/m^3 ($(15 \div 28) \cdot 10^{-5}$ Bq/m^3). Moreover, the enterprise has stationary sources of radioactive releases. These are two slipways where NSs are also repaired, SRW and LRW storage, an incineration furnace, an equipment decontamination bay, nuclear and radiation safety laboratories etc. The total annual release of radioactive airborne particles in 1998 was $3{,}41 \cdot 10^{-2}$ Ci.

Over the past 10 years the free air radioactivity for airborne particles at every checkpoint did not exceed the reference level of $5000 \cdot 10^{-5}$ Bq/m^3 set by local supervisory bodies according to the Minoboronprom's guidance document. The radioactivity of fall-out over the same period in the restricted entry area varied within $(5{,}1 \div 36{,}0)$ MBq/km^2 which was much lower than the reference level (RL) of 370 MBq/km^2. In the settlement this value was $(4{,}4 \div 11{,}9)$ MBq/km^2. All of this is proof of the high efficiency of protective actions provided at the enterprise in the performance of radiation-hazardous works.

2. Radioactive releases and effluents from the SNF unloading complex

Additional radioactive contamination sources of the free air will appear at the enterprise after the NS SNF unloading complex has been put into operation and become operational. The most important of these sources under normal operation are: RC ventilation systems during SNF unloading, the buildings for charging casks TK-18 and area at the accumulation site to store loaded casks. The first and the second sources release radioactive airborne particles into the air after cleanup and inert radioactive gases (IRG) in a small amount; the third source mainly contains IRG. The complex service is sized for 3000 hrs per year. Four NSs must be handled for that period.

Calculations show that about $3 \cdot 10^7$ m^3 of air will be annually released, after purification, from the RC the ventilation systems. An averaged specific activity of the air will be $5 \cdot 10^{-11}$ Ci/m^3, consequently, the total annual release of radioactive airborne particles will be $1{,}5 \cdot 10^{-3}$ Ci.

The ventilation systems of the cask loading buildings over the same period will release $1 \cdot 10^{-4}$ Ci/year. The accumulation site will be a IRG source, mainly ^{85}Kr whose release will not exceed $2{,}8 \cdot 10^{-6}$ Ci/hr at the permissible release of $1 \cdot 10^{-3}$ Ci/hr under the valid regulatory rules OPB-83 which comply with IAEA regulation.

Thus, it can be seen that the additional release of radioactive airborne particles under normal operation of the plant will rise relative to the reached figure ($3{,}41 \cdot 10^{-2}$ Ci/year) by only 4,7 % and will come to $3{,}57 \cdot 10^{-2}$ Ci/year. This rise in radioactive release will not impact the nearby residential area. The authority of the RF Goskomecology (GMP "Zvezdochka") set for the year of 2000 a maximum permissible annual release of radioactive airborne particles in amount of 5,14 Ci, i.e. the actual radioactivity release by the plant, including the SNF unloading operation complex will come to less than 1 % of the maximum permissible release.

The stated release values are calculated for air cleanup filter efficiency of 99 %. If the efficiency decreases, filters have to be replaced by new ones. No discharge of liquid radioactive waste from the unloading premises into the water area of the plants is performed.

Together with radioactive releases the unloading complex is a source of solid and liquid radwastes. These mainly arise in the preparation for SNF unloading from RCs since dismantling works are conducted at that time, the primary and tertiary circuit systems are drained, sorbents are unloaded etc.

The amount of solid radioactive wastes during the preparation of one NS for unloading is ~ 30 m^3, including metal of ~ 8 m^3; combustibles (clothing, boards etc/) – 2 m^3; compactibles (flexible PVC, rubber, hoses, cables etc.) – 20 m^3. All the wastes are in the low active category and have a total activity of about $2 \cdot 10^{-3}$ Ci. Consequently, up to 120 m^3 of these wastes can annually arise which will increase the amount of SRW annually arising at the plant by 26 %. An incineration furnace is in service at the plant, so combustible waste will be incinerated and the others will be forwarded to SRW storage. The existing SRW store (facility 162) of 1500 m^3 is practically completely filled. A new SRW store of 2500 m^3 must be put in service. Moreover, a compaction unit is envisaged with the accumulation of additional SRW.

About ~ 50 m^3 of LRW per one NS arise together with SRW at the unloading complex. Mainly this is water due to the NPU circuits drainage performed prior to SNF unloading from reactors and decontamination water. These liquid radioactive wastes are the low-active category and their specific activity is $1 \cdot 10^{-4} \div 1 \cdot 10^{-8}$ Ci/l. Annually up to 200 m^3 of such wastes can arise which is ~ 20 % of the existing LRW amount arising annually at the plant.

Waters of the change rooms, as a rule, have a low specific activity premising their discharge into the water area. But they are separately collected, monitored and discharged, if their activity does not exceed PAL. In case of an increased radioactivity they are directed to the LRW store. LRW arising at the plant are collected for temporary storage in tanks of facility 159 whose total volume is 2200 m^3 and which are currently half full.

Currently a LRW treatment plant is ready for operation at the enterprise which will create capabilities for LRW arising at the plant including the SNF unloading complex.

Thus the creation and normal operation of the land-based SNF unloading complex and further fuel packing and forwarding for reprocessing will not lead to ecological problems in this region.

Total data on the increased releases and radwaste at GMP "Zvezdochka" are given in Tables 1 and 2.

Table 1. Rise in radioactive airborne particles (RAP) at enterprise in SNF unloading complex (UC) operation

Current RAP release by enterprise, Ci	RAP release by UC premises, Ci	Total RAP release by enterprise with operating UC, Ci	Permissible RAP release for enterprise for 2000, Ci
$3{,}41 \cdot 10^{-2}$	$1{,}6 \cdot 10^{-3}$	$3{,}57 \cdot 10^{-2}$ (4,7 %)	5,14

Table 2. Rise in radwaste volume at the enterprise in SNF unloading complex operation

RW type	RW volume after unloading of 1 N/SM, m^3	RW volume after unloading of 4 N/SM, m^3	Portion of the all-enterprise RW volume, %
SRW	30	120	26
LRW	50	200	20

3. Changes of radiation situation under accidental conditions

The failure-free operation of the SNF unloading complex is provided by: the handling process, reliability of premises and equipment, high qualification of personnel, up to – date control means, reliable electric power supply etc. Nevertheless natural occurrences, failures of instruments and equipment, personnel mistakes etc must be included.

Accidents due to natural occurrences include earthquakes, floods, storms etc. Accidents caused by external impacts are: aircraft drop, terrorism, external explosion. The analysis shows that accidents caused by natural occurrence and external impacts will not result in a substantial release of radioactivity beyond boundaries of premises being constructed and accordingly, will not cause considerable damage to the environment.

For accidents due to technical faults the following are considered: power blackout, transfer container drop (a maximal design-basis accident), shielding cask drop, fire in the cash fire in the fuel loading building, and unauthorized LRW discharge into the water area. Also considered are two types of beyond design-basis accidents: guiding device drop into an opened TK-18 cask loaded with fuel and super critical reactor (SCR) occurrence in the reactor during its preparation for fuel unloading – the most severe accident. Let us consider this accident in more detail. According to the process developed by NI PTB "Onega", SNF unloading is performed under a strict adherence to two safety conditions:

- regular absorbing elements of the reactor are reliably fixed in their extreme lower position;
- the reactor is drained of water prior to fuel unloading.

The failure to violate these two conditions simultaneously assures criticality safety of fuel unloading. However an authorized reactor filling with water before lid opening can be considered possible. Then a mistaken extraction of one or more control rods due to failure to disengage the drive from the working unit dismantling the drive or seizure of one or several CG when dismantling the reactor lid can be assumed. A simultaneous failure of several safety systems and/or several personnel mistakes must occur to accomplish the mentioned events. The SCR can result from the combination of these mistaken actions and failures. The probability of such an event will be not more than $1 \cdot 10^{-6}$ according to estimates.

Radiation consequences of SCR are specified by the following key factors:

- power generation of the core;
- delay from the time of reactor shutdown to the SCR start;
- burst power (total fissions at SCR);
- the nature of fission products release into the free air;
- free air condition.

Analyzing the SCR development scenario one proceeds from some provisions:

- the burst occurs in the cooled down reactor with the unsecured lid; with the RC opened;
- the burst duration is short due to steam-water mixture release from the reactor vessel which facilitates SCR termination;
- due to a slow lift of absorber elements the SCR is contained in the bottom part of the core and close to its center, the portion of completely failed fuel elements is taken as 10 %.

FP release will continue after all the water has been evaporated but with a smaller intensity. The total activity of the release will be specified by Th quantity of FPs accumulated in the core and integral fissions at SCR. In this case the integral fissions are taken as $1 \cdot 10^{20}$. The steam-gas jet height can, by estimates, be 100 m. The time of intensive FP release with evaporating water is ~ 10 min. In which case about 3000 Ci of FP with half-life over 1 hr can release into the free air including: ^{85}Kr – 350; ^{90}Sr – 260; ^{137}Cs – 2300 and other FP in amount of about 3000 Ci (Table 3).

Radiation consequences of such an accident can be:

- the total individual effective exposure dose for the public (without food chains) will not be more than 3 mSv for the early and middle accident phases (up to 10 days) outside the Special Premises Zone (SPZ);
- soil contamination outside the SPZ can be $2{,}9 \cdot 10^{6}$ Bq/m^2.

Table 3. In-core FP content during fuel unloading, post – SCR FP formation (over 10 min.) and a complete FP release into the air

Radio-nuclide	Half-life	Content during unloading, Ci	FP formation during SCR, Ci	Release into the air, Ci
^{85}Kr	10,8 y.	3730	0,0003	338
^{89}Sr	52,7 d.		4,64	0,023
^{90}Sr	27,7 y.	51300	0,105	257
^{103}Ru	39,3 d.		32,7	0,05
^{106}Ru	368 d.	283	0,23	0,42
^{131}J	8 d.		2,55	0,115
^{133}Xe	5,29 d.		0,19	0,017
^{137}Cs	30 y.	55200	0,103	2298
^{140}Ba	12,8 d.		105	0,54
^{141}Ce	32,3 d.		0,208	$3 \cdot 10^{-5}$
^{144}Ce	284 d.	11300	4,54	0,17
^{239}Pu	24300 y.	140	-	0,021
^{240}Pu	6,6 y.	50		0,0075
^{241}Pu	13 y.	2400		0,36
^{241}Am	458 y.	220		0,033
^{143}Ce	33 hr.		354	0,053

Note: The Table contains radionuclides having half-life over 1.5 hr.

The anticipated individual dose for the public for the first year after an accident at the SPZ boundary (without food chains) can be 22 mSv. Consequently no public evacuation is needed, since the established limit for the public of 50 mSv is not reached; the same dose 12 km away will be 5 mSv and this zone is specified as a zone of planned protective actions. In keeping with the recommendations, the monitoring zone is to be limited to 30 km where the public dose for the first year will not exceed 3 mSv (Table 4).

The width of a contamination strip outside the SPZ with the soil activity level of $3,7(10^4 \div 10^5)$ Bq/m^2 can be 200-300 m wide and 3-12 km in length (for different weather conditions). Therefore the conduction of decontamination actions at the most contaminated places will be rather effective.

Within the plants 5 km zone soil is not used for agricultural purposes. Local agricultural products are not allowed for use within the contaminated strip without decontamination and radiation monitoring at a distance up to 30 km in case of a beyond design-basis accident with SPZ.

Thus the construction and operation of the land-based complex for unloading SNF from dispositioned NS reactors in the GMP "Zvezdochka" territory under normal operating conditions will not release radioactive materials into the environment and will not effect living conditions in the heighborhood.

The public exposure under the normal operation of the SNF unloading complex at the SPZ boundary (1 km) will be below 1 μSv/year or 2,5 % of the departmental dose.

Table 4. Maximal value of anticipated individual effective dose, which can be received by the public in case of a maximal beyond-design-basis accident with SCR in N/SM NPU, Sv

Distance, km	Category of atmosphere stability	Exposure route[A])					
		1	2	3	4	1+2+3 [B])	1+2+3 [C])
0,1	A	$2,2\cdot10^{-4}$	$9,5\cdot10^{-12}$	$2,3\cdot10^{-10}$	$6,1\cdot10^{-8}$	$2,2\cdot10^{-4}$	$8,8\cdot10^{-5}$
0,15	A	$2,2\cdot10^{-4}$	$3,2\cdot10^{-6}$	$7,8\cdot10^{-5}$	$2,0\cdot10^{-2}$	$3,0\cdot10^{-4}$	$1,2\cdot10^{-4}$
0,2	A	$2,2\cdot10^{-4}$	$1,8\cdot10^{-4}$	$4,3\cdot10^{-3}$	1,1	$4,4\cdot10^{-3}$	$1,8\cdot10^{-3}$
0,3	A	$2,2\cdot10^{-4}$	$1,9\cdot10^{-3}$	$4,6\cdot10^{-2}$	12	$4,8\cdot10^{-2}$	$1,9\cdot10^{-2}$
0,5	A	$2,2\cdot10^{-4}$	$3,4\cdot10^{-3}$	$8,1\cdot10^{-2}$	21	$8,5\cdot10^{-2}$	$3,4\cdot10^{-2}$
0,7	A	$1,6\cdot10^{-4}$	$2,7\cdot10^{-3}$	$6,5\cdot10^{-2}$	17	$6,8\cdot10^{-2}$	$2,7\cdot10^{-2}$
1,0	B	$1,2\cdot10^{-4}$	$2,1\cdot10^{-3}$	$5,2\cdot10^{-2}$	13	$5,4\cdot10^{-2}$	$2,2\cdot10^{-2}$
1,5	C	$9,5\cdot10^{-5}$	$2,0\cdot10^{-3}$	$4,9\cdot10^{-2}$	13	$5,1\cdot10^{-2}$	$2,0\cdot10^{-2}$
2,0	B	$7,2\cdot10^{-5}$	$1,7\cdot10^{-3}$	$4,1\cdot10^{-2}$	11	$4,3\cdot10^{-2}$	$1,7\cdot10^{-2}$
3,0	D	$5,0\cdot10^{-5}$	$1,4\cdot10^{-3}$	$3,4\cdot10^{-2}$	8,8	$3,5\cdot10^{-2}$	$1,4\cdot10^{-2}$
5,0	D	$2,9\cdot10^{-5}$	$9,5\cdot10^{-4}$	$2,3\cdot10^{-2}$	6,0	$2,4\cdot10^{-2}$	$9,6\cdot10^{-3}$
10,0	E	$1,2\cdot10^{-5}$	$5,6\cdot10^{-4}$	$1,4\cdot10^{-2}$	3,6	$1,5\cdot10^{-2}$	$6,0\cdot10^{-3}$
15,0	E	$8,7\cdot10^{-6}$	$4,6\cdot10^{-4}$	$1,1\cdot10^{-2}$	2,9	$1,2\cdot10^{-2}$	$4,8\cdot10^{-3}$
20,0	F	$7,2\cdot10^{-6}$	$4,2\cdot10^{-4}$	$1,0\cdot10^{-2}$	2,7	$1,0\cdot10^{-2}$	$4,0\cdot10^{-3}$
25,0	F	$6,0\cdot10^{-6}$	$3,7\cdot10^{-4}$	$9,1\cdot10^{-3}$	2,4	$9,5\cdot10^{-3}$	$3,9\cdot10^{-3}$
30,0	F	$5,0\cdot10^{-6}$	$3,3\cdot10^{-4}$	$7,9\cdot10^{-3}$	2,0	$8,2\cdot10^{-3}$	$3,3\cdot10^{-3}$

A) 1. External exposure from a cloud. 2. Internal exposure due to inhalation. 3. External exposure from the earth surface. 4. Internal exposure due to the consumption of contaminated food.

B) Corresponds to the dose ignoring food consumption.

C) Corresponds to the dose ignoring food consumption with the radiation buildings screening ratio introduced – 0.4.

It follows, from a review of emergency situations for caused by natural events, external impacts and technical faults that the most hazardous are accidents caused by technical faults. The most hazardous can be a beyond design-basis accident with SCR occurrence in the course of fuel preparation for unloading from the reactor. In the event of such an accident a considerable radioactive materials release into the air will take place. As a consequence, there will be contamination of an area for which protective actions must be carried out but with no public evacuation, since dose expenses established for these actions will not be reached.

References

1. *Infrastructure of SNF Unloading from N/SM Reactors at GMP "Zvezdochka". FS. Environmental impact assessment.* (1998) LI "VNIPIEY", ref. № 98-01504, SPb.
2. *Health Physics Regulation (NRB-99)* (1999), Moscow.
3. *Basis Sanitary Rules to Assure Radiation Safety (OSPORB-99)* (2000), Moscow.
4. *Radiation Monitoring of the Environment at Enterprises Performing the Construction, Repair, Disposition of Ships and Vessels with NPU.* (1999) RD5.AEISH.2946-99. SPb.
5. *Permit № 36 to Release Radioactive Materials into the Free Air Granted to GMP "Zvezdochka" by the RF Goskomecology Authority* (Nov. 01, 1999).

METHODOLOGICAL ASPECTS OF ASSESSING THE RADIOECOLOGICAL SITUATION AT CONTAMINATED TERRITORIES ADJACENT TO RUSSIAN NAVAL BASES TO PERFORM THEIR SUBSEQUENT REHABILITATION

V. ZHERNOVOY, S. NATKHA, A.PUSIKOV
Research Institute of Industrial and Marine Medicine
St. Petersburg, Russia

1. Introduction

From the standpoint of: -the Russian Federation (RF) Law "On the Radiation Safety of the Population ", -"The Radiation Safety Standards (RSS-99)" and -"The Basic Sanitary Radiation Safety Rules (BSRSR-99)" that toughen the radiation safety requirements, a demand arises presently on revising the existing management directives and creating new ones to regulate issues of environmental protection in the area of preventing radionuclide contamination of different environment media.

Under conditions of reducing the Russian Armed Forces personnel and transferring a part of the Navy facilities to other departments, the lack of both unified standards & methods of assessing the status of contaminated territories adjacent to the above facilities complicates the task of both their rehabilitation and following land use.

Investigations aimed at elaborating a unified methodology to assess the radioecological situation at territories adjacent to facilities of both the Russian Navy and nuclear shipbuilding enterprises (performing nuclear submarine utilization operations) are especially urgent [1].

After closing Russian naval bases or changing their status and, thus, eliminating sources of the environment contamination, their adjacent territories can be used for different purposes (e.g. house-building, industry, agriculture, forestry, etc.). But this is only possible if the levels of the environmental contamination do not exceed permissible limits. Such limits represent basic criteria to assess both the contaminated territory status and potentialities for subsequent use.

Obviously, these criteria should be established for every individual contaminated territory with consideration for its potential use in the future depending on the specificity of radionuclide contamination, paths of radionuclide intake into human organism and many other factors.

2. Radioactive Contamination at Territories Adjacent to Russian Naval Bases

At the Arctic Navy nuclear vessel basing areas within some restricted industrial workcells elevated concentrations of ^{90}Sr, ^{137}Cs and ^{60}Co in soils and vegetation are recorded (average excess by a factor of 5 through 10 as

A.A. Sarkisov and L.G. LeSage (eds.),
Remaining Issues in the Decommissioning of Nuclear Powered Vessels, 199–206.

compared to the background level) indicating local-scale radioactive contamination of the environment [2].

Within nuclear vessel waterborne storage areas the marine ecosystem is most affected. Because of the low quality of Radioactive Waste (RW) collecting devices and of temporary storage facilities, appreciable amounts of Liquid Radioactive Wastes (LRWs) are occasionally dumped into water. This results in a considerable radionuclide accumulation in bottom sediments and hydrobionts. Within some restricted areas an excess of ^{137}Cs and ^{90}Sr concentrations by a factor of 3 to 5, as compared to the background level, is recorded [2, 3, 4].

The Russian naval base operating experience demonstrates that the most hazardous situation regarding the environment contamination characterizes the Coastal Servicing Enterprise (CSE) in Andreeva Bay and its adjacent territories. Here, an excess (by a factor of 5 through 100) over the background value is recorded. The integral radionuclide disposal into the Andreeva Bay water area is about $1{,}5 \cdot 10^5$ Bq/year wherein the main contribution is due to long-lived radionuclides of Cs & Sr. ^{60}Co is also recorded in marine waters: its presence can be explained by unauthorized RW dumping out of tanks of Nuclear Submarines (NSs) stored afloat in the immediate vicinity to Andreeva Bay [2].

It is widely known that radionuclides are principally accumulated in hydrobionts, bottom sediments and suspensions. At the same time radionuclide concentrations in marine waters remain at a low level. From this a well-known fact follows that for long-term evolution of the radioecological situation (worsening evolution is caused by emergency dumping of contaminated waters) the volumetric activity of marine waters becomes a monitoring parameter containing little information and is unable to characterize the real situation.

There is practically no way of determining the integral radionuclide concentrations in soils within both the CSE site and its adjacent territories. Hydro-meteorological factors have contributed to the creation of a spotty soil-contamination structure. Maximum contamination levels are recorded in natural depression bottoms as well as within some areas of immediate RW dumping. The specific activity in soils at the CSE site reaches about $1{,}3 \cdot 10^6$ Bq/kg (220 Bq/kg on average), whereas its average value within the buffer area equals 14 Bq/kg.

An analysis of the available data characterizing the radioecological situation within territories adjacent to CSEs makes possible the following conclusions:

1) Operation over a long period of time of many installations of imperfect quality at Russian CSEs resulted in a considerable contamination of their sites and of a part of buffer areas by long-lived radionuclides. Thus, based on analysis of the available data, the radiation situation at individual CSE-adjacent territories can be considered as unfavorable;
2) Radioactive atmospheric aerosols and radioactive fallout, radioactive contamination of soils, plants, bottom sediments and hydrobionts are principal parameters determining the radiation impact of CSEs facilities on the environment;

3) Thanks to self-decontamination processes in natural hydrocenoses, their contamination in the CSE vicinity does not result in environmentally important implications even with consideration for eventual dumping of large amounts of LRWs. However in areas of nuclear vessel handling operations a considerable radionuclide portion has been accumulated within the upper layer of bottom sediments resulting in periodic increases in the volumetric activity in marine water.

3. Methods and Criteria of Assessing the Radiation Situation at Contaminated Territories Adjacent to Russian Naval Bases

Human beings are the key subjects of radioecological safety investigations. Therefore, to assess the radiation situation at contaminated territories, the radiation-hygienic method is used most frequently [5].

This implies the application of an equidosage approach with respect to human exposure, on the one hand, and other biota species exposure, on the other hand [6, 7, 8].

However, in the case of a potential severe radiation accident accompanied with considerable radionuclide releases into the environment, dose loads to biota can reach 100 Gy/year, whereas they would not exceed permissible levels to human beings [6].

Thus, to assess the radioecological situation, the radiation-hygienic criteria are only applicable if the radiation contamination values are close to permissible levels. In a case of emergency their application is incorrect and can result in a wrong assessment.

To implement into practice the RF Law "On the Environment Protection", the RF Ministry of Natural Resources (Ministry of Ecology) issued a document known under the name of "The Criteria of Assessing the Environmental Situation to Identify Territories of the Environmental Emergency Status and of the Environmental Disaster Status" (1992) [9]. As distinct to this document, our purpose consists in the development of a methodology to evaluate the actual environmental status of territories with consideration for different options for their potential future use.

A Methodical Directive "The Criteria & Methods of Assessing the Radiation Situation at the Contaminated Territories Adjacent to Russian Naval Bases" has been developed at the Russian Research Institute of Industrial and Marine Medicine (RIIMM). The Directive deals with both the methods of radiation-hygienic evaluations at territories adjacent to coastal facilities (CSEs) of the Russian Navy transferred under the authority of Ministry of Atomic Energy (Minatom) and the contamination due to operations performed at the concerned CSEs.

In keeping with this document, an assessment of the actual radioecological status of a contaminated territory is aimed at establishing some permissible levels needed to make a decision on the necessity/possibility of its rehabilitation as well as on future use.

When performing the radiation-hygienic assessment, the following tasks are solved:

1) Organization & carrying out of radiation monitoring at a territory under consideration with reference to a territory with background contamination;
2) Identification of all potential exposure paths with respect to all population groups within the studied territory;
3) Processing of the obtained information, calculations of radionuclide intake into the human organism and of individual effective exposure doses regarding "critical" population groups;
4) Comparative analysis of the radiation monitoring results between territories adjacent to CSE facilities and a reference territory with background contamination;
5) Establishing permissible radiation contamination levels for different environment media in the territory under investigation;
6) Comparing the obtained population & environment exposure levels with the RSS–99 limits of doses for the population or of annual exposure doses, permissible volumetric activities, intervention levels and the calculated values (based on the above limits) of permissible levels with respect to specific conditions.

When assessing the radiation situation in territories, one should take into account all detected radionuclides of technogenic origin [8, 9].

The proposed criteria to assess the radiation-hygienic situation in contaminated territories are illustrated in Table 1 and Table 2.

TABLE 1. Radiation-Hygienic Criteria to Assess the Status of Territories Contaminated by Technogenic-Origin Radionuclides

Territory status	**Individual effective dose (D) caused by technogenic radionuclides within the considered territory, mSv**
Normal	$D_{Bg} < D < D_{Bg} + 3\sigma$
Satisfactory	$D_{Bg} + 3\sigma < D < 1,0$
Unsatisfactory	$D > 1,0$

Comments: 1) DBg – annual effective population exposure dose formed by technogenic radionuclides in a reference background-contamination region outside the studied territory;

2) σ - integral standard deviation of all measured parameters of the radiation situation (caused by technogenic radionuclides) contributing to individual effective exposure doses with respect to individuals of "critical" population groups. Uncertainties of the dosimetric models (used in keeping with RSS-99) and of their parameters are not included in the inaccuracy.

TABLE 2.

Radiation-Hygienic Criteria to Assess the Status of Territories Using Values of Individual Parameters of the Radiation Situation (radionuclide concentrations in aerosols of near-surface air, atmospheric depositions, soils, marine & drinking water and bottom sediments at the studied territory as compared to a reference region with background radiation level)

Territory status	**Value of the radiation situation parameter at the territory to be assessed (P)**
Normal	$P_{Bg} < P \leq P_{Bg}+3\sigma$
Satisfactory	$P_{Bg}+3\sigma < P \leq 5P_{Bg}$
Unsatisfactory	$P > 5P_{Bg}$

Comments: A factor of 5 excess over a value measured in the reference background region (P_{Bg}) is used as a criterion of assessing the radiation contamination in keeping with "The Criteria... "[9] recommendations.

The assessment criteria, based on a comparison of the calculated effective doses with permissible doses (see Table 1), are used as principal ones, whereas those based on a comparison of the radiation parameters measured at a specific territory with their corresponding background values (see Table 2) are only applied if the contamination values do not exceed permissible levels regarding the population but are above the background values.

Within some territories the impact of many technogenic sources can result in the creation of dose loads at a level close to the dose limit. In such a case the dose quota falling at the territory under consideration can be too low. Here, one should begin with establishing a quota of the effective dose (caused by the contaminated territory effects) for all population groups.

The radiation situation is to be considered as "normal" if all its parameters correspond to background levels typical for the concerned region.

If the assessment results have demonstrated that no normalized value for the "critical" population group is exceeded, the territory status is considered as a "satisfactory" one.

The radiation situation is considered as "unsatisfactory" in the case that an excess of dose limits for the population or of annual exposure limits, of permissible volumetric activities, intervention levels or of calculated (based on these limits) permissible levels for the territory under consideration over the corresponding parameters regulated by RSS –99 is recorded.

4. Establishing Permissible Radiation Parameter Levels to Make a Decision on Rehabilitation of the Contaminated Territories

When closing or changing the status of the navy facilities, their buffer areas and radiation-control areas can also lose their destination and be reoriented for sanitary-hygienic, house-building, industrial, agricultural and forestry use.

From the viewpoint of the risk concept, decision-making on rehabilitation of the contaminated territories is related to a choice of one of the following options:

1) The population is living within the concerned territory and is performing its normal economic activities without any change (restriction) because the radiation impact was unimportant and did not affect appreciably the environmental & social situation in the area;
2) The population relocation and cessation of economic activities are necessary because of potential hazard for both the health of the population and the quality of products even with consideration for realizing countermeasures;
3) The population is not relocated and still continues its economic activities in a new social and environmental situation (after a radiation impact) under condition that some restrictions are observed which ensure a minimum risk for both the health of the population and the quality of products, as compared to Option (2) due to the implementation of some rehabilitative, protective and compensation countermeasures.

The choice of an option represents a multi-criteria procedure and includes two stages: estimation and substantiation of the territory rehabilitation potentialities; analysis and evaluation of the rehabilitation expediency.

If the performed assessment resulted in the identification of some areas with the "unsatisfactory" radiation status, these areas are to be rehabilitated. In such a case the land use reorientation represents one way of possible rehabilitation.

The choice of land use type within the contaminated territory depends mainly on radioinuclide concentrations in soils [6, 7, 8]. Using the radiation monitoring results, as applied to a planned (potential) type of future land use, permissible levels of radionuclide concentrations in soils and in other environment media are established.

Permissible levels concerning specific types of the land use are calculated taking into account all possible exposure paths and conditions (periodicity, rate and duration). These levels are calculated using the value of maximum effective dose for "most critical" population group.

Permissible levels are established for all radiation parameters (radiation contamination of air, soils, underground waters, foodstuffs, surface waters, bottom sediments and equivalent exposure dose rate) that can cause an exposure of the "critical" population group over a minimum-significant value of 10 μSv/year.

The integral calculated effective dose due to all radiation indices should not exceed a population exposure dose limit determined in RSS-99 with consideration for quoting.

For the "house-building" type of rehabilitation at a contaminated territory, the radiation level requirements are most strict.

For "agricultural rehabilitation" of a contaminated territory both agricultural workers (farmers) and consumers of the produced agricultural production will constitute the "critical" population group.

In case that the radiation situation within the concerned territory is considered as "normal" to perform house building, no measure of sanitary-hygienic rehabilitation is needed. In such a situation the land use re-orientation is possible without decontamination work.

If the territory under investigation obtains the "satisfactory" status, sanitary-hygienic rehabilitation is necessary. It includes:

1) Precision of both the buffer area and the radiation control area boundaries;
2) Restrictions for visiting the contaminated territories by the population;
3) Restriction (elimination) of radionuclide intake into human organism via the identified exposure pathways (with respect to "critical" population groups).

In such a situation the change in the land use type is only possible after partial decontamination. Here, the most suitable options for future land use are industrial or agricultural specialization.

The "unsatisfactory" status of a territory under consideration necessitates its sanitary and hygienic rehabilitation that comprises:

1) Changes of the buffer area & radiation control area boundaries;
2) Restrictions for visiting the contaminated territory by the population;
3) Restriction (elimination) of radionuclide intake into human organism via the identified exposure pathways (with respect to "critical" population groups).
4) Construction/reconstruction of protective and waste disposal facilities.

The type of land use within a territory of the "unsatisfactory" radiation & hygienic status can be changed after decontamination operations only. Note that in this case the most suitable is industrial specialization.

In conclusion it may be said that to ensure the uniformity of methodical approaches for assessing the status of contaminated territories and to improve both the trustworthiness & reliability of such assessments needed for further use of the territories under consideration, it is advisable to continue investigations and apply the proposed radiation and hygienic assessment method on other territories surrounding radiation-dangerous facilities of the Russian Navy and nuclear shipbuilding enterprises.

5. References

1. Gorigledzhan, E. (1995) International Advanced Research Workshop "Issues of the Decommissioning and Utilization of Nuclear Submarines", *Monitoring. Life Safety* **3** (in Russian)

2. Lissovskiy, I., Petrov, O., Belikov A. (1996) Radiological Monitoring at Naval Bases of the Arctic Navy, *Monitoring. Life Safety* **2**, 7-9 (in Russian)

3. Danilian V., et al. (2000) Radioecological Situation within Coastal Servicing Enterprises of the Far East Region of Russia, *Atomic Energy*, **89**, **2**, 160-166 (in Russian)

4. Chajkovskaya E., et al. (2001) Mechanisms of Forming the Radioactive Contamination Pattern in the Primorskij Kraj, Russia, *Atomic Energy* **91**, **3**, 223-237 (in Russian)

5. Assessment of Doses to the Public from Ingested Radionuclides (1999) *IAEA Safety Series Reports*, **14**

6. Begak, O., Nechaev, A., Projaev, V. (1992) *Introduction to the Radioecology*, Saint Petersburg (in Russian)

7. Bykov, A., Ushmaeva T. (1994) *Methods of Analyzing the Impacts of Industrial Facilities on the Population Health,* VZPI Publishers, Moscow (in Russian)

8. *Risk Assessment: Guidance for Superfund.* (1989) EPA/540/1-89/001, Washington, DC

9. *The Criteria of Assessing the Environmental Situation to Identify Territories of the Environmental Emergency Status and of the Environmental Disaster Status* (1992) The RF Minprirody Publishers, Moscow (in Russian).

RADIOECOLOGICAL MONITORING OF DEFUELLING OF DAMAGED SPENT FUEL FROM STORAGE FACILITIES OF FLOATING SHOPS

V. VYSOTSKY, N. POKIDYUK, D. SALKO. A. MAXIMOV, AND A. BORISSOV
Russian Navy

V. BULYGIN, YU. SDERZHIKOV, AND E. STEPANOV
SPA "Ecoatom"

V. DANILYAN
IBRAE RAS
52 Bolshaya Tulskaya, Moscow 115191, Russia

G. NEZHDANOV
RSC "Kurchatov Institute"
1 Kurchatova sq., 123182 Moscow, Russia

A. LUK'YANETS
FSUE "Dal'RAO"

N. RUBTSOV AND V. IVANOV
FSUE "30 Navy SRE"

A. GODNEV
FSUE "SVRTs MOD RF"

The concept of national security of Russia considers protection and enhancement of the environment in the national interest. Stable development of industries and enterprises managing fissible and radioactive materials requires versatile and purposeful activities in all main directions. However, up to the moment, there is a certain disproportion in the scales of the scientific and technological tasks to be solved and appearing ecological problems. The growth speed of ecological problems exceeds the speed of their solution. The situation is aggravated by the continuing accumulation of radioactive waste owing to decommissioning and utilization of the large number of spent nuclear objects [1]. One of the directions of interest is the removal of the accumulated damaged spent nuclear fuel from the emergency floating workshops.

The spent nuclear fuel (SNF) of the state enterprise "Atomflot" and joint stock company "MPP" has been in storage for many years at the floating maintenance bases (FMB) "Lotta" and, partially, "Imandra" and "Lepse". The Navy SNF is stored at the coastal maintenance bases and floating workshops (FW); the damaged SNF is stored at FWs PM-80 and PM- 32 (Pacific Fleet). Unloading of the SNF from the

A.A. Sarkisov and L.G. LeSage (eds.),
Remaining Issues in the Decommissioning of Nuclear Powered Vessels, 207–231.

the international project. Since 2000, there are 640 spent fuel assemblies (SFA) at "Lepse"; 430 of them are damaged and not treatable under customary extraction technology (the SFA acquired corrosion-caused cracks and changed their geometry) [2]. During the same period, PM-80 and PM-32 (decommissioned from Navy in 1994 and 1996, respectively) stored approximately 250 damaged SFA with total activity up to 9 PBq [3].

The most unfavorable is the situation with PM-80, which operated over 30 years (Primorsky region). In the 80-s, the Navy specialists attempted to extract the damaged SNF, but failed. As a result, the radiation situation grew worse. The estimate of the radiation situation in the SFA storages (SFAS) at PM-80 and PM-32 prior to the start of the main operations showed that it was out of the permissible limits and should be identified as hazardous or, for certain individual sectors, as extremely hazardous [4, 5].

In spite of the mentioned difficulties, the scientific and production association "Ecoatom" (Sosnovy Bor city), in collaboration with a number of leading institutes and organizations, developed and later improved a technology for removing the damaged SFA [6]. The operations at PM-80 lasted from 1999 to 2002, while at PM-32 they were started and completed within 2002. With respect to the potential hazard, the operations face the highest rate of complexity. These unique operations were carried out under marine conditions, close to the dwelling sites, and were completed in Russia for the first time completely and with minimum consequences for the environment.

According to the developed technology, the SNF from PM-80 (Primosky region) was transported by special vehicles from the Konyushkova bay to the coastal maintenance base (CMB) in the Sysoeva bay. In Kamchatka, the SNF from PM-32 was reloaded to the FW PM-74 and shipped to Primorsky region with further reloading to the storages of the FSUE "Dal'RAO".

Radiation and radioecological monitoring of the damaged SNF removal operations included: state assessment of the radioactive contamination sources; forecasting of the situation at the preparation stage and further improvement of predictions in the course of operation implementation allowing for the actual state of sources and meteorological data; permanent control over radiation fields in the SFAS, discharge of the radioactive materials in the storage atmosphere and outside vessels, including the surface layer of the atmosphere, water area, and surrounding territories. We also carried out monitoring of the activity changes of air, soil, fresh and sea water, bottom sediments, flora and fauna; the meteorological situation was carefully observed. At all stages, we carried out mapping of gamma-fields of the territories and water area before the start of the operations, in course of the operations, and after their completion. The process of the SNF removal was corrected according to the obtained radioecological data.

In the Primorsky region, the radioecological studies were performed at the FW PM-80 Konyushkova bay (internal bay of the Strelok gulf and Peter the Great gulf), at the transportation trunk and the surrounding part of the Dunai settlement. In the Kamchatka region, the studies included PM-32, Sel'devaya bay (internal bay of the Krasheninnikova bay), at the territories of the ship-repairing plant FSUE "SVRTs MOD RF" and the neighboring settlement Sel'devoi (Fig. 1).

The main goal of monitoring consisted in minimizing the unjustified exposure of the personnel and population, and man-caused radioactive contamination of the environment at all stages of the SNF removing technology: extraction of the SFA from the cells; SFA loading into cases and transportation containers; container transportation to the CMB (shipping to another FW); SFA reloading to the coastal storage. Monitoring procedures united the forecasting of possible radiation and radioecological consequences. It used the basic code packages from the specialized geoinformational system "TRAVE-C" and "NOSTRADAMUS" [7], natural studies [8-10], assessment of the radiation and radioecological situation [11, 12], radiation risk and hazards for the personnel and population [13, 14].

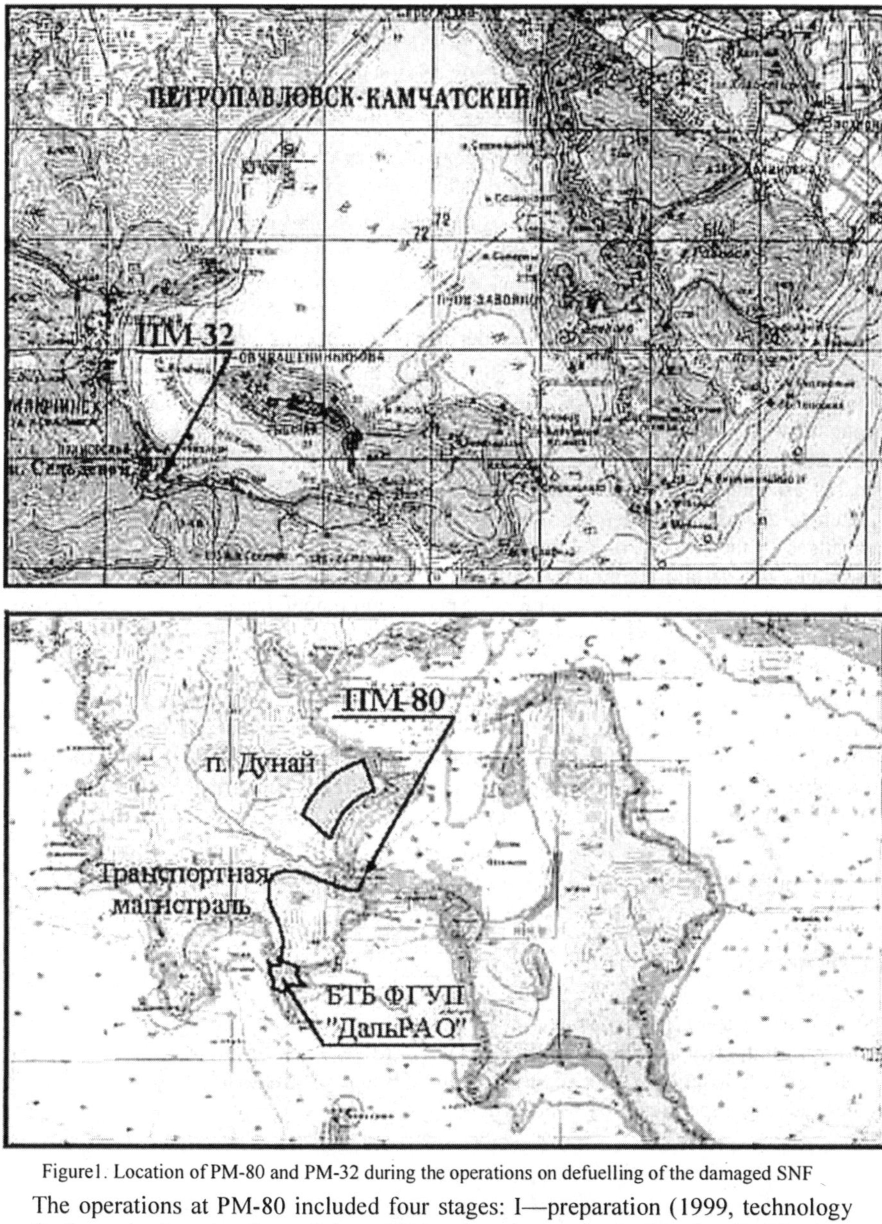

Figure1. Location of PM-80 and PM-32 during the operations on defuelling of the damaged SNF

The operations at PM-80 included four stages: I—preparation (1999, technology perfection, check extraction of three SFA using the method of hydraulic blasting;

II—main stage (2000, removal of 84 SFA by blasting); III—completion of the main stage (2001, removal of 21 SFA in cases by cutting); IV—completion of operations (2002, cutting and removing of the cases with the SFA fragments). At PM-32, all operations were completed in two months—130 damaged SFA in cases were trimmed and reloaded to the operating FW PM-74.

The radiation situation at the floating workshops is determined by the activity of the damaged SFA and liquid radioactive waste (LRW); radioactive contamination of SFAS and upper deck of the vessel; type of the performed operations. The main types of the performed operations were: assessment of the radiation situation, radiometric and dosimetric control, installation of the biological shield, adjustment of the equipment for blasting the SFA and cutting the case bottlenecks, assembling of the equipment for LRW recycling, SFA blasting, case cutting, case inspection for possible fragments of SFA and LRW, LRW recycling, SFA and SRW removing, surface covering by polymeric materials to fix and localize radioactive materials for further dry deactivation.

The auxiliary vessels were used to provide sanitary inspection, dosimetric and radiometric control facility, special treatment of personnel, cargo-handling operations, SRW storing, radiation and radioecological monitoring. The radiation situation at these vessels was determined by the SNF radiation, aerosol discharge from the FW, radioactive contamination of sanitary inspection rooms, and SRW radiation.

In order to refine the scope and character of the required operations, the status of nuclear and radiation safety by telemetry and remote radiometry methods, we carried out inspection of all cases in the nose and stern SFAS of both FW and control measurement of the exposure rate (ER) in individual cases (see Fig. 2, [15]). In some cases assumed to be empty, we found SFA fragments of various sizes. The failed attempt at case cutting on PM-80 in the 80-s resulted in their deformation and changes in their initial dimensions. In preceding years, the SFA were placed in the storage with certain technology violations, which resulted in bending of the trailing parts of some SFA and other defects. Special equipment was developed to remove such cases and SFA; it took additional operation time.

The operations aimed at removing the damaged SNF were justified and well-timed taking into consideration aggravation of the radiation situation in the SFAS and degradation of the technical state of the vessels. Long-term observation of the activity variation of the SNF cooling contours showed that by the middle of 90-s, the irreversible process of depressurization in individual fuel elements of the SFA started growing in an unpredictable impulse way (Table 1).

Table 1. Time variation of the volumetric activity of water media in the tanks of the SFAS in PM-32, 10^{-2} Ci/l

Sampling point	1993	1994	1996	1997	1998	1999	2000	2001	2002
Nose storage	0.002	0.05	0.2	2.3	0.1	0.1	4.2	1.0	1.0
Stern storage	0.002	0.03	4.2	7.9	1.8	13.0	26.0	9.5	9.5

Nuclear fuel (U) was found in the LRW, which indicates significant violation of cladding integrity of individual fuel elements. Migration of radionuclides from contour A (SFA location) to the external contour B and intertank space indicates loss of the cladding impermeability (Tables 2, 3).

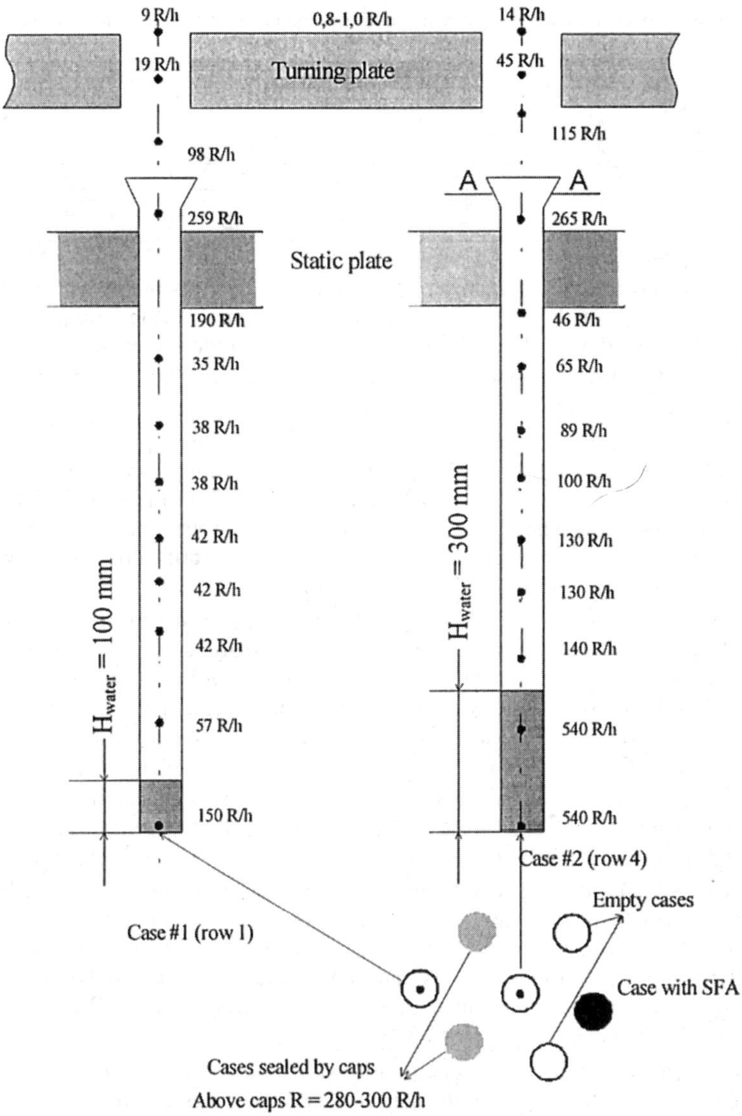

Figure 2. Gamma-radiation levels in the cases with LRW, surrounded by other cases

Table 2. Characteristics of water media of the SFAS at the PM-32 before cutting and extraction of cases with SFA

Sampling location	Volumetric activity, 10^{-2} Ci/l			Concentration, U, mg/l
	$A_{vol-\beta}$,	^{137}Cs	$^{90}Sr+^{90}Y$	
Nose SFAS	1.1 ± 0.3	0.6 ± 0.2	0.4 ± 0.1	0.1±0.04
Stern SFAS	9.5 ± 3.2	6.3 ± 2.1	3.1 ± 0.7	0.1±0.04

Table 3. State of the liquid radioactive waste in the storages of spent fuel assemblies of PM-80 at the I-III stages of work

Control period	Sampling location	$A_{vol-\alpha}$, 10^{-6} Ci/l	$A_{vol-\beta}$, 10^{-2} Ci/l	^{137}Cs, 10^{-2} Ci/l	$^{90}Sr+^{90}Y$, 10^{-2} Ci/l	U, mg/l
Start of the I stage of work	Nose SFAS	0.5±0.2	0.7±0.1	0.3±0.1	0.3±0.1	1.5±0.5
	Stern SFAS	41±15	2.9±0.7	2.1±0.5	0.7±0.3	15.3±3.6
Finish of the III stage of work	Nose SFAS	0.5±0.2	1.5±0.3	0.7±0.3	0.7±0.3	1.5±0.4
	Stern SFAS	44±15	14.0±3.5	8.3±2.5	5.1±1.7	15.5±3.5
	Intertank space	0.02±0.005	1.7±0.5	1.0±0.2	0.6±0.3	0.4±0.2

Comment: Content of ^{134}Cs, ^{60}Co and other nuclides does not exceed 1%.

The data in Tables 2 and 3 show that the dominating nuclides in the water media of the SFAS are ^{137}Cs and $^{90}Sr+^{90}Y$. In the actual situation, cleaning is required for the medium-active LRW of the SNF cooling contours to reduce their activity from 10^{-1} Ci/l to 10^{-6} Ci/l using specially developed technology[16] and thus reduce the radioactive contamination of surfaces and the environment.

Preliminary estimates and further studies showed that the main paths of the man-caused nuclide release in the environment are: at the territory—migration in the atmosphere (aerosols, atmospheric fallouts) and discharge from the surfaces of the SNF-transportation equipment along the traffic path; at the water area—migration in the atmosphere, washing from the vessel deck by rain, wind drift, dropping of varnish-and-paint coverings and corrosion products into the water (SFAS region).

A hazardous and extremely hazardous radiation situation at PM-80 was determined by the exposure rate 0.03-0.35 R/h inside the compartments on the surface of the deck and turning plates; for certain individual points, the ER amounted to 1-2 R/h (at PM-32, it did not exceed 0.15 R/h). When the technological aperture in the turning plate is registered with the case bottleneck, the ER amounted to 10-15 R/h (at PM-32, the corresponding value is 2-6 R/h). The flux densities of α-particles and β-particles varied within the intervals 10^1-10^3 decays/cm^2·min and 10^4-10^6 decays/cm^2·min, respectively. The ER on the deck in the region of the SFAS amounted to 0.1-1.5 mR/h, the α- and β-flux densities were within the intervals 1-5 decays/cm^2·min and 10^1-10^3 decays/cm^2·min, respectively. The situation in the dwelling compartments satisfied the permissible limits. In the SFAS region, "holes" in the physical shield were found on both boards of the vessel; the ER near the "holes" amounted to 5-7 mR/h. Later, when SFA and LRW, which played the role of the passive shield, were removed, the radiation intensity at these points increased and changed to within the interval 10-25 mR/h.

The maximum radiation contamination at PM-80 was observed in 2000, upon unloading of the main part (85%) of the damaged SFA. The reason is that some parts of nuclear fuel from two fuel elements were spilled under the turning plate (the SFA were extracted by the hydraulic blasting method under telemetric and radiometric control). The ER at the height of 1 m in the compartment increased to 0.5-1 R/h; in individual points—to 2-6 R/h (at the level of 0.1 m—to 6-9 R/h). The surface contamination density amounted to 10^2-10^4 decays/cm^2·min for α-particles and 10^5-10^8 decays/cm^2·min for β-particles. The ER on the upper deck in the SFAS region increased to 70 mR/h, in other parts of the vessels, the ER remained unchanged. β-contamination at the pier at the place of container loading on to special trucks increased to 10^3-10^4 decays/cm^2·min, that on the upper deck of the maintenance vessels remained within the permissible limits (upon deactivation, the situation came back to normal).

The radiation situation at PM-32 was substantially better, because the operation technology accounted for the experience gained at PM-80. In order to unambiguously provide absence of fuel discharge from the damaged SFA to the SFAS compartment, it was decided to extract the SFA only with containment cases using preliminary cutting technology. The approach turned out to be completely justified. The radiation situation in the SFAS was within the initial state during the whole operation period (Fig. 3).

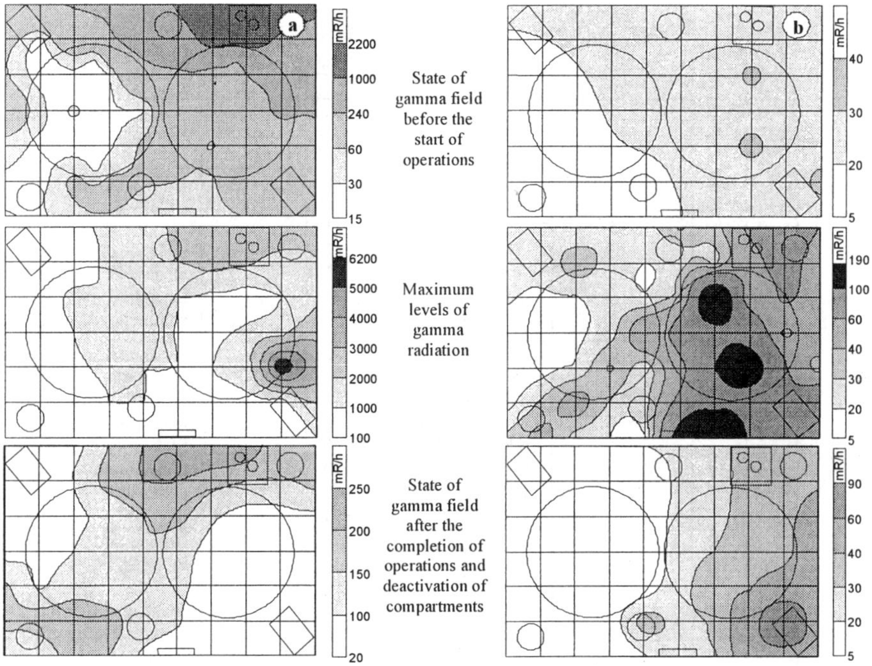

Figure 3. Maps of gamma radiation exposure dose rates in SFAS of PM-80 (a) and PM-32 (b) at various stages of operations

During the whole time of the operations, the radionuclide content in the atmosphere in the SFAS, at the upper deck of the FW, and in the observation zones (Dunai and Seldevoy settlements) was under permanent control. Depending on the technological operation, the aerosol concentration in the SFAS compartment varied within the interval 10-3,000 Bq/m^3 (SNF spilling did not result in the abrupt growth of the aerosol content in the atmosphere, because active mechanical operations were immediately paused, and upon preliminary deactivation, the surface was covered by the additional polymeric protection layer). The concentration of radioactive materials in the atmospheric aerosols increased usually between 1-3 hours. Maximum concentration was observed after the SFA extraction, cutting or blasting of the cases with SFA. The concentration decrease to the initial value consumed more time and lasted 2-6 hours, sometimes up to 8-12 hours (the operations at PM-80 were carried out with closed or 25% opened SFAS caps, at PM-32—in a completely pressurized compartment). In case of successive extraction, cutting, or blasting of several SFA or SFA cases, the aerosol concentration increased by the factor of 2-5 and amounted to

10-32 maximum permitted concentration (MPC) with respect to ^{90}Sr, but within 1 MPC with respect to ^{137}Cs (α-activity of aerosols at maximum concentration did not exceed 0.3 MPC). Owing to more intensive radioactive surface contamination of the SFAS in PM-80, similar operations resulted in the aerosol concentration inside compartments several dozens or hundreds times higher than at PM-32 (see Tables 4, 5).

Table 4. Dynamics of the aerosol concentration in the atmosphere of the storage of the spent fuel assemblies at PM-80 during various types of operations

Time of the day	Volumetric activity, Bq/m^3	Radionuclide activity, Bq/m^3 and factor of the MPC excess				Type of operation
		^{90}Sr	MPC	^{137}Cs	MPC	
08.30-09.30	2	0.8	< 0.01	0.6	< 0.01	No operations
10.30-11.30	15	10	0.2	2	< 0.01	Preparation
11.31-12.31	5	2	0.04	90	< 0.01	
12.32-13.32	250	160	3.0	16	0.05	LRW
13.30-14.30	61	42	0.8	6	0.01	recycling
14.40-15.40	14	6	0.1	51	< 0.01	
15.40-17.00	117	56	1.0	37	0.03	Case
17.25-18.25	86	44	0.8	10	0.02	inspection
18.26-19.26	19	6	0.1	4	< 0.01	Operation
19.27-20.27	12	5	0.1		< 0.01	completion
10.00-11.00	7	2	0.04	2	< 0.01	No operations
11.05-12.05	180	86	1.6	82	0.05	Preparation
12.10-13.10	1470	830	15.7	630	0.37	Extraction and
13.15-14.15	2670	1440	27.2	1100	0.16	loading of 7
15.15-16.15	3010	1670	31.5	1260	0.64	SFA in the
16.15-17.15	1340	710	13.4	480	0.28	transportation
17.20-18.20	370	190	5.9	160	0.09	container
18.25-19.25	260	130	2.6	115	0.07	
11.00-12.00	2	0.7	< 0.01	0.7	< 0.01	No operations
12.00-13.00	11	5	0.1	4	< 0.01	Preparation
13.20-14.20	87	53	1.0	27	0.02	Cutting
14.30-15.30	185	110	2.0	64	0.04	Blasting
15.45-16.45	350	260	4.9	59	0.03	Cutting
16.55-17.55	18	11	0.2	4	< 0.01	Preparation
18.10-19.10	423	250	4.7	143	0.08	Blasting
19.15-20.15	42	24	0.4	9	< 0.01	Completion
09.50-10.50	5	1	< 0.01	2	< 0.01	No operations
12.30-13.30	206	130	2.4	59	0.03	Preparation
13.40-14.40	1120	730	13.8	310	0.20	Cutting
14.50-15.50	660	410	7.7	190	0.10	of three cases
16.00-17.00	64	43	0.8	11	< 0.01	
10.00-11.00	2	0.7	< 0.01	0.5	< 0.01	No operations
11.10-12.10	14	6	0.1	6	< 0.01	Preparation
12.15-13.15	37	24	0.2	12	< 0.01	
13.20-14.20	184	106	2.0	67	0.04	
14.25-15.25	1350	870	16.4	380	0.20	Blasting
15.30-16.30	750	456	8.6	198	0.10	of three cases
16.40-17.40	86	50	0.9	24	0.01	
10.10-11.10	5	2	0.04	2	< 0.01	No operations
11.30-12.30	152	76	1.4	67	0.04	Preparation
12.30-13.30	680	330	6.2	328	0.19	Unloading of
14.00-15.00	540	240	6.5	280	0.16	six cases with
15.15-16.15	110	67	1.3	36	0.02	SFA
16.15-17.15	264	110	2.1	134	0.08	
17.20-18.20	47	19	0.4	16	0.01	

Time of the day	Volumetric activity, Bq/m^3	Radionuclide activity, Bq/m^3 and factor of the MPC excess				Type of operation
		^{90}Sr	MPC	^{137}Cs	MPC	
10.00-11.00	3	1.3	< 0.01	1.0	< 0.01	Operations stopped
12.30-13.30	1	0.6	< 0.01	0.3	< 0.01	
14.00-15.00	1	0.4	< 0.01	0.3	< 0.01	
15.30-16.30	3	1.5	< 0.01	1.1	< 0.01	
16.30-17.30	3	0.9	< 0.01	0.6	< 0.01	

Comment: MPC in air for the personnel is 53 Bq/m^3 for ^{90}Sr and 1700 Bq/m^3 for ^{137}Cs [12]. The measurement accuracy is within 10-20%

Table 5. Dynamics of the aerosol concentration in the atmosphere of the storage of the spent fuel assemblies at PM-32 during various types of operations

Time of the day	Volumetric activity, Bq/m^3	Radionuclide activity, Bq/m^3 and factor of the MPC excess				Type of operation
		^{90}Sr	MPC	^{137}Cs	MPC	
08.20-09.20	1.2	0.8	0.02	0.2	< 0.01	Mounting of the cutting
11.00-13.00	25.3	5.1	0.1	20.0	0.01	
14.00-16.00	3.8	2.5	0.05	1.2	< 0.01	
09.20-10.20	0.3	0.1	< 0.01	0.2	< 0.01	Cutting and blasting of 20 cases
11.00-12.00	3.8	2.5	0.05	1.2	< 0.01	
14.30-16.00	10.7	4.3	0.08	6.1	< 0.01	
11.30-12.30	2.9	1.2	0.02	1.6	< 0.01	Reloading of 29 cases with SFA to PM-74
15.15-16.15	4.1	1.6	0.03	1.8	< 0.01	
17.15-18.15	2.4	0.8	0.02	1.0	< 0.01	
12.15-13.30	96.4	23.1	0.4	72	0.04	Mechanical deactivation
15.00-16.00	1.1	-	-	-	-	
09.30-10.30	19.1	9	0.17	10	< 0.01	Deactivation by chemical combinations
11.00-12.00	19.6	10	0.2	8	< 0.01	
13.30-15.00	6.5	0.8	0.02	5.6	< 0.01	

The aerosol concentration at the upper deck of the FW in the region of the SNF unloading was always lower than in the SFAS by the factor 10^4-10^5, but exceeded the background by the factor 30-1000 (2500 at the maximum) at PM-80 and 2-5 at PM-32. The concentration was reduced by taking special measures: enhanced ventilation and daily sputtering of quick-drying polymeric coverings on internal surfaces. The atmospheric fallout exceeded the background values by the factor of several hundreds (thousands) at PM-80 and 1.5-2 at PM-32. Despite this, no increase in aerosol concentration or density of the atmospheric fallout above the background was observed in the Dunai settlement at a distant 1,000-2,000 m from the pier. During operations with PM-32, the aerosol concentration decreased to the background value already at 200 m from the vessel, and, therefore, no excess above the background was observed at the territory of the plant and in the Seldevoy settlement at the distances of 200-800 m from the pier (see Table 6).

Table 6. Daily mean aerosol activity of air and atmospheric fallouts in the radiation safety regime zone at the upper deck of PM-80 and in the observation zone in the Dunai settlement

Date	Volumetric aerosol activity, 10^{-4} Bq/m^3				Fallout density, Bq/m^2·day			
	PM-80*			Dunai settlement	PM-80*			Dunai settlement
	A_{vol}	^{90}Sr	^{137}Cs	A_{vol}	A_{sq}	^{90}Sr	^{137}Cs	A_{sq}
20.10.	9	6	3	2	-	-	-	-
22.10.	38	22	16	-	-	-	-	-
24.10.	13	8	5	-	2600	1300	1260	3
25.10.	23	16	7	3	-	-	-	-
27.10.	610	370	160	-	-	-	-	-
29.10.	1690	920	640	-	-	-	-	-
01.11.	13	7	5	-	12040	8400	3910	3
05.11.	12	10	2	-	-	-	-	-
09.11.	124	100	24	2	218	160	46	4
10.11.	40	25	15	-	-	-	-	-
11.11	46	24	12	-	-	-	-	-
15.11.	6	2	3	3	-	-	-	-
16.11.	12	7	4	-	46	31	10	3
18.11.	9	6	3	1	-	-	-	-
28.11.	8	4	4	2	80	56	21	5
30.11.	740	460	274	-	-	-	-	-
01.12.	465	340	123	-	-	-	-	-
04.12.	24	14	9	1	1400	880	490	3
12.12.	1030	654	483	1	2700	1200	1480	2
13.12.	106	61	44	-	-	-	-	-
17.12	2190	1200	970	-	-	-	-	-
18.12	1790	1100	680	-	-	-	-	-
19.12	6	3	2	1	540	320	217	2

Comment: *air inlet and pan for sampling atmospheric fallouts were located on the upper deck of PM-80 5 m from the SFAS in the direction of the Dunai settlement. Air inlet pumped 48,000 m^3 of air daily; atmospheric fallouts were collected for a week. In the Primorsky region, the volumetric aerosol activity is $(2\pm1)\cdot10^{-4}$ Bq/m^3; the atmospheric fallout density is 2±1 Bq/m^2·day. [17]. The MPC for the population is 2.7 Bq/m^3 with respect to ^{90}Sr and 27 Bq/m^3 with respect to ^{137}Cs [12]. The measurement accuracy is 10-30%.

During the whole operation period, the volumetric aerosol activity in the SFAS compartment of PM-80 varied within the limits (depending on the type of performed operation): no operations (background)—1-6 Bq/m^3; preparation, LRW recycling—30-250 Bq/m^3; blasting, extraction, and loading of the SFA into transportation container—1000-3000 Bq/m^3; cutting and blasting of cases with SFA—250-1300 Bq/m^3. The same figures for PM-32 are: background—0.1-0.3 Bq/m^3; preparation, LRW recycling, and deactivation—3-20 Bq/m^3 (mechanical deactivation—up to 100 Bq/m^3); cutting and blasting of cases with SFA—10-25 Bq/m^3.

Respectively, aerosol release to the atmosphere during the operations at PM-80 amounted to: background—$(5\text{-}10)\cdot10^{-4}$ Bq/m^3; preparations, including LRW recycling—$(10\text{-}50)\cdot10^{-4}$ Bq/m^3; blasting, extraction, and loading of the SFA into transportation container—$(2000\text{-}5000)\cdot10^{-4}$ Bq/m^3; cutting and blasting of cases—$(500\text{-}2000)\cdot10^{-4}$ Bq/m^3. The same for PM-32: background—$(0.6\text{-}0.7)\cdot10^{-4}$ Bq/m^3; preparation—$(1\text{-}2)\cdot10^{-4}$ Bq/m^3; cutting and blasting of cases—$(1\text{-}4)\cdot10^{-4}$ Bq/m^3; reloading of cases with SFA to PM-74—$(1\text{-}3)\cdot10^{-4}$ Bq/m^3 (only once there was a release with the volumetric activity $(217\pm60)\cdot10^{-4}$ Bq/m^3 at the distance 10 m from the vessel).

The density of the atmospheric fallouts at PM-80 amounted to: no operations—10-30 Bq/m^2·day; preparations—100-500 Bq/m^2·day; radiation-dangerous operations—2,000-8,000 Bq/m^2·day (12,000-18,000 Bq/m^2·day maximum). The same for PM-32 never exceeded 1 Bq/m^2·day for all types of operations. As a result, the largest radioactive contamination of the land and water area during operations at PM-80 through the air path occurred in the zone of strict regime (radius 30 m), a lower level of contamination was observed in the zone of radiation safety regime (100 m), and only insignificant contamination occurred in the sanitary protection zone (300 m, see Table 7).

Table 7. Contents of man-caused radionuclides in the external environmental objects of Konyushkova bay before and after the operations at the distance up to 300 m from PM-80

Konyushkova bay	Concentration, Bq/kg	
	Before operations	After operations
Soil*, ^{137}Cs, Bq/kg		
- 70 m ;	72±18	120±25
- 120 m;	29±5	88±4
- 150 m;	49±15	62±18
- 200 m;	44±13	63±18
- 300 m.	36±13	43±13
Sand, ^{137}Cs, Bq/kg		
- 50 m;	< 3	6±2
- 70 m;	3±1	4±2
- 100 m;	< 3	< 3
- 130 m.	< 3	4±2
Sea water (10 m):		
- ^{137}Cs, Bq/m^3;	4.5±0.9	9.9±1.5
- ^{90}Sr, Bq/m^3;	2.2±0.3	4.3±0.4
- ^{60}Co, Bq/m^3	< 1.4	< 1.4

Pier point # (2-20 m)	Density β, decays/cm^2·min	
	Before operations	After operations
1	8±3	80±10
2	6±3	64±6
3	14±3	20±5
4	12±3	30±7
5	14±3	360±30
6	10±3	44±5
7	10±2	144±13
8	12±3	80±9
9	10±2	8±2
10	12±2	1440±110
11	14±3	14±3
12	12±3	14±3
13	16±2	120±12
14	8±2	8±3
15	10±2	12±2
16	14±3	2800±200
17	12±3	1180±10
18	15±3	20±5

Comment: Background soil content of ^{137}Cs in Primorsky region amounts to 4-65 Bq/kg [17].
*concentration of artificial radionuclides in the grass did not exceed 3 Bq/kg before and after operations.

The situation is confirmed by the monitoring of the surface layer of sea water (up to 2 m depth) near PM-80 by the complexes "Arakhis (Na (Tl) receiver, size 63x250 mm, 5 m from the vessel, Navy) and "Fortep'yano" (Na (Tl), 200x400 mm, 50 m from the vessel, RSC "Kurchatov Institute" [18]). Both complexes simultaneously registered natural and man-caused events, The comparative analysis showed that the intensity of gamma radiation resulting from washing out of the natural U-Th fission products from the atmosphere with rain exceeds that of the man-caused nuclides released from PM-80 (see Fig. 4).

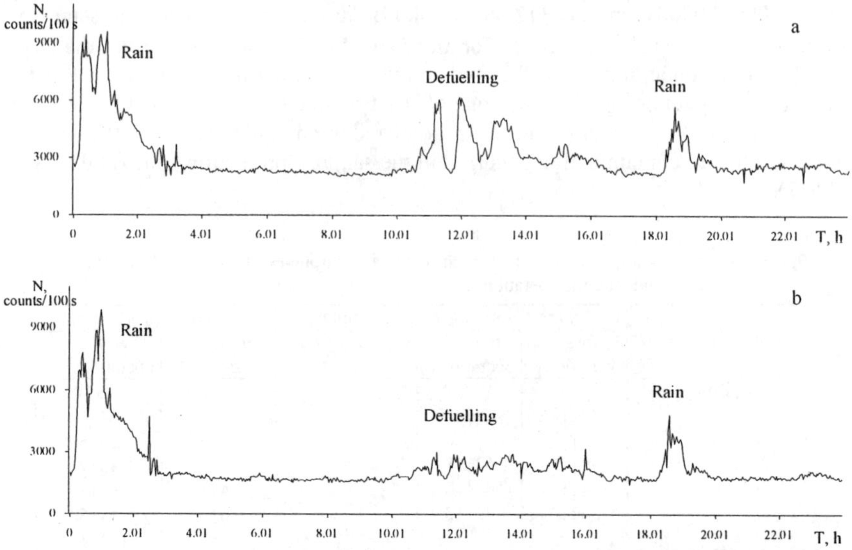

Figure 4. Daily variations in the counting rates of the scintillation detectors of the complexes "Arakhis" (a) and "Fortep'yano" (b) located 5 and 50 m far from PM-80 at the depth 1 m during the defuelling operations with the damaged SNF

^{137}Cs (E_γ = 662 keV) concentration in water during the operational day might vary sometimes from 10 to 300 Bq/m^3 (at 50 m distance from the vessel up to 10 Bq/m^3), while natural ^{214}Bi (E_γ = 609 keV) concentration over the whole water area of Konyushkova bay varies from 10 to 1,000 Bq/m^3. In the Kamchatka region, near PM-32, the increase in radionuclide concentration in water was observed only once, upon completion of operations on unloading of the radioactively contaminated equipment from storage—the ^{137}Cs concentration near the vessel amounted, according to the data of the "Arakhis" complex, 80±20 Bq/m^3 (0.04 MPC). In other cases, the concentration was always at the background level.

As a result, the activity of the surface level of atmosphere in the Dunai settlement remained unchanged during the whole period of operations with PM-80. The volumetric activity of aerosols and the density of the atmospheric fallouts were permanently at the levels of (1-3)$\cdot 10^{-4}$ Bq/m^3 and 1-5 Bq/m$^2\cdot$day, respectively, that is, at the background level. The month-average volumetric fission-fragment activity of the fresh and sea water also remained unchanged; no pattern contamination of soil was observed. The forecast about the insignificant hazard of the defuelling operations with the damaged SNF for the population of the Konyushkova and Seldevaya bays (see Tables 8-10) was conect.

Table 8. Month-average volumetric activity of the air aerosols and density of the atmospheric fallouts in the Dunai settlement before and after radiation-hazardous operations at PM-80

Before operations with PM-80	Dunai settlement $A_{vol.}$ 10^{-4} Bq/m^3	Dunai settlement $A_{sq.}$ Bq/m^2·day	During operations with PM-80	Dunai settlement $A_{об.}$ 10^{-4} Бк/м^3	Dunai settlement $A_{sq.}$ Bq/m^2·day
January	2.3±0.6	3.0±0.7	September	0.7±0.3	1.8±0.4
February	2.0±0.5	2.5±0.6	October	1.9±0.4	2.3±0.6
March	1.1±0.3	2.0±0.5	November	2.4±0.6	3.9±0.9
April	2.0±0.5	2.2±0.6	December	1.7±0.4	2.6±0.6
May	2.0±0.5	3.3±0.8	-	-	-
June	2.0±0.5	2.5±0.6	-	-	-
July	2.4±0.6	3.4±0.8	-	-	-
August	0.5±0.2	2.0±0.5	-	-	-
Average	1.8±0.5	2.6±0.7	Average	1.7±0.5	2.7±0.9

Comment: Year average volumetric aerosol activity and atmospheric fallout density in the Far East region in 1990-2000 amounted to (3±1)·10^{-4} Bq/m^3 and 2±1 Bq/m^2·day, respectively [17].

Table 9. Month-average fission-fragment activity of fresh and sea water in the Dunai settlement before and after radiation-hazardous operations at PM-80, Bq/m^3

No operations: Month	PM-80 Pier*	Dunai settlement: Dunai river	Dunai settlement: Drinkable	During the operations: Month	PM-80 Pier*	Dunai settlement: Dunai river	Dunai settlement: Drinkable
January	171±8	162±8	124±6	September	160±8	152±7	112±5
February	149±7	150±7	122±6	October	158±8	158±8	122±6
March	159±7	146±7	123±6	November	168±8	156±7	119±6
April	163±8	154±7	120±6	December	164±8	154±7	121±6
May	180±9	157±8	120±6	-	-	-	-
June	162±8	158±8	123±6	-	-	-	-
July	160±8	153±7	140±7	-	-	-	-
August	158±7	151±7	112±5	-	-	-	-
Average	163±8	154±7	123±6	Average	162±8	155±7	120±6

Comment: *in the background region (6 km far from PM-80), the average fission-fragment activity of sea water during the year amounted to 160±10 Bq/m^3.

Table 10. ^{137}Cs content in the soil of the Dunai settlement at various distances from PM-80, Bq/kg

Distance, m	A_{sp}, Bq/kg	Distance, m	A_{sp}, Bq/kg	Distance, m	A_{sp}, Bq/kg
450±50	20-50	900±50	< 9-52	2200±200	< 9-56
600±50	< 10-57	1300±100	< 9-60	2800±300	< 9-45
600±50	< 10-157*	1700±150	< 7-30	3500±500	16-51

Comment: We inspected the sector 60° with respect to PM-80, where the settlement is located. The ^{137}Cs background in the soil of the Primorsky region is 4-65 Bq/kg [17].

The analysis of the results of full-scale studies and the forecast drive us to the conclusion that the air path of the man-caused nuclide migration under the most unfavorable conditions of defuelling the damaged SNF from the floating workshops does not result in radioactive contamination outside the zone of the radiation safety regime (100 m) and sanitary protection zone (300-500 m); in the observation zone (> 500 m), it does not affect the background level.

At the same time, it is necessary to notice that the radiation protection zone turned out to be penetrable for partial migration of radionuclides by special trucks. In particular, it grew evident upon removal of those SFA, which featured partial spillage of fuel in the SFAS of PM-80. The radioactive contamination of the compartment resulted in the contamination of the transportation container, in spite of the performed partial deactivation. The container, in turn, turned into the source of the secondary radioactive contamination of the road (there was no possibility to change the container, because the contaminated unit was the only one adjusted to carry the damaged SFA. The surface of the special truck was covered by the quick-drying polymeric covering prior to shipping it from the Konyushkova bay to the coastal maintenance base. The covering sealed a significant part of the radioactive materials but failed to protect the truck against losses along the traffic path.

Owing to vibration, a certain amount of radioactive particles was dropped on the roadsides. At the beginning of the path, in the zones of the radiation safety regime, the ER at individual points amounted to 100-20 μR/h (Fig. 5). Soil sampling in the zones of maximum contamination at various distances from the source showed that the atmospheric events, including the most powerful typhoon for last 30 years, moved man-caused nuclides distances of 7-9 m from the roadside. However, most of the nuclides stay within 1 m from the primary contamination point (Table 11).

Table 11. Contents of the man-caused nuclides in soil along the transportation path at various distances from the contamination sources, Bq/kg

Distance from the source, m	Point #1		Point #2		Point #3	
	^{90}Sr	^{137}Cs	^{90}Sr	^{137}Cs	^{90}Sr	^{137}Cs
0	$3.7 \cdot 10^4$	$(3.9\pm1.1)\cdot10^4$	$(2.6\pm0.7)\cdot10^5$	$(2.1\pm0.7)\cdot10^5$	$(2.1\pm0.7)\cdot10^5$	$(1.9\pm0.5)\cdot10^5$
1	439±60	437±60	-	600±90	2600±250	2200±220
2	-	18±6	-	3370±650	370±50	116±20
3	150±30	151±30	1060	1080±320	40±120	13±5
5	-	44±12	-	780±100	125±30	79±25
7	23±6	21±6	-	125±30	-	-
8	-	-	-	-	32±8	23±6
10	-	4±4	-	< 3	5±2	4±2

The roadside contamination decreases along the direction of motion of special trucks. The pedestrian gamma survey showed that the ER distribution density along the transportation path decreases to the values comparable with the ER distribution in the settlement where radioactive materials were never transported. However, soil sampling indicated the presence of man-caused nuclides with a density that amounts in individual points to 10 MPC (Table 12).

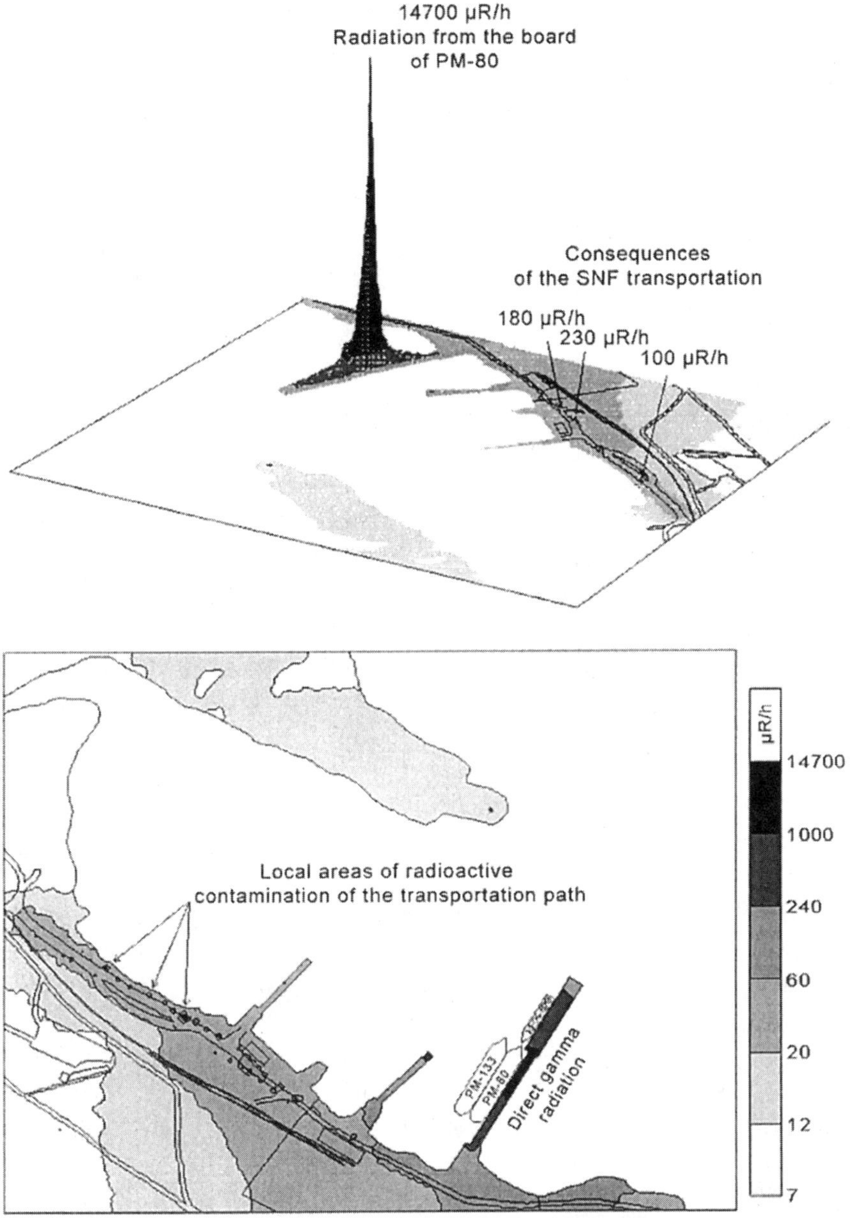

Figure 5. 3D and topographic view of gamma-field levels of the shore line at the Konyushkova bay during the defuelling of the damaged SNF from PM-80

Table 12. ^{137}Cs contents in soil along the transportation path of nuclear fuel from the Konyushkova bay to Sysoeva bay, Bq/kg

Point #	A_{sp}	MPC	Point #	A_{sp}	MPC	Point #	A_{sp}	MPC
1	38	0.01	32	187	0.06	62	10	< 0.01
2	117	0.04	33	8920	3.4	63	< 9	< 0.01
3	988	0.4	34	100	0.04	64	< 9	< 0.01
4	< 9	< 0.01	35	134	0.05	65	< 9	< 0.01
5	5960	2.3	36	21	< 0.01	66	< 9	< 0.01
6	1980	0.8	37	59	0.02	67	< 9	< 0.01
7	260	0.1	38	303	0.1	68	37	0.01
8	< 9	< 0.01	39	< 9	< 0.01	69	198	0.08
9	< 7	< 0.01	40	< 9	< 0.01	70	< 9	< 0.01
10	21	< 0.01	41	25	< 0.01	71	< 9	< 0.01
11	< 9	< 0.01	42	< 9	< 0.01	72	39	0.01
12	28	0.01	43	< 9	< 0.01	73	54	0.02
13	< 10	< 0.01	44	< 9	< 0.01	74	< 9	< 0.01
14	145	0.05	45	< 9	< 0.01	75	< 9	< 0.01
15	< 10	< 0.01	46	27	0.01	76	19	< 0.01
16	< 9	< 0.01	47	< 9	< 0.01	77	25	< 0.01
17	14	< 0.01	48	< 9	< 0.01	78	18	< 0.01
18	25	< 0.01	49	< 9	< 0.01	79	< 9	< 0.01
19	21	< 0.01	50	< 9	< 0.01	80	44	0.02
20	< 9	< 0.01	51	22	< 0.01	81	64	0.02
21	< 9	< 0.01	52	12	< 0.01	82	33	0.01
22	10	< 0.01	53	60	0.02	83	37	0.01
23	< 9	< 0.01	54	24	< 0.01	84	41	0.01
24	< 9	< 0.01	55	< 9	< 0.01	85	30	0.01
25	< 9	< 0.01	56	< 9	< 0.01	86	49	0.02
26	277	0.1	57	62	0.02	87	48	0.02
27	< 9	< 0.01	58	41	0.01	88	< 9	< 0.01
28	2453 0	9.5	59	207	0.08	89	17	< 0.01
29	8070	3.1	60	106	0.04	90	792	0.4
30	414	0.2	61	744	0.3	91	650	0.3
31	655	0.2	-	-	-	-	-	-

Comment: According to RKVS-90 (Navy) [11], the maximum permissible concentration (MPC) of ^{137}Cs in soil is 2,590 Bq/kg.

The analysis of the pointlike radioactive contamination along the path showed that in 90% of the cases it was located within 1-3 m from the road edge. Average specific activity of ^{137}Cs amounted to 1,000±500 Bq/kg (0.4 MPC). At the distance of 5-100 m from the road edge, caesium concentration remained unchanged within the limits < 9-50 Bq/kg, that is, at the background level. Therefore, it was decided to perform partial deactivation of the road in the inhabited region and to proceed with the SNF unloading up to complete removal of the damaged SFA to the CMB

"Dal'RAO" followed by deactivation of the transportation path. The decision was implemented.

Before operations at PM-80 started, it was found that ^{137}Cs from global fallout and man-caused sources is present in the soil at the coast line of Konyushkova bay. The concentration varied within the limits 10-100 Bq/kg. The concentration of the natural ^{40}K in the same territory varies within the limits 200-400 Bq/kg. Closer to the shore line, the concentration of ^{137}Cs decreased to 3-5 Bq/kg (^{40}K in sand increased to 600-800 Bq/kg); this level is the same along the whole water area except local zones of radioactive contamination near the reloading pier, which were formed in preceding years. In these zones, ^{137}Cs concentration amounted to 1,300 Bq/kg, ^{60}Co concentration—to 500 Bq/kg. The artificial radionuclide content of algae and aquatic life did not exceed 3 Bq/kg.

In addition to PM-80, there were other sources of the radioactive contamination of Konyushkova bay during the period 1999-2002: PM-74 (multiple unloading of the SNF in the normal mode) and the emergency technical tanker TNT-72 (anchored at the pier for two years filled with the LRW). The intensity of the radioactive contamination from all sources was approximately the same, but the radionuclide content differs: PM-80 was mainly the source of ^{90}Sr and ^{137}Cs, while PM-74 and TNT-27 also released ^{60}Co. Under this condition, the monitoring task additionally included the problem of classification of the radioactive anomalies, their attribution to the vessel type, and assessment of their contributions to the radioecological situation.

The investigations showed that bottom sediments at the point of the long-term location of PM-80 at the next pier (approximately a year) feature anomalies with ER amounting to 2200 μR/h; the anomalies are not related with the SNF removal, but rather determined by the wind drift and atmospheric wash-out of the radioactive materials from the deck together with corrosion products and varnish covering. However, the SNF removal also resulted in formation of anomalies with the ER amounting to 2400 μR/h. Owing to various reasons, radioactive anomalies appeared also from TNT-27 (ER up to 7-16 μR/h) and PM-74 (ER up to 1400 μR/h) (see Fig. 6).

The radioactive contamination of the bottom sediments did not lead to the changes in the radionuclide content of aquatic life during the first three years following the start of the operations at PM-80; at the beginning of the first year, the ^{137}Cs content in dry algae slightly increased, but in the alive algae (near the pier ground) it remained at the background level, see Table 13.

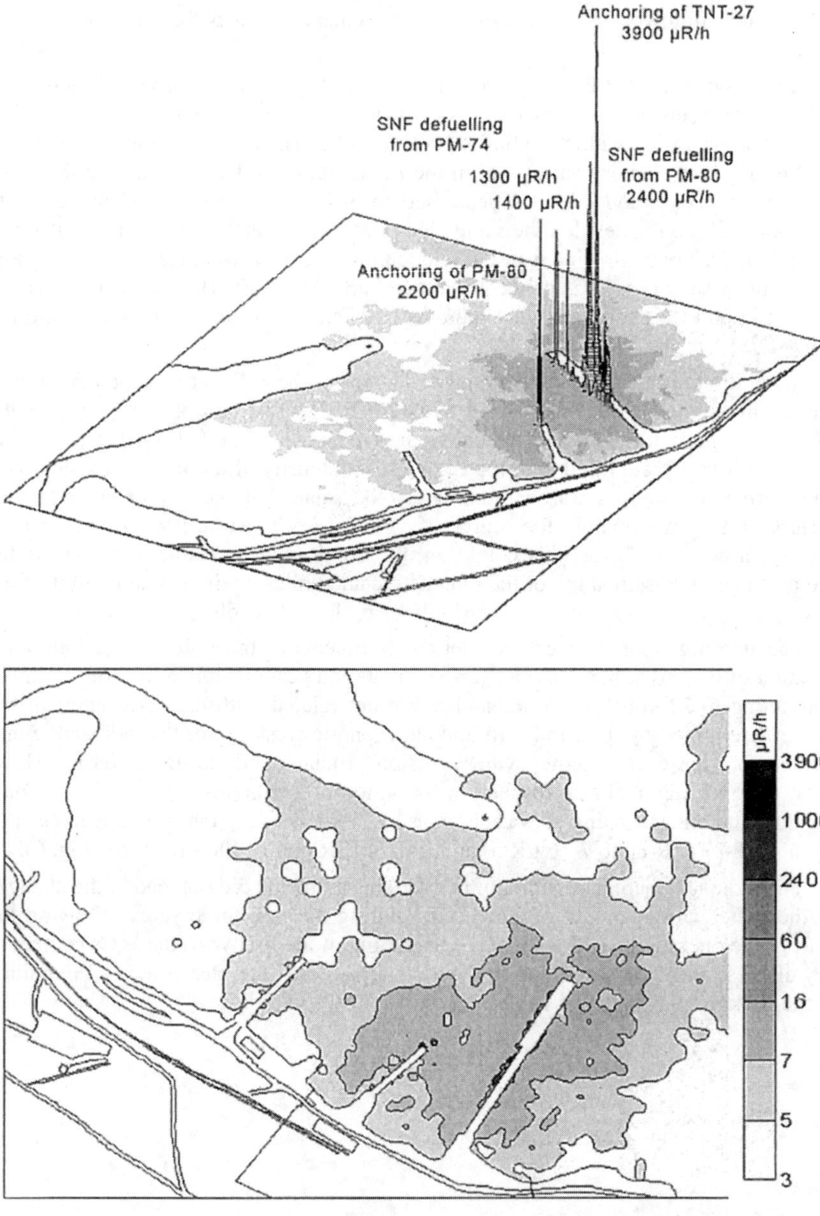

Figure 6. 3D and topographic view of gamma-field levels in the bottom sediments after the completion of defuelling operations of the damaged SNF from PM-80 and RW management at nuclear maintenance vessels (Konyushkova bay, 2002)

Table 13. Variation of the specific activity of natural and man-caused radionuclides at the water area of the Konyushkova bay resulting from the defuelling of the damaged SNF from PM-80 and SNF and LRW management by other vessels of nuclear maintenance in 1999-2002, Bq/kg

Objects of marine environment	Before SNF removal		After SNF removal		Natural ^{40}K
	^{137}Cs	^{60}Co	^{137}Cs	^{60}Co	
Bottom sediments					
p. 1	6 ± 3	5 ± 3	2550±250	150±30	313±30
p. 2	1310±160	65 ± 20	3960±290	70±20	290±30
p. 3	195 ± 60	< 5	151500±150	< 5	534±50
p. 4	< 5	< 5	< 5	< 5	307±30
p. 1	11 ± 5	< 5	8880±900	70±20	516±50
p. 2	< 5	< 5	1650±160	150±30	690±70
p. 3	202 ± 40	494 ± 80	12280±1200	1510±50	376±40
p. 4	6 ± 3	< 5	9240±900	< 5	234±30
p. 1	39 ± 15	< 5	< 5	< 5	807±80
p. 2	6 ± 3	< 5	1440±140	790±40	617±60
p. 3	< 5	< 5	6050±600	316±50	844±80
p. 4	< 5	< 5	17790±1700	< 5	634±60
p. 1	165 ± 50	< 5	2210±220	366±50	606±60
p. 2	< 5	< 5	2970±300	1510±150	885±90
p. 3	6 ± 3	< 5	< 5	< 5	700±70
p. 4	< 5	< 5	48±10	< 5	550±60
p. 1	6 ± 3	< 5	126±30	< 5	550±60
p. 2	204 ± 60	28 ± 13	700±80	26 ± 15	384±40
Sand					
p. 1	4 ± 1	< 3	4 ± 1	< 3	630±70
p. 2	5 ± 2	< 3	6 ± 2	< 3	550±60
p. 3	4 ± 1	< 3	5 ± 1	< 3	880±90
p. 4	4 ± 1	< 3	4 ± 1	< 3	740±80
Algae:					
- mortified;	< 3	< 3	< 3	< 3	150±20
p. 1	6 ± 2	< 3	157 ± 20	< 3	356±35
p. 2	3 ± 1	< 3	15 ± 5	< 3	230±25
p. 3	4 ± 3	< 3	20 ± 5	< 3	246±25
- alive					
p. 1	< 3	< 3	< 3	< 3	407±40
p. 2	< 3	< 3	< 3	< 3	210±20
p. 3	< 3	< 3	< 3	< 3	230±25
p. 4	< 3	< 3	< 3	< 3	330±35
Fish:					
- perch;	< 3	< 3	< 3	< 3	125±20
- bullhead;	< 3	< 3	< 3	< 3	113±20
- flounder.	< 3	< 3	< 3	< 3	105±20
Squids	< 3	< 3	< 3	< 3	63±20
Shrimp	< 3	< 3	< 3	< 3	68±20
Crabs	< 3	< 3	< 3	< 3	80±20
Starfish	< 3	< 3	< 3	< 3	51±15

Comment: According to [10], MPC with respect to ^{137}Cs is 2,590 Bq/kg, with respect to ^{60}Co— 370 Bq/kg (maximum ^{137}Cs concentration is 59 MPC, maximum ^{60}Co concentration is 4 MPC).

The space-time analysis of the radiation maps of bottom sediments obtained in various years and of the state of gamma field in sea water at the level 2 m from the bottom made it possible to reveal the migration direction of the radioactive materials with respect to the pier. Currently, the area of the most intensive radioactive contamination occupies the space 30-100 m with respect to the pier. The migration of radionuclides is rather complicated. There is branching in the directions of the coastline and the central part of the bay. At individual points, separations of the small-sized (several meters) radioactive anomalies from the continuous zone are observed;

the separated anomalies are later scattered away (it is worth noting that this process is additionally induced by frequent approaches of vessels to the pier). The motion velocity of the external edges is not large; depending on the exact location, it varies within the limits 5-40 m/year (Fig. 7).

During defuelling operations with PM-32 in the Kamchatka region, no nuclide migration to environmental objects was observed in the territory of the ship-repairing enterprise or in the water area of the Seldevaya bay. The chosen technology of the SNF removal together with cases was the safest one with regard to the radioecological aspect; the radiation situation in the SFAS of PM-32 was substantially better, and the complete duration of the defuelling operations was rather short. As a result, the radiation characteristics of the environmental objects remained unchanged (Table 14).

Table 14. Specific activity of artificial radionuclides in the environmental objects at the territory of the ship-repairing enterprise, in the Seldevoy settlement, and water area of the Seldevaya bay before and after the operations on the defuelling of the damaged SNF from PM-32, Bq/kg (m^3)

Date	Sampling region	Sampling point #	Sample type	Operation period, nuclide			
				Before operations		After operations	
				^{137}Cs	^{60}Co	^{137}Cs	^{60}Co
13.06-29.08.	Ship-repairing enterprise	1	Soil	24 ± 3*	BT	23 ± 3*	BT
			Grass	BT	BT	BT	BT
		2	Soil	5 ± 1	BT	4±1	BT
			Grass	BT	BT	BT	BT
13.06 - 29.08	Settlement	3	Soil	4±1	BT	3 ± 1	BT
			Grass	BT	BT	BT	BT
		4	Soil	2 ± 1	BT	BT	BT
			Grass	2 ± 1	BT	BT	BT
		5	Soil	BT	BT	BT	BT
			Grass	BT	BT	BT	BT
14.06-29.08	Water area	1	Water	< 20	< 20	< 20	< 20
		1	Bottom	25±10*	25±10*	27±10*	23±10*
		1	Bottom	40±15*	36±12*	38±15*	34±12*
		1	Benthos	BT	BT	BT	BT
		1	Flounder	BT	BT	BT	BT

Comment: *residual radioactive contamination of bottom sediments occurred in 80-s; BT—below the detection threshold 1.5 Bq/kg.

The analysis of the radiation situation outside the PM-80 leads to the conclusion that the residual radioactive contamination of the vessel hull under conditions of long-term anchoring produces an area 50 m in radius of increased content of man-made radionuclides in air with the concentration of ^{137}Cs and ^{90}Sr in aerosols exceeding the background values by the factor of 2-3 (for PM-32, at the background level), which corresponds to 0.0001 MPC for the population [12]. During preparation operations related to the defuelling of the damaged SNF, the radionuclide concentration in aerosols

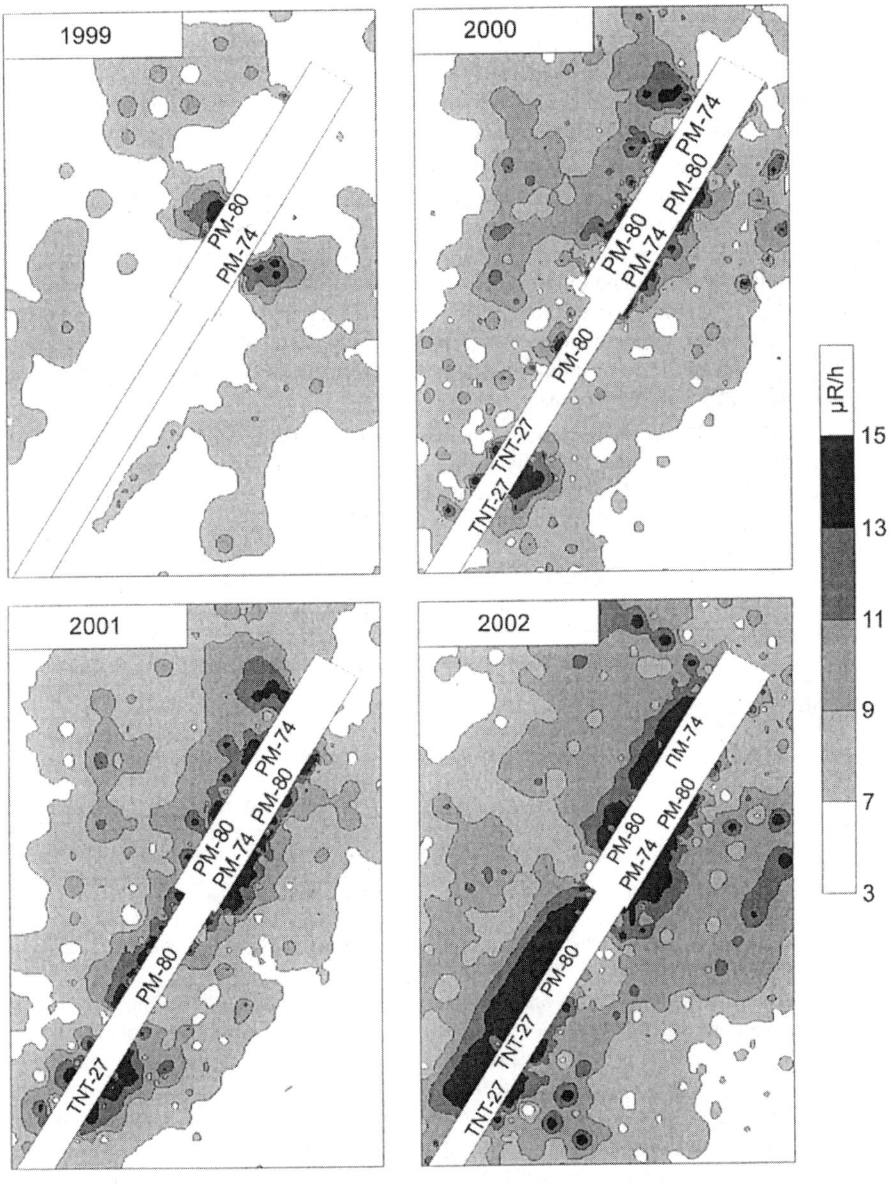

Figure 7. Variation dynamics of the external edges of the radioactive contamination of bottom sediments in the pier region for the period of the SNF defuelling from PM-80 and RW management by the Navy nuclear maintenance vessels (1999-2002, Konyushkova bay)

and atmospheric fallouts at PM-80 increase by a factor of 10-30 (0.001 MPC for the population). The SNF extraction, case cutting and blasting, and LRW recycling increase radionuclide content in aerosols by a factor of 1,000-5,000 with respect to the background, but this amount does not exceed 0.1 MPC for the population. At the same time, the aerosol concentration in the SFAS compartment in the absence of operations corresponds to 0.01 MPC, during preparations—to 0.1-2 MPC, and during the main operations amounts to 32 MPC (for PM-32—below 1 MPC).

The exposure rate of gamma radiation and the density of the beta-particle flux on the work surface of SFAS at PM-80 exceed the permissible level by factor of several hundreds or thousands (the same at PM-32 is lower by an order of magnitude). No excessive radioactive contamination was observed in the habitation zones of the floating workshops and maintenance vessels.

The concentration of man-caused nuclides in sea water, flora, and fauna in the pier region, at the territory, and water area of Konyushkova bay does not exceed 1 MPC according to the Navy regulations [11] (for PM-32, the corresponding figure is at the background level); in the drinkable and fresh water, the concentration does not exceed the background. The only exceptions (before deactivation) are the local spots of radioactive contamination of the roadsides and the bottom sediments in the anchoring region of PM-80, where concentrations of ^{137}Cs, ^{90}Sr, and ^{60}Co amount to 1-60 MPC (in the Seldevaya bay, no additional contamination of the bottom sediments in the anchoring region of PM-32 was observed).

Upon completion of operations, the dose burden and radiation risk (R) for the main personnel were substantially higher than for the auxiliary personnel and the population, but still did bot exceed the annual permissible level (Table 15).

Table 15. Radiation risk of the operations on the damaged SNF removal from PM-80

Personnel	Doze, rem (10^{-2} Sv)		Radiation	
	Daily average	Total	Risk, 10^{-6}	Hazard*
Management	0.005–0.015	0.05–0.15	30-80	Moderate
Engineers	0.05–0.1	0.15–0.8	80-500	Moderate
Main personnel	0.1–0.2	0.7–1.8	400-1200	Moderate/High
Auxiliary personnel	0.005–0.05	0.05–0.7	30-400	Moderate

Comment: *the hazard was estimated according to the step dependence of the form: $\leq 10^{-6}$—insignificant; 10^{-6}-10^{-5}—low; 10^{-5}-10^{-3}—moderate; 10^{-3}-10^{-2}—high; $\geq 10^{-2}$—excessive (limits $\geq 10^{-2}$ and $\leq 10^{-6}$ correspond to the probabilities of the natural death and death in result of the natural disaster, respectively).

Relatively low dose levels are provided by the application of the shadow shielding, replacement of the personnel, and optimization of the operator disposition in the SFAS during the operations. The latter method made it possible to reduce the exposure doses by a factor of 1.5-3 with respect to the regime with arbitrary motion of operators. The optimization is based on the choice of the most appropriate route of motion (disposition) of operators with minimum radiation levels. For this purpose, we carried out radiation monitoring of the compartment while mapping the intensity of gamma fields at various heights (01 and 1 m) and developed the corresponding recommendations.

The personnel of the maintenance vessels and the part of the population (B cohort) that stood in the region of PM-80 anchoring during the period of radiation-hazardous operations for 100 days per year were exposed to an external dose from 100 to 500 μSv, which corresponds to the radiation risk $(7\text{-}40)\cdot 10^{-6}$. The increased content of man-caused radionuclides in the surface layer of the atmosphere also contributed to the additional radiation burden, but this contribution is insignificant (Table 16).

Table 16. Maximum exposure doses and radiation risks resulting from the atmosphere contamination by the man-caused radionuclides during the defuelling operations with the damaged SNF at PM-80

Object	Doze, 10^{-6} Sv			Radiation risk, 10^{-6}		
	^{90}Sr	^{137}Cs	ΣD	^{90}Sr	^{137}Cs	ΣR
PM-80 (SFAS)*:						
- preparation;	300	2	302	17	0.1	17
- SNF removal.	400	2	402	22	0.1	22
Konyushkova bay**	0.3	0.01	0.31	0.02	0.001	0.02
Dunai settlement ***	< 0.1	< 0.1	< 0.1	< 0.001	< 0.001	< 0.001

Comments: *personnel (cohort A),
**population (cohort B),
***population of the Dunai settlement

At PM-32, in the Seldevaya bay, and in the Seldevoy settlement, all figures are lower by 2-3 orders of magnitude.

The data in Table 16 show that the total annual dose for the personnel formed by the man-caused radionuclides released to the atmosphere during the defuelling operations does not exceed 700 μSv, which corresponds to a radiation risk of $40 \cdot 10^{-6}$; for the population of the cohort B, the corresponding figures are 0.3 μSv ($R < 1 \cdot 10^{-7}$); for the population in the Dunai settlement, the place of permanent residence—< 0.1 μSv ($R < 1 \cdot 10^{-8}$). The maximum contribution to the radiation burden comes from ^{90}Sr.

The next source of radiation risk is the consumption of the contaminated marine products by the population. A short-term appearance of man-caused radionuclides in the surface water layer was observed during the operations on the SNF removal; during the last year, they appeared in algae for a longer time period. Taking into account that the accumulation coefficients of ^{137}Cs and ^{90}Sr by algae for the considered region are 70±30 and 100±50 [19], we can assume that their content in the bottom water layer in the region of the radioactive contamination of the bottom sediments (up to 1 m from the bottom) will vary between 200-1500 Bq/m^3 (< 1 MPC [11]; the concentration in the upper layers is lower by 1-2 orders of magnitude). This concentration determines the contamination of the aquatic life. The investigations showed that the possible variation intervals of the specific nuclide activities are: ^{60}Co—2-100 Bq/m^3, ^{90}Sr—6-700 Bq/m^3, ^{137}Cs—8-1,500 Bq/m^3; the nuclide content in fish and benthos amounts to 0.05-0,3 Bq/kg and 10-100 Bq/kg, respectively.

If a person consumes 100 kg of benthos annually (and unlimited amount of fish) the additional dose of internal exposure (allowing for the activity reduction after thermal processing) can amount to 10-100 μSv/year, which is equivalent to a risk of $(1\text{-}7) \cdot 10^{-6}$, or a low radiation hazard. However, such consumption is unlikely, because the Konyushkova bay does not produce benthos in the considered amount, and all fishery in the pier region is banned. In the Seldevaya bay, this contamination factor is insignificant, because fishery is completely excluded, while benthos usable for feeding is absent.

Under the impact of all negative effects, the radiation risk of defuelling of the damaged SNF from PM-80 and PM-32 for the population did not exceed $1 \cdot 10^{-6}$, while the additional radiation burden was within 1% of the radiation background. The gained "profit" is high, because a radiation accident might result in severe consequences for the environment, worsening of the psychological status and social state of the residents.

In conclusion, it is worth noting that during the period 1999-2002, the operations on removing the damaged spent nuclear fuel from two emergency floating workshops were, for the first time, completed in Russia. In this way, the existing prerequisites for an accident with severe radioecological consequences for the south of the Primorsky region and the eastern shore of the Kamchatka peninsula were eliminated. The monitoring showed that the SNF removal did not result in a massive release of radioactive materials outside the zones of the strict safety regime of the special vessels and did not violate the radioecological balance in the environment. The radioecological situation in the places of temporary anchoring of PM-80 and PM-32 in the Konyushkova bay and Seldevaya bay can be characterized as normal; the situation in the neighboring settlements is indistinguishable from background.

The developed complex approach to the monitoring problem was approved in complicated radiation conditions. The approach involved elements of technological control; complete reproduction of information in the form of mapped radiation fields with marked hazardous zones; study of the dynamics of the space-time motion of the man-caused radionuclides in the environment; assessment of the risk and radiation hazard for the population. The approach, however, is beyond the existing regulations for the control over the contamination of the environment. The expense of its implementation amounts, according to the experience gained, to 5-15% of the total project cost. These expenses are sometimes disregarded, when requirements to provide complete and reliable information about the radioecological hazard of preparation and utilization of the emergency objects are posed, but the provided financial support covers only 1% of the expenses.

References

1. Lebedev, V.A. and Akhubov, V.D. (2002) Ecological problems of utilization of nuclear submarines, *Voprosy utilizatsii APL* **1**, 6-12.
2. Suggestions of theMinatom of Russia to the state report anout the state of the environment of the Russian Federation in 1999 (2000) *Obrashchenie s radioactivnymi otkhodami i otrabotavshim yadernym toplivom* **3**, 40-48.
3. *Development of the Requirements to Provide Radiation and Nuclear Safety of Long-Term Storage of the Damaged SFA* (1994), Nizhnii Novgorod (in Russian)
4. SPA "Ecoatom" (1998) *Technological Regulations for Operations on the Damaged SNF Removal from the Storages of PM-80 and PM-32, Stage 4,* Sosnovyi Bor (in Russian).
5. Navy (1986) *Manual for Radiaton Safety in Operation of the Navy Buclear-Powered Vessels (NORB-83)*, Voenizdat, Moscow (in Russian).
6. SPA "Ecoatom" (2000) *Technological Process of the Damaged SNF Removal from the Storages of PM-80 and PM-32*, Sosnovyi Bor (in Russian).
7. *Development of Recommendations for Users of the Computer-Advised Desk of an Operator Analysing the Situation at the Sites of Anchoring and Repair of Vessels in Case of an Emergency Discharge of Radioactive Mateials in the Atmosphere* (1996), IBRAE RAS, Moscow.
8. Vysotsky, V.L., Bulygin, V.K. et al. (2001) *Impact of Defuelling of the Spent Nuclear Fuel from the Floating Workshop PM-80 on the Radioecological Situation at the Temporary Anchoring Site in the Konyushkova Bay*, Pacific Fleet, SPA "Ecoatom", Vladivostok (in Russian).
9. Vysotsky, V.L., Bulygin, V.K. et al. (2001) *Study of the Impact of Defuelling of the Damaged Spent Nuclear Fuel from the Decommissioned from the Pacific Fleet Emergency Floating Workshop PM-80 on the Radioecological Situation at the Temporary Anchoring Site in the Konyushkova Bay*, Pacific Fleet, SPA "Ecoatom", Vladivostok (in Russian).
10. Vysotsky, V.L., Bulygin, V.K., Maximov A.A. et al. (2002) *Impact of Defuelling of the Damaged Spent Nuclear Fuel from Emergency Floating Workshop PM-32 on the RAdioecological Situation at the Seldevaya Bay*, Pacific Fleet, SPA "Ecoatom", Vladivostok (in Russian).

11. *Manual for Control Over the Radioactive Contamination of the Environment and Internal Exposure of Staff of the Nuclear-Powered Vessels (RKVS-90)* (1991), Voenizdat, Moscow (in Russian).

12. *Regulations for Radiation Safety (NRB-99). SP 2.6.1. 758 — 99*, (1999), Minzdrav of Russia, Moscow (in Russian).

13. ICRP Publication 45 (1989) *Quantitative Substantiation of the Unified Damage Index*, Energoatomizdat, Moscow (in Russian).

14. ICRP Publication 60 (1994) *Limits of Annual Intake of Radionuclides by Staff Organisms Based on 1990 Recommendations*, Energoatomizdat, Moscow (in Russian).

15. Bulygin, V.K., Sderzhikov, Yu.A. et al. (2000) *Inspection of the SFAS cases on PM-80 for Presence of Residual Nuclear Fuel Upon Completion of the Year 2000 Operations*, Pacific Fleet, SPA "Ecoatom", Sosnovyi Bor (in Russian).

16. Bulygin, V.K., Sderzhikov, Yu.A., Vysotsky V.L. et al. (1999) *Development of the Technological Solutions and Carrying out Experimental Operations Aimed at the Removal of the LRW from the Floating Workshops*, SPA "Ecoatom", Sosnovyi Bor.

17. Chaikovskaya, E.L., Vysotsky, V.L., and Gichev, D.V. (2001) Regularities in fornation of the radiation situation at the territory of the Primorsky rregion *Atomnaya Energiya*

18. Sivintsev, Yu.V. (1985) *Development of the High-Sensitive Spectromtry Hardware Complex for Measuring Activity of Sea Water*, Kurchatov Institute, Moscow.

19. Danilyan, V.A., Vysotsky, V.L., Kryshev, I.I. et al. (2002) Doze burden on aquatic life and population in the regions of operation of nuclear-powered vessels *Atomnaya Energiya* **92**, 158-168.

ISSUES OF NOVAYA ZEMLIA BAY REHABILITATION DUE TO THE DUMPING OF RADIATION-DANGEROUS FACILITIES

Y.SIVINTSEV
RRC "Kurchatov Institute"
Moscow, Russia

Present-day quantitative radioecology allows for ranking unresolved radiation-related tasks depending on their potential environmental hazard. Successive stages of such assessments are widely known. They comprise estimates of:

- source term activity and radionuclide composition;
- radionuclide release rate from the source term;
- radionuclide transfer and spreading in the environment;
- radionuclide impact on the anthroposphere;
- radionuclide intake by human beings with air, water and food;
- exposure doses with respect to critical population groups;
- comparison of the above doses with admissible levels of radiation impact.

If the chosen quantitative criterion does not exceed an admissible limit in force, the radiation situation is usually considered as "satisfactory" or "normal." If such a criterion (e.g., radionuclide concentrations in air or water) exceeds the admissible level by only few times, no real radiation hazard arises. In this case the term "the environment contamination" is applied, and both administrative and technical measures are undertaken to normalize the radiation situation. In a case that the contamination level exceeds by many times the admissible values but the exposure dose with regard to critical population group is below the annual admissible dose, the term of "radiation incident" is recommended for use. And it is only in rare super-exposure cases when the population exposure dose approaches an extreme value (admissible only once in a life-long) and a real hazard regarding human beings arises that the notion of "radiation accident" is employed. In the case of long-term contamination of air, water of food above admissible values, one is dealing with "an environmental catastrophe". When eliminating its consequences, the task of rehabilitating the contaminated and/or polluted (above admissible levels) territories can be set. Unfortunately, even in radioecology no *quantitative* criteria for such a gradation are elaborated yet. Because of political factors, an attempt to create such a scale, when mitigating the Chernobyl accident consequences, 1986, has failed.

If neglecting at least one of the above factors or using only a part of them, both wrong estimations of the radiation situation and incorrect conclusions can be reached. A dramatic example of such an error is an estimation of the radioecological consequences resulting from Radioactive Waste (RW) dumping including that of Nuclear Submarines (NSs) with Spent Nuclear Fuel (SNF) in water areas close to the Novaya Zemlia (New Land) archipelago.

Similar to the stage of creating new-type devices & facilities in other industries, the very first stage of NS construction was accompanied by a number of radiation accidents. Some of them had so serious radiation consequences that the restoration of

A.A. Sarkisov and L.G. LeSage (eds.),
Remaining Issues in the Decommissioning of Nuclear Powered Vessels, 233–240.

NS Power Reactor Facilities (PRFs) was considered as impossible. Note that all such-type accidents took place in rather imperfect NSs of the first & the second generation. A severe radiation accident on NS K-19 (serial number 901) occurred first and resulted in the death of some members of the crew. After elimination of the accident the NS came back to its basing area in Andreeva Bay and was here afloat for a long period of time. The radiation measurements peformed concluded that the optimum solution for restoring K-19 consisted of replacing the damaged Reactor Compartment (RC) by a new one. Thus, the cut out RC of the two-reactor facility was dumped in 1965 in Ambrosimova Bay on the east coast of Novaya Zemlia. It comprised both the damaged reactor and the reserve one with intact cores. At the time of the dumping the activity of fission products, actinides and activation radionuclides within these reactors was 80 kCi or 2,9 TBq. These and similar-type data were obtained using initial information on nuclear fuel mass, enrichment and burnup: for NSs with water-moderated reactors in RRC "Kurchatov Institute" (RRC KI) and in case of liquid-metal coolant NSs in SRC "Physics & Power Engineering Institute" (FEI) [1-3]. Prior to dumping, both reactors were drained of water and filled with hardening radiation-resistant mixture based on furaldehyde (below furfurol) aimed at creating an additional protective barrier between SNF radionuclides and marine water. In keeping with initial expert estimates, the furfurol lifetime was taken equal to 500 years; based on the results of more recent calculations it was reduced to 100 years.

In 1965 another NS with SNF, order 285, was also dumped in Ambrosimova Bay with similar precautionary measures. Due to a shorter waterborne storage period between the time of accident and that of RC dumping, the radionuclide activity in the NS reactors reached its maximum by the dumping time and equaled about 314 kCi or 11 600 TBq. Altogether in 1965 through 1988 in bays close to Novaya Zemlia Archipelago 13 vessel reactors were dumped including 6 facilities with SNF & furfurol and 7 facilities without these components.

The history of the first OK-150-type PRF of the "Lenin" nuclear icebreaker is worthy of special consideration. This facility was dumped as two separate units: -a central compartment comprising three reactors without SNF & furfurol and –a special container with a part of SNF from reactor №2 filled up with furfurol. As distinct from widely disseminating information on the special container loss during hauling by a barge towards Novaya Zemlia, reliable evidences of the "Lenin" crewmembers can be cited (among them, the testimony of captain B. Sokolov should be mentioned first) on dumping both the RC and special container in Tsyvol'ki Bay at a distance of about 1 km from each other. In keeping with our estimates, the above container presents the main source of potential radiation hazard in the Arctic Ocean containing about 60% of the activity of environmentally important radionuclides dumped in the vicinity of the Novaya Zemlia archipelago.

Including the icebreaker, during 1965 through 1988 16 vessel nuclear reactors were dumped around the Novaya Zemlia archipelago: -10 facilities without SNF; -6 facilities with SNF; and -a container with a part of SNF of reactor n° 2 of "Lenin" icebreaker (see Table 1). Data on their activities published in a report by A. Jablokov Commission (the so-called "1993 White Book" (WB-93) [4]) are unreliable because, first of all, erroneous assumptions on SNF burnup in NS reactors were used. As a result, the activity of facilities dumped in the Arctic indicated in WB-93 was above actual values by factors of 3 through 27. In keeping with WB-93 the maximum integral radionuclide activity in the Kara Sea due to reactor dumping reached 10 MCi; in reality it did not exceed 0.77 MCi with consideration for the process of radioactive decay [5] (Figure 1). According to the data of both RRC KI and FEI, fission products were dominant among all radionuclides in the dumped facilities: their fraction reached 86% (mainly ^{137}Cs & ^{90}Sr). Activation radionuclides ranked second (12%) among which, as of 1994, not ^{60}Co but longer-lived ^{63}Ni was dominant. Minimum contribution (2%) was due to actinides with the maximum activity of ^{241}Pu.

TABLE 1. Radionuclide Activity within Vessel Reactors in the Kara Sea when Dumping, kCi

Facility name	Dumping year	Activity when sinking		Activity, as of 2000	
		RRC KI	WB-93	RRC KI	NNC
"Lenin" Nuclear Icebreaker					
RC and container with a part of SNF	1967	528	150	49,7	53,3
Six NS Reactors Dumped with SNF					
NS order 285 (starboard)	1965	312	800	15,2	15,2
NS order 901	1965	80	400	16,9	17,6
NS order 421	1972	28	800	6,8	7,6
NS order 601	1981	37	200	15,3	22,8
Seven NS Reactors Dumped without SNF					
NS order 254	1965	2,5	Total ~ 50	0,22	-
NS order 285 (port side)	1965	1,3		0,13	-
NS order 260	1966	1,2		0,1	-
NS order 538	1988	0,15		0,1	-
Integral		990	2 400	104,5	116,6

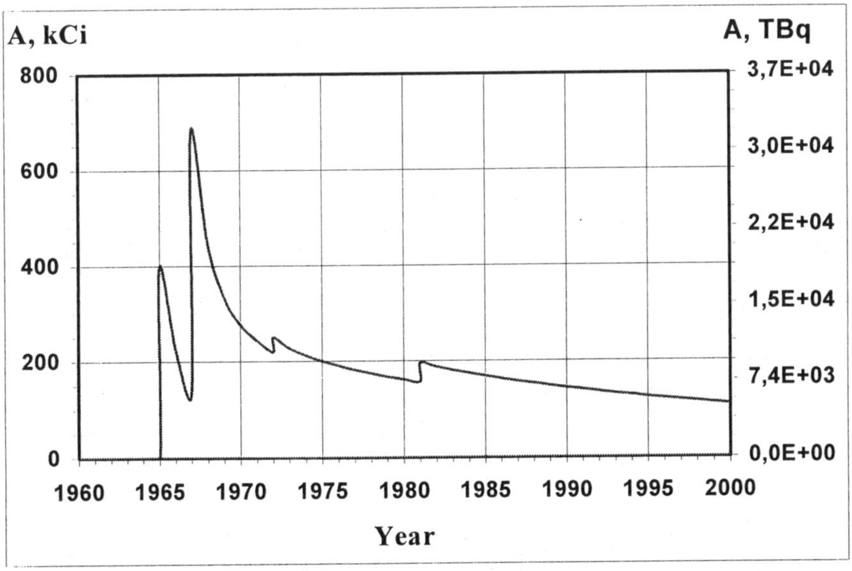

Figure 1. Changes in the Integral Activity of Solid Radioactive Wastes Dumped in the Arctic

To assess the radiation hazard to the marine environment and, first of all, to the aquatic life, the next investigation step should comprise calculations of the rate of radionuclide release from the dumped wastes. For this purposes the data on composition and properties of protective barriers (including furfurol), which prevent the contact of marine water with environmentally important radionuclides in SNF, reactor internals and reactor vessel, should be used. Such calculations were performed by a group of experts within the framework of International Arctic Seas Assessment Project (IASAP Project) under the aegis of IAEA. The project was aimed at:

- assessing real radiation situation in the Arctic seas in order to find out whether a radionuclide release from the dumped wastes (first of all from nuclear vessel reactors) took place;
- calculating the radiation situation in the Arctic for the foreseeable future taking into account potential consequences of long-lived and radioecologically-important radionuclide release from the dumped reactors;
- if necessary, making proposals on the rehabilitation of the Novaya Zemlia bays due to dumping of nuclear reactors of the former NSs & "Lenin" icebreaker.

Over 50 specialists of 14 countries participated in the project. In the course of Russian-Norwegian research expeditions of 1992/94 reliable data were obtained confirming the lack of radionuclide leaks from the dumped reactors (containing SNF and without SNF). A few individual samples taken in the immediate vicinity of the sunken facilities were the only exceptions (note that one of possible reasons of an elevated activity of the above samples could be due to radionuclide washing down from the Novaya Zemlia nuclear testing ground). In keeping with estimates of the Norwegian Radiation Protection Authority, annual individual exposure dose due to artificial radionuclides in both the Kara Sea and the Barents Sea varied within 1-20 μSv, the natural background dose being equal to 2400 μSv (2.4 mSv) per year.

The calculations performed within the framework of the IASAP Project demonstrated that radionuclide release from all dumped facilities would reach its maximum of 3000 GBq (81 Ci)/year in the coming century; after 300 years it could also increase up to 2100 GBq (57 Ci)/year. The rest of the time this parameter would vary within 2 - 20 GBq (from 54 mCi to 540 mCi) per year. In keeping with some exotic scenarios, a "volley" release could reach 110 000 GBq (3 kCi) per year (case of reactor destruction by an iceberg in 2050) and 6600 GBq (178 Ci) a year (case of global climate warming by 3000) [6]. It should be especially stressed that each of these values was considerably below the activity of persistent radionuclide dumping into the Irish Sea by the radiochemical plant in Sellafield, the UK, and into the English Channel by the similar-type plant in La Hague, France. Thus, both calculated and experimental investigations performed during many years proved the radiation safety of the concerned dumped wastes with respect to both the aquatic life of the Arctic and the critical population groups.

At the following stage of work on the IASAP Project the results obtained were used to estimate the radiation situation in the Novaya Zemlia bays and in the Kara Sea using multi-compartment and hydrodynamic models. Considered were: -normal conditions (static ones save for ebb/flow phenomena) and -some exotic scenarios including global warming of the Earth's climate and the collapse of dumped facilities due to the impact of an iceberg. The radiation consequences for hydrobionts and the population were considered separately. Taking into account the mode of life of both the critical population groups in the Kara Sea coastal zone and the personnel of the Novaya Zemlia testing ground, exposure doses due to different radiation intake paths were assessed with consideration for radionuclides issuing from reactors with SNF (as units of most potential hazard).

To calculate radionuclide concentrations in the Kara Sea, IAEA experts within the framework of the IASAP Project used the following 9 models:

- three 3D baroclinic models of the Barents Sea and the Kara Sea with a possibility of 10-year forecast (these models were developed by scientists of the International Marine Environmental Laboratory (MEL) of IAEA);
- three regional multi-compartment models of the Arctic, as a whole, and of the Kara Sea, in particular, with consideration for both trophic chains and a 100-year forecast (these models were generated by Russian, Netherlands and Japanese specialists);
- three global multi-compartment models taking into account seawater stratification & insoluble substance sedimentation as well as the possibility of a 1000-year forecast (these models were generated by specialists of MEL, the UK, and a Denmark-Norway team of scientists).

It should be mentioned that the Kara Sea hydrodynamic model created by Japanese scientists possessed the best spatial resolution.

The results of the performed simulations appeared to be close to the real situation in the Novaya Zemlia bays and on the open waters of the Kara Sea (Table 2) [7]. The only exception was provided by a few above-mentioned samples taken by Russian-Norwegian scientific expeditions of 1992/94 in the immediate vicinity of the dumped facilities. Good agreement between the calculated results (with the application of different models) and sample measurements proved the validity of the models used to assess the Kara Sea radiation situation.

Using some models, exposure doses for hydrobionts in Ambrosimova Bay (possessing the highest activity of the dumped wastes) were calculated. It was established that the additional hydrobiont exposure dose did not exceed 0.1 μGy/h. This value is thousands of times below minimum levels at which radiobiological effects become apparent; therefore it can be neglected.

TABLE 2. *Concentrations of Technogenic Radionuclides in Marine Water and Bottom Sediments in Different Areas of the Kara Sea*

Kara Sea region	^{137}Cs			^{90}Sr			$^{239+240}$Pu			^{60}Co
	In water, Bq/m^3		Bottom sediments, Bq/kg, dry weight	In water, Bq/m^3		Bottom sediments, Bq/kg, dry weight	In water, Bq/m^3		Bottom sediments, Bq/kg, dry weight	Bottom sediments, Bq/kg, dry weight
	Surface layer	Near-bottom layer		Surface layer	Near-bottom layer		Surface layer	Near-bottom layer		
Ambrosimova Bay	4 – 7	4 – 9	9 – 8400	2 – 4	2 – 4	0,3 – 3500	3 – 7	3 – 5	1 – 18	< 1 – 66
Stepovoy Bay	3 – 9	6 – 32	7 – 193000	2 – 7	3 – 26	0,4 – 300	2 – 5	2 – 18	0,1 – 15	0,1 – 3100
Tsyvol'ki Bay	4 – 6	6 – 14	4 – 30	4 – 6	3 – 4	0,4 – 1	4 – 10	5 – 8	0,03 – 0,5	< 1 – 4
NovoZemel'skaya Sea Trench	4 – 7	7 – 14	7 – 30	2 – 3	2 – 4	0,8	3 – 4	7 – 12	1	< 2
Open Sea	3 – 8	8 – 20	2 – 3	3 – 11	4 – 6	0,3 – 0,8	2 – 8	5 – 16	0,4 – 1,3	

In addition, exposure doses were estimated regarding critical groups of the population living in the lower courses of the Ob & the Yenisei river as well as in the Kola Peninsula, on the one hand, and to the military personnel of the Novaya Zemlia testing ground, on the other hand. Generalized results of these calculations (Table 3) demonstrated no radiation hazard due to dumping of the considered radiation facilities either for the military personnel at the Novaya Zemlia testing ground (Group 2) or for the seafood consumers (Groups 1 and 3). Note that the critical groups 1 and 3 receive exposure doses of 500 μSv and 100 μSv per year, correspondingly, due to natural-origin 210Po. It is of special interest that in all cases considered the maximum contribution to human exposure dose was due to radionuclides of actinides.

TABLE 3. *Maximum Exposure Doses for Critical Population Groups Due to Radiation-Dangerous Facilities Dumped in the Vicinity of the Novaya Zemlia Archipelago*

Scenario	Seafood consumers (groups 1 and 3), μSv/year	Military personnel (group 2), μSv/year
Most probable scenario	< 0.1	700
Worst scenario	< 1	4000
Global climate warming scenario	0.1	3000

Thus, the presented data prove no need of either extracting/deepening the dumped radiation-dangerous facilities (in particular, nuclear vessel reactors with SNF) or creating additional protective barriers around them. Based on the above data, one can conclude that such a task is groundless.

Taking into account:

- minimum (relative to other seas) radionuclide concentrations measured in samples of marine water, bottom sediments and hydrobionts during international research expeditions to radioactive waste dumping areas in the Kara Sea;
- results of forecast estimates using some extreme scenarios (performed, in particular, by the IASAP Project regarding the facilities dumped in the vicinity of the Novaya Zemlia archipelago); and
- insignificant exposure doses for the Kara Sea hydrobionts and the lack of resident population in the Novaya Zemlia archipelago, as a whole, and within the testing ground, in particular, a conclusion can be drawn that the problem of the studied territory & water area rehabilitation has no need for further consideration from the viewpoint of ensuring the radiation safety of the population, hydrobionts and other environmental components. Such a problem does not exist at all!

As an additional remark, the very high cost for extracting the dumped facilities, especially with regard to actual economic situation in Russia, is worthy of mention. Thus, the expenses for sea-work only are estimated at >13 million USD; by this we mean that the cost of 1 man·Sv of prevented exposure dose would exceed 200,000 USD [7]. Nevertheless, if such funds were assigned, in our opinion, they should be rather spent for scientific expeditions to Tsyvol'ki Bay where the container with SNF of

"Lenin" icebreaker, in stored and to the NovoZemel'skaya trench housing the reactor of NS, order 421.

References

1. Sivintsev, Yu. (1994) Study of Nuclide Composition and Characteristics of Fuel in Dumped Submarine Reactors and Atomic Icebreaker 'Lenin'. *IAEA-IASAP-1. Part 1 - atomic icebreaker. Working Material IAEA.* IAEA, Vienna.
2. Sivintsev, Yu. (1995) Study of Nuclide Composition and Characteristics of Fuel in Dumped Submarine Reactors and Atomic Icebreaker 'Lenin'. *IAEA-IASAP-5. Part 2 - nuclear submarines. Working Material IAEA.* IAEA, Vienna.
3. Efimov, E. (1994) Radionuclide Composition, Characteristics of Shielding Barriers and Analysis of Weak Points of Dumped Reactors of Submarine No. 601. *IAEA-IASAP-6. Working Material IAEA.* IAEA, Vienna.
4. *Facts and Problems Related to Radioactive Waste Dumping in Seas Surrounding the Russian Federation Territory. Materials of the Report by Government Commission* (1993) Administration of RF President, Moscow (in Russian).
5. Sivintsev, Yu, Kiknadze, O. (1999) Quantitative Analysis of Radioecological Consequences Due to Radioactive Waste Dumping in the Arctic, the North Atlantic and in Russian Far East. *Materials of the 5th Conference in Memory of Academician Alexandrov*, RRC "Kurchatov Institute" (in Russian).
6. *Predicted Radionuclide Release from Marine Reactors Dumped in the Kara Sea. Report of the Source Term Working Groupe of the International Arctic Seas Assessment Project (IASAP)* (1997) IAEA-TECDOC-938, Vienna.
7. Sjoblom, K.L., Salo, A., Bewer, J.M., et al. (1999) International Arctic Seas Assessment Project. *Science of the Total Environment*, **237/238**, pp. 153 - 166.

ENSURING SAFETY OF NUCLEAR SUBMARIMES STORED AFLOAT AFTER DECOMMISSIONING

S. PETROV
1st Central Research Institute of Russian Ministry of Defense
Saint Petersburg, Russia

An analysis of the information on managing the decommissioned Russian and foreign Nuclear Powered Vessels (NPVs) allows the following statements.

In the past 50 years of active development of the nuclear industry, about 500 Nuclear Submarines (NSs) and over 30 other type Nuclear Powered Vessels (NPVs) equipped with approximately 800 nuclear reactors were built in the developed countries.

At the actual stage of the nuclear industry development issues related to the decommissioning of transport Power Reactor Facilities (PRF) constructed in 1950-70s, because of the termination of their lifetime, become rather topical. By the beginning of the 21st century the integral number the decommissioned NPVs in the USA, United Kingdom and France reached about 150 units, whereas in Russia it exceeded 190 NSs. The Russian rate of NS and PRF decommissioning exceed those of other countries by a factor of 1.3 and 2.5, correspondingly.

As of the beginning of 2002, over 50% of the decommissioned NSs were defueled and about 1/3 of those NSs were dismantled: their reactor compartments were cut out and prepared for long-term waterborne storage in the form of one-, three- and multi-compartment reactor units.

The decommissioned NSs and the cut out reactor compartment units are stored afloat within water areas at their temporary storage centers and at plants of both the Northwest and the Far East regions of Russia. They present a potential environmental hazard. Important funds are necessary to: -realize organizational and technical actions aimed at preventing the initiation/development of potential emergencies; -construct additional wharfs; and -maintain such vessels afloat. The latter task necessitates considerable human resources (personnel) to ensure proper operation of general-purpose vessel systems.

The decommissioning of an important number of NSs requires rapid and efficient resolving the issues of their utilization. The topicality and complexity of the problem results from the following reasons:

- concentration of many decommissioned de-fueled & with fuel NSs (including NSs with damaged cores) within basing, repair & temporary afloat storage areas. This creates potential nuclear/radiation hazards for both the environment and the population;
- exhaustion of the design resources and/or termination of the lifetime of both the equipment and systems ensuring buoyancy, explosion protection and fire safety for most of the decommissioned NSs;
- need to ensure nuclear, radiation and environmental safety during waterborne storage of the decommissioned NSs, their cutting out, transportation,

A.A. Sarkisov and L.G. LeSage (eds.),
Remaining Issues in the Decommissioning of Nuclear Powered Vessels, 241–246.

storage and/or utilization of their Reactor Compartments (RCs) and also when managing their Spent Nuclear Fuel (SNF);

- need to unify and improve both the technologies and facilities involved in NS utilization and SNF handling operations for subsequent application to other nuclear vessel types, et al.

Under actual utilization rates in Russia forced temporary waterborne storage of NSs with fuel in the reactor cores is inevitable.

When considering NSs awaiting utilization, the following issues are of prime importance:

- ensuring nuclear & radiation safety of PRFs; ensuring NS explosion protection & floodability as well as of the environmental safety within NS waterborne storage areas;
- NS transportation to enterprises performing cutting operations;
- reactor de-fueling, temporary storage & removal of important amounts of NS SNF for reprocessing.

There are some distinctive peculiarities in waterborne storage of NSs with SNF, such as:

- reduced size of the NS crews (maintenance workers) and, thus, a reduced program of both environmental monitoring and control over safety systems, as well as inadequate possibilities to perform reciprocal follow-up actions;
- A particular vulnerability to NS sinking, losses in stiffness, freezing of water (liquid) channels and circuits due to non-operability of some important systems and also to natural ageing of basic structure elements.

To substantiate all issues related to safe management of the decommissioned NS, their RCs & SNF, with consideration for economic efficiency of these activities, the following top-priority actions are necessary:

- development (revision) of standards & rules related to management of the decommissioned NSs and their RCs with consideration for up-to-date nuclear, radiation and environmental safety requirements;
- elaboration and implementation of the relevant requirements for centers of RC waterborne & coastal storage and also for their engineering equipment;
- development and realization of proposals on priorities for defueling decommissioned NSs;
- establishing a list of top-priority actions (including technical and economical issues) aimed at improving SNF and RC management;
- performing technical justification of safety when managing the decommissioned NSs, their RCs and SNF;
- developing information technologies to perform explanatory and elucidative work with the concerned population on issues of creating storage centers for RCs and SNF casks.

The results of some already realized actions are discussed below.

1. Ensuring both nuclear safety of PRFs and radiation safety of NSs is the top-priority task is. To resolve these tasks, a complex of actions is being performed at every decommissioned NS in order to ensure reliable & safe PRF storage through its transferring into the following condition:
   - reactors cooled down and all standard absorber rods inserted, shim lattices being on bottom supports;
   - drivers of actuating mechanisms for the control & protection system mechanically secured, cables for their power supply made dead (disconnected from distribution devices) and cut off;
   - pressure in the primary circuit lowered, needed water regime in the primary circuit maintained;
   - periodic radiation monitoring and (if necessary) normalization of the radiation situation are performed.
2. Waterborne storage of the decommissioned NSs necessitates additional safety measures.

   When handling such NSs (including non-de-fueled NSs), as well as their RCs and SNF, a probability of initiating an emergency situation exists despite preventive measures. If it progresses, such an emergency can result in losses in NS flotation ability and, in case of external impact of both natural and technogenic origin, in the subsequent release of radonuclides beyond safety barriers, unpredicted exposure of personnel and contamination of the environment.

   It should be stressed that the list and parameters of possible RC accidents, as well as recommendations to be introduced into the emergency plans (aimed at protecting NS crews and the population living close to nuclear vessel basing areas) should take into account extreme natural & technogenic impacts, as applied to specific geographical conditions.

   The document known, as "The Technical Justification of Safety" should include information needed by governmental bodies involved into the use of atomic energy as well as by nuclear safety authorities, their basic organizations and commissions of experts when assessing the completeness and relevance of safety standards, rules and requirements to be introduced into practice in order to ensure safe management of the decommissioned NSs, their RCs and SNF.

   Measures on substantiating safety requirements are especially pressing in the context of reinforcing antiterrorist activity, in particular, in the Russian Federation.
3. To realize the optimum sequence of NS de-fueling, the decommissioned NSs are arranged (classified) depending on potential radioecological consequences due to a potential radiation accident the progression of which can result in NS sinking. Such a classification takes into account the probability of radionuclide release into the environment and water penetration into the NS strong hull when the NS sinks.

   The safe period of SNF storage within reactors of the decommissioned NSs depends on the status of both the SNF and its package (here "package" means the NS hull and structures of the primary circuit).

Thus, the duration of safe SNF storage in NS reactors and, indirectly, the priority for SNF unloading are estimated depending on the leaktightness characteristics of the three safety barriers: NS strong hull, primary circuit and fuel element claddings.

Based on an analysis of a representative amount of data, a classification (ranking) of the decommissioned NSs housing SNF via combining the most important indices regarding the status of the strong hull, PRF and the reactor core is proposed. The weight of every considered index is to be determined by experts. In this case every concerned NS will receive a resulting sum of indices. An example of the decommissioned NS ranking depending on the strong hull technical condition is presented in Table 1.

TABLE 1. Quantitative Estimate of NS Strong Hull Indices Needed to Determine De-fueling Priorities

Strong Hull Technical Status Characteristics	Score
NS kept afloat using special techniques or NS considered as "damaged" in keeping with annual report data	1
NS lunched > 30 years ago	2
NS dock repair and installation of protective barriers performed > 12 years ago	3
NS mid-life repair carried out > 15 years ago	4
NS decommissioned > 12 years ago	5

Another type of NS ranking depending on PRF technical condition of the reactor unit can be proposed with consideration for the following positions: - the NS was decommissioned during PRF repair; -NS with depressurized PRF primary circuit; -NS decommissioned because of the poor state of PRF; -last PFR shutdown performed > 15 years ago; -actual PRF storage mode introduced > 10 years ago, et al.

One can continue the proposed ranking depending on the technical condition of NS reactor cores.

The proposed method makes it possible to group NSs of similar technical condition in order to determine the potential (admissible) duration of SNF storage in the NS reactors and also to develop recommendations on NS de-fueling priorities.

When assessing the general condition of NS, taking into account the sum of all three indices (NS hull, PRF and reactor core), the worst index should be considered first. Note that, if only one parameter of the sum obtains "score 1", the NS in question should be placed into the "urgent de-fueling" category.

If 2 or 3 parameters have a "score 1," a "damaged" NS condition should be understood and, thus, such a NS should be de-fueled immediately. Further NS ranking should be performed with consideration for a successive increase in the score sum. In a case that the sum exceeds 9 scores, or NS indices remain out of consideration, the NS de-fueling queue should depend on the technical potentialities of the relevant enterprise.

The information presented in the paper taken together with the assessment scores (Table 1) makes it possible to propose the following recommendations regarding NS de-fueling priorities (Table 2).

TABLE 2. Recommendations on Priorities for defueling Decommissioned NSs.

Serial number of order		Sum of scores regarding the considered indices	Group number	Recommendations on NS de-fueling
Northwest region	Far East region			
.....	.....	< 5	1	Top-priority de-fueling ("damaged" status)
.....	.....	6-7	2	NS waterborne storage prior to de-fueling < 3 years
.....	.....	8-9	3	NS waterborne storage prior to de-fueling 3-7 years
.....	.....	> 9	4	De-fueling queue depends on technical potentialities of the enterprise

4. It is obvious that, when ensuring the continuity of both the NS utilization process and transfer of cut out/dismantled equipment, materials and nuclear fuel to processing/reprocessing, the task of minimizing the integral expenses (calculated for a specific time), with consideration for both non-simultaneous charges for the whole process realization and the inflation rate, is rather important. Calculating a minimum sum of expenditures makes it possible to substantiate an economically balanced annual rate of both NS utilization and their RC & SNF management. These data can be used to substantiate an acceptable structure and values of basic parameters regarding the whole SNF handling system.

The Efficiency Function (EF), as an optimization criterion regarding the whole NS utilization process, can be calculated as follows:

$$EF = (1+\alpha)^{t_0}\left\{\sum_i\sum_j\begin{pmatrix} C_{wbst}^{NS} + \\ + C_{houl}^{NS}\end{pmatrix} + \sum_i\sum_j\begin{bmatrix}(1+\alpha)^{t_{defuel}}(C_{defuel}^{core} + C_{stor}^{core} + C_{stor}^{SNF}) + \\ +(1+\alpha)^{t_{defuel}+t_{stor}}C_{RC}^{NS}\end{bmatrix}\right\} +$$

$$+(1+\alpha)^{t}\sum_f\sum_s C_{maint}^{f} + \sum_{jx}(1+\alpha)^{t_\mu}C_{jx}^{stor} + (1+\alpha)^{t}\sum_q\sum_k C_{maint_k}^{q} -$$

$$-\sum_i\sum_j(C_{met}^{NS} + C_{reuse}^{NS} + C_{reuse}^{stor_j}) \rightarrow \min$$

where: C signifies expenses for:

- waterborne storage of the decommissioned NSs and their haulage to the de-fueling center;
- de-fueling of NS reactors and performing attendant dismantling/mounting operations;
- RC cutting out, haulage and storage within a temporary (or long-term) storage center;
- construction or/and maintenance of the facilities and elements of the RC & SNF management system including the costs for transportation & storage as well as those for personnel wages;

- NS SNF chemical reprocessing.

The formula also comprises:

- integral cost of: -potentially reusable NS equipment and the metal of both NS rostrum & stern parts and –recovered nuclear fuel that can be reused in nuclear fuel cycle;
- a – inflation rate;
- (1+a) – inflation index, i.e. an average inflation rate with respect to individual operations.

Actually the problem of multi-purpose NS utilization is realized on the basis of special programs of the RF Government aimed at ensuring the nuclear and radiation safety of the State (see for example the RF Governmental Regulations № 518 of May 28, 1998, № 149 of February 22, 2000, et al.).

But, the Program of multi-purpose NS utilization (in its part concerning operations prior to NS cutting out) is far from considering in full the radiation and environmental hazards associated with the decommissioned NSs. It should be especially stressed that in most cases relatively new strategic-missile NSs with a "normal" state for their PRFs are actually utilized first. These NSs pose no special hazard when considering the potential damages which could result from their sinking.

The "oldest" NSs (principally, of the first-generation) are the most serious environmental hazard; this is especially true regarding NSs stored afloat in temporary storage centers far from coastal complexes. De-fueling and the following utilization of such NSs should be considered as a top-priority task.

Unfortunately, the international agreements ensuring the largest part of the financing within the framework of the Russian NS utilization Program are concentrated on the disarmament issues in Russia rather than on the problem of reducing/eliminating the environmental threat issuing from the decommissioned NSs.

The actual regime of NS utilization in Russia allows concluding that the Russian Navy still performs a part of NS utilization operations including RW handling and NS maintenance up to the stage of three-compartment unit creation. And these activities require important human resources from the Navy Forces.

In our opinion the Navy personnel should perform only special types of work, such as:

- unloading of reserve ammunition and of special counteracting engines;
- unloading of: spare parts, tools, devices, and chemical regeneration supplies;
- NS haulage to the shipyard (but only after making a special contract).

POSSIBILITIES OF USING A MULTIFUNCTIONAL SYSTEM OF SANOGENETIC MONITORING TO PERFORM A PHYSIOLOGICAL AND HYGIENIC SURVEY OF DIFFERENT WORK-STATUS PERSONNEL AT NUCLEAR SHIPBUILDING ENTERPRISES

V. DOVGUSHA, L. IVANINA, S. SAENKO
Research Institute of Industrial and Marine Medicine
St. Petersburg, Russia

1. Introduction

Working conditions at the large industrial facilities performing repair & reconstruction of Russian Nuclear Submarines (NSs) cannot exclude fluctuations in the effects of different technogenic factors during some process operations. However the detection of biological effects resulting from these technogenic impacts seems to be a very difficult task since they affect the human organism as a whole depending on the individual sensitivity level only [2, 3, 4, 10].

To detect the individual sensitivity of a human organism, instrumental methods of multi-system analysis, regarding the functional sufficiency of the systems responsible for the processes of adaptation to risk factor effects, were proposed in recent years [6]. The given investigation is aimed at approving the information sensitivity of the multi-system sanogenetic monitoring to detect professional risks for different categories of personnel involved in multipurpose NS repair & utilization operations at the Federal State Unitary Enterprise “State Machine-Building Enterprise "Zvezdochka" (SMBE “Zvesdochka”) in Severodvinsk-town. The given examination of the actual status of the principal life-support systems for “Zvezdochka” workers using the proposed method of express diagnostics was performed within the framework of the project "Studying the Possibilities of Using the Method of Sanogenetic Monitoring to Assess Harmful Effects of Industrial Factors at Atomic Shipbuilding Enterprises." This project was performed by the Research Institute of Industrial and Marine Medicine of the Russian Federation (RF) Ministry of Health.

2. Investigation Data & Methods

In 2000 through 2001 a two-phase examination of 100 men-workers was performed at SMBE "Zvezdochka" at one-year intervals. The workers were selected from virtually all shipyard professional groups, such as:

- engineers and technicians – a conventional control group of 15 persons (average age: 44,3±6,3 years; average record service: 19,3±7,5 years);
- 26 welders (gas welders, gas collectors, cutters and electric welders: average age 40,0±5,1 years; average record service: 17,1±6,2 years);
- 20 mounters and pipeliners (average age: 50,3±5,8 years; average record service: 24,0±8,3 years);

A.A. Sarkisov and L.G. LeSage (eds.),
Remaining Issues in the Decommissioning of Nuclear Powered Vessels, 247–254.

- 17 rubberizers (average age: 45,0±7,8 years; average record service: 18,3±8,1 years);
- 18 dosimetrists including meter men, inspectors, decontaminators with consideration for their individual radiation exposure risk (average age: 46,4±8,1 years; average record service: 18,1±7,7 years).

Workers of all the above-listed groups are exposed to different effects of their working environment including physical (microclimatic conditions, noise, local vibrations, dynamic & static loads) and chemical (asbestos and glass tissue dusts, condensation & disintegration welding aerosols, organic solvents, paints, sorbents, heavy metals, etc.) factors and also low-dose contacts with different Ionizing Radiation Sources (IRSs). Every examination included:

- transcutaneous bilirubinometry to detect the bilirubin level in peripheral cutaneous vessels [6];
- studying the state of metabolism & humoral immunity regulation using the method of blood plasma Laser Correlation Spectrometry (LCS) in keeping with the Methodic Recommendations [5]. To analyze biological solutions, an LCS spectrometer developed at "Intox" LS, Saint Petersburg (certification of approval the type of measurement devices RU.C.39.003.A №5381 of July 28, 1998, license №577 of June 15, 1999) was used;
- Studying the vascular & cardiac rate regulating systems using the multi-parameter cardio-arterio-rhythmography method in keeping with the criteria recommended by the European Cardiology Association regarding the standards of analyzing cardiac rate variability [12];
- Examining the systems of central, subcortical and peripheral regulation of psychomotor functions using the Computerized Movement Investigation (CMI) method according to the Methodic Recommendations [8].
- Studying systems responsible for respiratory apparatus regulation using the multi-parameter spirometry method in keeping with the Methodical Directions [9].
- The results of multi-system examinations were analyzed via establishing the levels of intersystem balance for individual functional indices in keeping with the Methodic Recommendations [7].

3. Results and Discussion

The dynamic examination of the functional conjugation of sanogenetic systems regarding 'Zvezdochka" workers comprised: -an initial-state assessment of individual functions and -a dynamic assessment of the functional balance characteristic variability throughout the survey year.

The initial state of individual functions for health-preserving systems was determined with respect to three centile-ranked levels [7]:

- 1st "sufficient" level wherein quantitative ranks of individual parameters did not exceed the 2nd centile values;
- 2nd "stress" level wherein quantitative ranks of individual parameters varied from the 2nd to the 3rd centile values;
- 3rd "pronounced stress" level wherein quantitative ranks of individual parameters varied from the 3rd to the 4th centile values.

Dynamic characteristics of the functional balance variability throughout the survey year were assessed for three categories:

1. Negative dynamics – increase of centile values assessing all of the health-preserving systems by ≥1 centiles during the survey year.
2. Positive dynamics – replacement of "more stressed" centile values of the corresponding functional systems by "less stressed" ones (by ≥1 centiles).
3. No dynamics – stability (changes within 1 centile) of indices assessing the State of individual functions.

The highest occurrence of the "pronounced stress" state (Level 3) among all considered sanogenetic systems (Table 1) with respect to "Zvezdochka" workers independent of their work-status characterizes the disintoxication system (44% of all cases). At the same time "positive dynamics" cases are also recorded here most frequently (57%). Regarding "negative" kinetics, the rubberizer group is worthy of mentioning (35% of patients have "negative dynamics" in functional balance of the system under examination). This professional group is especially affected by chemical risk factors. There negative effects on the disintoxication hepatic system are widely known. It is also worth noting that 35% of the patients of the given professional group trend toward improving the functional balance of the studied system. In other professional groups "positive dynamics" prevailed throughout the survey year.

TABLE 1. Annual Dynamics of the Disintoxication System Balance Depending on Work Status of "Zvezdochka" Workers

Balance level and its dynamics	All members of workers	Professional groups:				
		Engineers & Technicians	Dosimetrists	Mounters and pipeliners	Rubberizers	Welders
Pronounced unbalance	44%	53%	36%	50%	48%	40%
Stress balance	26%	20%	32%	25%	22%	25%
Sufficient balance	30%	27%	32%	25%	30%	35%
Negative dynamics	11%	7%	0%	10%	35%	8%
Positive dynamics	57%	73%	45%	60%	35%	69%
No dynamics	32%	20%	55%	30%	30%	23%

It is worthy of notice that the established facts correlate well with the results of studying plasma homeostasis characteristics using LCS investigations [11] of blood plasma (Table 2). In most cases dystrophic-like semeiotic characteristics were recorded (55% of the whole studied cohort of "Zvezdochka" workers). The annual dynamics for every professional group indicates that the given LCS-deviation type is rather stable and does not depend on professional status: the occurrence of dystrophic-like deviations in all groups varies within 42-62%.

TABLE 2. Occurrence of different semeiotic-linked LC spectra of blood plasma regarding the whole studied cohort of "Zvezdochka" workers

Type of blood plasma LC spectrum deviation	Occurrence of different LC spectra in professional groups (%)					
	All members of workers	Engineers & Technicians	Dosimetrists	Mounters and pipeliners	Rubberizers	Welders
Normokinesis	8	4	8	0	9	18
Allergo-like	1	0	0	2	0	0
Intoxicational-like	16	13	25	15	12	13
Catabolitic-like	5	8	8	4	6	2
Autoimmune-like	0	0	0	0	0	0
Dystrophic-like	55	62	42	61	57	58
Allergo-intoxicational-like	2	4	3	2	0	2
Autoimmune-intoxic.-like	12	9	11	15	16	7
Allergo-dystrophic-like	1	0	3	0	0	0

It is obvious that only a stress in the desintoxication system of human organism can compensate for the established catabolic-type metabolism. Thus, when considering the given sanogenetic system, the studied cohort of “Zvezdochka” workers seems to be adequate for the trend formed in the system of the tissue & plasma homeostasis regulation and testifies to the presense of an adaptive response to the actual type of the metabolism in the disintoxication function.

Principally another situation characterizes functioning of both the cardiac and the vascular systems (Tables 3 and 4). Thus, levels 2 and 3 of the initial stress are recorded for 66-67% of workers, and, if considering the annual dynamics, the stress degree increases in both cases (41%). By this we never that these are namely the cardiac and vascular systems that are most vulnerable when taking into account the stress degree. But, as a matter of fact, such a result is quite predictable for the cohort with an average age of 40-50 years. However, if considering dynamic characteristics independent of professional status, the functional stress indices for almost 41-48% of the patients remains at a stable level throughout the survey year, whereas in 11-18% of the cases positive dynamics is recorded. Thus, when considering the functional stress of both the cardiac and vascular systems, the cohort in question should be placed into a “safe” category, taking into account its age structure.

TABLE 3. Annual dynamics of the balance of the cardiac system state depending on the professional status of "Zvezdochka" workers

Balance level and its dynamics		All members of workers	Professional groups: Engineers & Technicians	Dosimetrists	Mounters and pipeliners	Rubberizers	Welders
Levels 2 and 3	Pronounced and stressed unbalance	66%	60%	64%	60%	41%	46%
Level 1	Sufficient balance	34%	40%	36%	40%	59%	54%
Negative dynamics		41%	33%	27%	40%	40%	58%
Positive dynamics		18%	27%	14%	10%	30%	15%
No dynamics		41%	40%	59%	50%	30%	27%

TABLE 4. Annual dynamics of the balance of the vascular system state depending on the professional status of "Zvezdochka" workers

Balance level and its dynamics		All members of workers	Professional groups: Engineers & Technicians	Dosimetrists	Mounters and pipeliners	Rubberizers	Welders
Levels 2 and 3	Pronounced and stressed unbalance	67%	40%	73%	70%	76%	69%
Level 1	Sufficient balance	33%	60%	27%	30%	24%	31%
Negative dynamics		41%	33%	32%	40%	41%	54%
Positive dynamics		11%	13%	9%	10%	12%	10%
No dynamics		48%	54%	59%	50%	47%	36%

Let us consider the state of the cardiac system regulation with respect to workers of individual groups implying different types of professional risk. The data of Table 3 demonstrate that, regarding the functional sufficiency of the cardiac system the less favorable groups are: mounters & pipeliners (stress levels 2 and 3 are recorded in 60% studied cases with 40% of negative dynamics and only 50% of "no dynamics" cases) and -welders (stress levels 2 and 3 in 46% cases with 58% cases of the "negative dynamics" regarding the cardiac system regulation). These facts illustrate that within the professional groups of "mounters & pipeliners" and of "welders" both vegetative cardiac regulation and the heart muscular contractility are most affected by the working environment factors.

The data of Table 4 shows that the criteria characterizing the state of the vascular system are similar to those of the cardiac system. If considering the vascular system stress, the same groups of "mounters & pipeliners" (stress levels 2 and 3 are recorded in 70% of cases with 40% cases of "negative dynamics") and "welders" (stress-levels 2 and 3 in 69% cases and 54% cases of the "negative dynamics") constitute the group of highest professional risk.

Table 5 represents the results of examining the respiratory system of "Zvezdochka" workers, which can be (within certain limits) considered as positive. Thus, in only 11% of the cases stress-levels 2 and 3 are identified, whereas no stress dynamics in the studied system was recorded in most cases within the studied cohort (62%) throughout the survey year. Based on these results, a conclusion can be drawn that most of "Zvezdochka" workers have adapted to the potential impacts of the working environment factors on the functional balance of their respiratory system. If, considering the state of pulmonary regulation in different professional groups, one should state the following: when comparing the relevant characteristics for all concerned professional groups, negative contrasts are recorded in the case of "mounters & pipeliners" only. Here, stress-levels 2 and 3 are identified in 40% of all cases (although regarding kinetics indices this professional group does not stand from of other groups of the cohort). Most likely, the results obtained are due to unfavorable microclimatic conditions specific to the working environment of mounters and pipeliners.

TABLE 5. Annual dynamics of the balance of the respiratory system state depending on the professional status of "Zvezdochka" workers

Balance level and its dynamics		All members of workers	Professional groups: Engineers & Technicians	Dosimetrists	Mounters and pipeliners	Rubberizers	Welders
Levels 2 and 3	Pronounced and stressed unbalance	25%	7%	23%	40%	20%	27%
Level 1	Sufficient balance	75%	93%	77%	60%	80%	73%
Negative dynamics		21%	20%	18%	20%	24%	23%
Positive dynamics		17%	13%	23%	30%	10%	8%
No dynamics		62%	67%	59%	50%	66%	69%

As viewed from psychomotor function regulation (see Table 6), stress-levels 2 and 3 are recorded in < 1/3 cases of all "Zvezdochka" workers. In dynamic examinations the trend to stability of the studied characteristics of psychomotor functions is obviously dominant ("no dynamics" in 46% cases) with a considerable share of positive dynamics (38% cases). Most likely, such a pattern is due to good organization of most laborious work operations influencing namely the coordination of the psychomotor processes. If considering every professional group, the "rubberizers" and "welders" stand out.

Here, on a relatively favorable background, as compared to other professional groups, the number of cases with stress-levels 2 and 3 in the psychomotor system is rather high (up to 40-38%), though throughout the survey year positive dynamics are predominant in the considered two groups (46% and 39% respectively). With consideration for the dynamic characteristics of the functional balance variability, the group of "mounters & pipeliners" stands out. This is the "oldest" professional group wherein no deviation dynamics is recorded in most cases (80%). This is testimony to the fact that there is sufficient compensation of psychomotor stresses during long labor activity.

TABLE 6. Annual dynamics of the balance of the psychomotor system state depending on the professional status of "Zvezdochka" workers

Balance level and its dynamics		**All members of workers**	**Professional groups:**				
			Engineers & Technicians	**Dosimetrists**	**Mounters and pipeliners**	**Rubberizers**	**Welders**
Levels 2 and 3	Pronounced and stresses unbalance	29%	20%	17%	25%	40%	38%
Level 1	Sufficient balance	71%	80%	83%	75%	60%	62%
Negative dynamics		16%	14%	27%	0	30%	11%
Positive dynamics		38%	53%	36%	20%	46%	39%
No dynamics		46%	33%	36%	80%	24%	50%

4. Conclusions

1. The checkup of SMBE "Zvezdochka" workers (Severodvinsk-town) performed using the method of sanogenetic monitoring has demonstrated the following. Highest stresses in functioning of health-protective systems are recorded in the metabolism regulation systems having catabolitic trend that, in its turn, results in functional stresses within the disintoxication system.
2. The results obtained regarding functional sufficiency of cardiac, vascular, respiratory and psychomotor systems using the sanogenetic monitoring method allow a definite conclusion that, among the above-mentioned risk factors of the working environment, the effects of microclimate characteristics stand out exerting the largest impact on the professional group of "mounters and pipeliners".
3. The compatibility of LCS deviations in regulation of the metabolism processes with rather favorable dynamic characteristics in other sanogenesis systems makes it possible to place the recorded metabolism-regulation type into an adaptive-sufficient category.
4. The sanogenetic monitoring is a noninvasive method that does not require additional materials or reagents and seems to be a rather economical, fast, automated, and standardized procedure.

5. References

1. Akleev, A, Kisselev, F., Pashkov, I., Klopov, N., Noskin, B. and Noskin L. (1999) Laser Correlation Spectrometry (LCS) of Blood Plasma to Diagnose Post-Radiation Consequences, *Radiobiology*, 39, 1, 64-78 (in Russian).
2. Anokhon, P. (1980) *Key Issues of the System Functioning Theory,* Nayka (Science), Moscow (in Russian).
3. Baevskiy, R. (1989) Assessment and Classification of Health Levels from the Standpoint of the Adaptive Theory, *Bulletin of the USSR Academy of Medical Sciences*, **9**, 73-79 (in Russian).
4. Dmitrieva, N., Glasachev, O. (2000) *Individual Health and Multi-parametric Diagnosis of Human Functional States*, Gorizont, Moscow (in Russian).
5. Kisselev, M., Akleev, A., Polischuk, V., Dovgusha, V., Semenov, A., Noskin, V., Noskin, L. and Landa, S. (2000) *Monitoring of Risk Groups Regarding Individual Diseases and Intoxications Using Laser Correlation Spectrometry of Blood Plasma within Territories of Both Natural and Technogenic Anomalies. Methodic Recommendations*, Russian Ministry of Health ("Medbioextreme" Federal Department), Moscow (in Russian).
6. Komarov, G., Karganov, M., Kisselev, M., Kuindzhi, N., Kurneshova, L., Kuchma, V., Noskin, L., Landa, S., Noskin, V., Pivovarov, V., Stepanova, M. and Sukhareva, L. (2001) *Multisystem Sanogenetic Monitoring,* MIPKRO, Moscow (in Russian).
7. Kuchma, V., Komarov, G., Sukhareva, L., Kuindzhi N., Stepanova, M., Teksheva, L., Noskin, L., Noskin, B. and Pivovarov, V. (2001) *Evaluation of Adaptive Reserves of Human Organism in the System of Physiologic-Hygienic Standardization of Different Activities. Methodic Recommendations № 11-1/282-09,* Appr. by SSED Department of Russian Ministry of Health on 22.11.2001, Moscow (in Russian).
8. Kuchma, V., Sukhareva, L., Kuindzhi N., Stepanova, M., Komarov, G., Siniakova, D., Karganov, M., Noskin, L., Bogoliubova, A., Tsyrenova, N., Noskin, V., Landa, S., Pivovarov, V., Milushkina, O., Issaeva, D. (2001) *Investigation of Psychomotor Activity to Assess the Effects of Educational Technologies on the Health of Children and Teenagers. Methodic Recommendations № Mos MU 2.4.8.002-01*, Appr. SSED on 27.08.2001, Moscow (in Russian).
9. Kuchma, V., Komarov, G., Sukhareva, L., Kuindzhi N., Stepanova, M., Siniakova, D., Karganov, M., Noskin, L., Bogoliubova, A., Tsyrenova, N., Noskin, V., Landa, S., Pivovarov, V., Milushkina, O., Issaeva, D. (2001-2002) *Investigation of Sanogenetic Status of Children and Teenagers During Education. Methodic Recommendations № Mos MU 2.4.8.004-01*, Appr. By the Health Committee of Moscow Government on 22.01.2002, Moscow (in Russian).
10. Sudakov, K. (1993) *The Health Diagnosis,* Sechenov MMA, Moscow (in Russian).
11. Ternovoy, K., Kryzhanovskiy, G., Musijchuk, Y., Noskin, L., Klopov, N. and Noskin, V. (1998) Classification of the Results of Blood Plasma Investigations using Laser Correlation Spectrometry on the Basis of the Semiotics of Preclinical and Clinical States, *Ukrainian Biochemical Journal*, **70**, **2**, 53-85 (in Russian).
12. Heart rate variability. Standards of measurements, physiological interpretation, and clinical use, *European Heart Journal*, 1996, **17**, 354-381.

SPENT NUCLEAR FUEL ISSUES

ACTUAL STATUS AND PROBLEMS OF SPENT NUCLEAR FUEL MANAGEMENT AT COASTAL FACILITIES OF THE NORTHWEST REGION AND THE FAR EAST REGION OF RUSSIA

A. PIMENOV, V. MAZOKIN, N. GONTSARIUK
FSUE Russian Design Institute of Power Engineering (RDIPE)
Moscow, Russia

The Russian Nuclear Submarine (NS) coastal supporting infrastructure was constructed in the early sixties of the last century during the initial period of active NS fleet construction in Russia. Created first were Coastal Servicing Enterprises (CSEs) in Andreeva Bay and Gremikha settlement (the Northwest region) and those of Sysoeva Bay and Gorbushechja Bay (the Far East region). These CSEs were principally designed to ensure temporary storage of Spent Nuclear Fuel (SNF) prior to its transportation for reprocessing.

By now the SNF handling cycle at the above CSEs still remains "unclosed"; and, because of four-decades of operation, these CSEs are today a technological problem and an environmental hazard. With time many buildings and installations of the concerned CSEs have become worn out, whereas some facilities of SNF-supporting infrastructure are inoperative.

Actually all of the four concerned CSEs have been transferred to be under the authority of Minatom of Russia, and do not receive new SNF for storage within so-called Dry-Storage Blocks (DSBs). The CSE in Gorbushechja Bay is not considered below because it is not presently involved into the Transport-Technological Flowsheet (TTF) for SNF management.

It is obvious that, when performing rehabilitation operations at CSEs in Andreeva Bay, Gremikha-Settlement and Sysoeva-Bay, issues of the collected SNF management are of prime importance, and their proper solution is one of high-priority tasks. No doubt that this task should be resolved with consideration for all requirements & standards of nuclear, radiation and environmental safety. Taking into account the problem topicality, complexity and capital intensity, on the one hand, and possible environmental implications, on the other hand, both potentialities and resources of the international society are needed, as much as possible, to solve the problem properly.

The TTF of handling SNF unloaded directly out of the decommissioned NSs prior to their dismantling is described in detail in the "Concept of Multi-Purpose Nuclear Vessel Utilization" developed by the Dollezhal Russian Design Institute of Power Engineering (RDIPE) with the participation of other Russian research institutes and design offices. The Concept was agreed to by the Ministry of Defense of Russia, Ministry of Economic Development, Ministry of Industry & Science, Ministry of the Environment and Natural Resources, Ministry of Health, Gosatomnadzor (Federal Nuclear & Radiation Safety Authority), Rossudostroenie (Shipbuilding) and approved by the Science & Technical Panel of Minatom of Russia. Thereafter the Concept underwent an environmental expert examination by the State authorities and was ratified by Minatom leaders.

A.A. Sarkisov and L.G. LeSage (eds.),
Remaining Issues in the Decommissioning of Nuclear Powered Vessels, 257–260.

During the last three years the concerned TTF of SNF management was improved in such a way as to ensure removal of all unloaded SNF for reprocessing rather than its accumulation within the CSEs. This was reached through the realization of a number of organizational, technical and technological actions with the attraction of foreign investors. By way of example, one can mention putting into operation of two coastal complexes known as "The Infrastructure of Fuel Unloading from Reactors of NS Reactor Units" at the Federal State Unitary Enterprise (FSUE) "State Machine-Building Enterprise "Zvezdochka" (the Northwest region) and FSUE Far East Plant "Zvezda" (the Far East Region), as well as the commissioning of the second-stage facilities for SNF transportation.

The actually employed TTF for SNF management is realized in "undertime" conditions, wherein all elements are linked with each other in a day-to-day fashion. Thus, any failure within any element would result in malfunctions throughout the TTF. To solve this problem, special buffer sites for SNF cask temporary storage are being constructed at FSUE SMBE "Zvezdochka" and FSUE FEP "Zvezda"; enlarging of a site to be used for the same-purpose in the CSE in Sysoeva Bay is also planned.

However a buffer site within the Service and Repair Enterprise (SRE) "Atomflot" in the Murmansk region (for which commissioning was planned in 1999) is still not constructed despite its very high importance in ensuring SNF handling operations in the Northwest region. Thus, the SRE "Atomflot" is the only enterprise in the Northwest region where SNF packages are reloaded directly from Nuclear Maintenance Support Vessels (NMSV) to transportation railcars. Other enterprises performing SNF unloading from the decommissioned NS reactors do not have railroad terminals, and their SNF can be removed by sea only. The construction of such a buffer site at SRE "Atomflot" would allow for avoiding downtimes in the operation of both the marine and railroad TTF elements when waiting for each other.

Due to the above circumstances and taking into account of a full workload at all elements of the actual TTF, one has no possibility for initiating SNF removal from repositories and sites of the CSEs in Andreeva Bay, Gremikha-settlement or Sysoeva Bay. However this is not the key reason explaining delays in solving this problem.

Many issues related to SNF handling require both technical and economic assessments & investigations, delivery of new devices and reconstruction of the available infrastructure. Thus, the CSE in Sysoeva Bay still ensures operations of SNF removal from the decommissioned NSs in the Far East region. With consideration for SNF handling operations, about 1200 packages with SNF are stored at the CSE including over 1100 packages in Building 30A and about 80 packages within TK-18 Transportation Casks (TCs).

To ensure safety of SNF handling operations using TK-18 TCs, capital construction plans provided for:

- construction of a reloading site to perform SNF loading into TK-18 casks (Building 227);
- construction of a railroad section between "The Danube" railroad station and Building 227 at the CSE;
- construction of a stationary pier.

But up to now no installation is started and only the site of the TK-18 cask reloading is mounted.

The CSE under consideration possesses a site to place (store temporarily) about 20 TCs of TK-18-type or of TK-108/1-type.

In 1997-1998 a concrete-surfaced road section connecting the Sysoeva Bay CSE and a similar-type road between Koniushkova Bay and "The Danube" railroad station was commissioned that made it possible to perform SNF handling operations using a temporary flowsheet. In August 1998 the concerned temporary SNF handling flowsheet at the Sysoeva Bay CSE underwent both interdepartmental testing and certification.

With consideration for the actual economic difficulties in Russia even planned construction of different facilities at CSEs of both the Northwest region and the Far East region in the near future is problematical. Thus, a demand arises for additional consideration of the construction expediency issues.

After decreasing the intensity of operating the actual TTF for SNF management (the peak is forecasted for 2005-2006), SNF removal from repositories of the Sysoeva Bay CSE can be realized via the actual TTF using:

- TK-18-type (TK-108/1-type) transportation casks,
- a road-train and
- an area of the railcar train makeup at a special site of the Far East Railroad.

Enlarging the capacity of the actual buffer site up to 30 TCs at the Sysoeva Bay CSE with an additional possibility to store a supplementary lot of TK-108/1-type casks would make it possible to decrease the actual TTF intensity. Furthermore, in such a case SNF package unloading from the repository as well as preparative operations for sending SNF for reprocessing can be started.

The actual situation within the CSEs in Andreeva Bay and Gremikha-settlement is much more difficult.

At present at the following are stored at the Gremikha CSE:

- 6 Spent Extractable Parts (SEPs) dismantled form Liquid-Metal Coolant (LMC) NSs;
- Spent Fuel Assemblies (SFA) taken from Pool №2 of Building №1 and placed into 116 packages of ChT-22-type (including 9 units arranged in casks of 11-type and 107 units in casks of 6-type stored at an open-air SRW storage site);
- metal casks (21 units), reinforced concrete casks (14 units) and concrete-lead casks (20 units) with SFA fragments.

SEPs of LMC reactors containing SNF are to be stored within special cells of Building 1B because their removal out of the Gremikha CSE in not planned. No procedure of their handling has been elaborated yet. Thus, a comprehensive technical and economic assessment is necessary the results of which can be used as a basis for making a decision related to further management of the concerned SEPs.

There is a particular problem with the old-type cask handling, the transportation of which is forbidden. Neither procedure nor technique of managing SNF stored within the old-type casks (type 6 and type 11) as well as in SRW containers are actually available and need special development.

Thus, to solve issues of the removal all-types of SNF stored presently at the Gremikha CSE, one needs to develop special techniques, perform technical and economical assessments and, possibly, create new types of specialized transport ships.

About 3000 packages with SNF, the integral activity of which exceeds 10^6 Ci, are stored at the CSE in Andreeva Bay.

Because of cracks at the ceiling panel interface of building 3a, high humidity and atmospheric precipitations affect channels housing SNF packages. According to preliminary examinations, 122 SFA packages in the “3A” storage facility and 5 SFA packages within the “2B” repository fall into the “damaged” category. Elevated specific activity values (up to 10^{-2} Ci/l) are recorded within individual cells. The results of different-year examinations do not allow for considering the radiation situation as “favorable” within the 2A, 2B and 3A buildings & repositories.

In addition, both the CSE infrastructure and technical equipment are worn out.

The above facts demonstrate that to normalize SNF storage conditions at the CSE facilities prior to SNF removal for reprocessing and to prevent further destruction of both packages and SFAs housed therein, urgent multi-purpose work aimed at preventing water penetration into the storage facility of Building 3A and dewatering of all repositories is necessary.

Among rehabilitation tasks to be performed at the Andreeva Bay CSE, SNF removal is one of the most topical and pressing issues. One needs to develop special technologies and find optimal solutions to ensure maximum efficiency when using the required investments and minimizing risks for the involved personnel, the population of surrounding territories and the environment. For these purposes, in Russia, basic conceptual approaches have been developed and principal solutions determined which presume a three-stage work to resolve SNF handling problems at the Andreeva Bay CSE:

- preliminary insulation of SFA Dry Storage Blocks (DSBs) from the external environment (this task is to be realized simultaneously with the reconstruction of supporting infrastructure);
- SNF inventory, preparation for removal and removal from DSBs for reprocessing;
- DBS preservation.

To realize SNF removal from the CSE site one needs: -first, determine SNF technical status for the purpose of considering possible options for its further management and -next, investigate different alternatives for SNF handling & removal for reprocessing from both the technical & economical viewpoint and -finally, update the CSE infrastructure to ensure SNF management operations are in keeping with the chosen TTF option.

Urgent solution of the above-considered tasks would make possible reliable operation of the TTF for SNF handling and would ensure nuclear, radiation and environmental safety in the involved regions contributing to further development of fruitful international cooperation.

TRANSPORT AND TECHNOLOGICAL FLOWSHEETS FOR MANAGEMENT OF SPENT NUCLEAR FUEL FROM NUCLEAR SUBMARINES UNDER UTILIZATION IN THE NORTHWEST REGION & THE FAR EAST REGION OF RUSSIA: PROBLEMS AND SOLUTIONS

V. SHISHKIN, V. MASOKIN, N. GONTSARIUK, A. PIMENOV
FSUE Russian Design Institute of Power Engineering (RDIPE)
Moscow, Russia

Unprepared both legally nor technically, the decommissioning of a large number of Nuclear Powered Vessels (NPVs) during a rather short period of time (peaked in 1993 - 1997) resulted in a sharp aggravation of the problem of nuclear & radiation safety for the Russian Navy NPVs and for their basing areas. The lack of a worked over concept regarding NPV utilization gave birth to many problems requiring urgent solutions.

At that time the infrastructure of the existing shipbuilding and ship repair enterprises in the Northwest & the Far East regions of Russia was principally involved in Nuclear Submarine (NS) construction and repair operations and, thus, was unprepared to ensure their mass decommissioning.

When solving multi-dimensional problems of NPV utilization and performing environmental rehabilitation of radiation-dangerous installations of the Navy, the issues of Spent Nuclear Fuel (SNF) management are of prime importance.

Once a decommissioned NS is defueled:

- first, its status can be changed from the "nuclear-dangerous" to "radiation-
- dangerous" due to the removal from the NS of fissile materials capable of a self-sustaining chain reaction; in such a case the possibility of initiating radiation-dangerous incidents aboard the decommissioned NS is ruled out;
- second, the radioecological hazard resulting from a defueled NS decreases by about an order of magnitude, because, in keeping with the results of both theoretical and experimental investigations, SNF concentrates about 90% of the activity in the Reactor Compartment (RC).

Thus, ensuring a fast pace of SNF unloading from NS reactors is one of high-priority tasks when utilizing NSs.

The earlier used Transport-Technological Flowsheet (TTF) for NS SNF management was employed to reload reactors of NSs in operation and, therefore, was unable to ensure mass NS decommissioning. At the beginning of 1998 the TTF for handling SNF from both NSs in operation and NSs after the decommissioning comprised:

- Three recharge Floating Servicing Vessels (FSVs), design 2020 (one FSV in the Pacific and 2 FSVs in the Arctic);
- 8 FSVs, design 326 and 326 M, of earlier construction, but their lifetime was exceeded; because of many technical failures their participation in SNF-unloading operations was disallowed by the supervising authorities;
- One special train of 4 railcars (TK-VG-18 – type) to remove SNF;

A.A. Sarkisov and L.G. LeSage (eds.),
Remaining Issues in the Decommissioning of Nuclear Powered Vessels, 261–266.

- 52 up-to-date protective casks (design TK-18) to transport SNF for reprocessing;
- regular SNF coastal storage facilities in Andreeva Bay and Gremikha-settlement (the Northwest region) and in Sysoeva Bay (the Far East region) decommissioned because of water leaks from their storage tanks;
- off-design "dry" storage facilities in Andreeva Bay equipped in empty underground repositories: by that time these facilities were full with packages of NS Spent Fuel Assemblies (SFAs) and, because of the poor technical status of the handling machinery, wharfs, industrial- & economic-purpose installations, no SNF acceptance/removal from these facilities was (and is) carried out;
- an off-design "dry" storage facility in the Far East region partly used as a "buffer" facility to house NS SNF (it is from this storage that SNF was (and is) removed for reprocessing);
- Worthless (because of long operation and insufficient financing to perform repair/reconstruction) infrastructure for SNF & Radioactive Waste (RW) storage in Andreeva Bay and Gremikha-settlement. This infrastructure was unable to ensure the needed safety during SNF & RW handling operations.

Thus, because of the above reasons and, as distinct from non-nuclear vessel utilization procedures, a new (non-regulated) lifecycle stage for decommissioned NPV management related to their long-term waterborne storage with SNF was faced.

Considering the previous SNF unloading rate of 4-8 NS a year, the task of defueling over 100 decommissioned NSs would require many years. Taking into account the poor technical status of the decommissioned NSs, this circumstance could have inevitably resulted in nuclear & radiation incidents.

In 1998 by a special regulation of the RF Government, RF Mimatom was appointed "The State Customer-Coordinator of Work on Multi-Purpose Utilization of NPVs Decommissioned from the Russian Navy" and also on "the Environmental Rehabilitation of the Russian Navy Radiation-Dangerous Facilities".

On the basis of the above Governmental Regulation, principal lines of investigations were formulated in order to develop & improve the existing NS SNF management infrastructure at enterprises performing NS utilization-related operations.

First, the actual status of both the SNF-handling TTF and the enterprise infrastructure was analyzed. On this basis a Concept and a Program of Multi-Purpose Utilization of NPVs were developed. Both documents were agreed upon by all interested Russian ministries and departments and underwent State Environmental Expert Examination.

So far the RF Minatom approved the Concept, whereas the Program was included (as a subproject) into a project of "The Program of Industrial Utilization of Weapons and Armaments for the period 2002 through 2005".

In keeping with the "Concept…", SNF from decommissioned NSs is to be unloaded from NS reactors and, then, removed from regions in special protective transport containers for reprocessing. This is one of the top-priority NS utilization tasks of the Concept, because NS defueling makes it possible to reduce (eliminate) the risk of radioecological incidents related to NS waterborne storage.

To develop further the Concept positions, many investigations have been performed in order to substantiate improvements in the SNF TTF flowsheet and determine optimum rates of both SNF unloading from NS reactors and the creation of reactor compartment units.

As the result, in 1999 proposals were substantiated providing for acceleration of SNF unloading rates by a factor of 2-2,5 (as compared to 1999). It is worthy of noting that in the 1990s only 2 to 6 NSs a year were defueled on average. In 1999 8 NS reactors were defueled.

To ensure acceleration of SNF-unloading rates, a number of actions was proposed and realized on the basis of performed investigations. This made it possible to accelerate SNF unloading rates in 2000 – 2002 by more than 2 times, as compared to 1999. The work was (and is) coordinated by the RF Minatom and other bodies of the Russian Federal executive authorities.

An acceleration of the planned SNF unloading rates required additional SNF unloading–transportation facilities within the concerned TTF. Optimum scope was determined and 3 FSVs, design 326M (their service life was over) were renewed; as a result the operational time of these vessels was prolonged by 5 years. Consequently, a twofold increase (from 3 FSVs to 6 FSVs) in the number of FSVs involved into the SNF TTF was reached.

Based on the results of examination/inspection of the reloading equipment at OK-300PB and OK-300PBM (including equipment whose service life was up in 2000-2001), 2 sets of OK-300PB reloading equipment were renewed and re-commissioned; in addition, a set of OK-300PBU equipment was constructed to ensure relevant operations in the Murmansk region.

2 sets of OK-300PBU reloading equipment were constructed for coastal SNF accepting complexes (one set to be used in Arkhangelsk region in the Russian Northwest, another one in the Far East region).

Two FSVs of the Murmansk Sea Navigation Enterprise (“Imandra” and “Lotta”) were also involved into the NS SNF management work (i.e. NS defueling and SNF transportation to a reloading center). During 1999-2001 “Imandra” FSV defueled 7 decommissioned NSs, another three NSs are to be defueled in 2002. FSV “Lotta” ensured SNF transportation from the NS defueling area to a reloading center in the Murmansk region as well as the placing of SNF into special transportation casks and railcars.

In the second half of 2000 after repair operations, an additional processing line was commissioned at plant RT-1 of PA “Mayak” to reprocess SNF from NSs. This event, taken together with placing into operation of the second special SNF transporting train, made it possible to augment SNF reprocessing by a factor of 1,5 and, thus, to accelerate SNF removal from coastal SNF accepting centers.

Using funds from international technical assistance the second special train (TK-VG-18A) for removing SNF from transport reactors was commissioned in September 2000. This allowed for increasing the number of hauls by a factor of 2.

In 2002 the construction and commissioning of the facilities known as “The Infrastructure of SNF Unloading from NPV Reactors” at SMBE “Zvezdochka”, the Northwest region, and Far East Plant “Zvezda”, are to be completed. Both facilities are being constructed using funds from international assistance programs and from the RF State Budget.

Thus, an improved TTF for managing SNF of the decommissioned NSs is under implementation. The facilities of "The Infrastructure of SNF Unloading from NPV Reactors"comprise sites (30 cask each) for temporary (buffer) SNF storage in radiation-protective casks TK-18 and TUK-108/1. Putting into operation "The Infrastructure..." facilities will ensure the planned rates of SNF unloading (defueling of 18 to 20 NS a year in both the Northwest and the Far East regions of Russia); simultaneously the workload on out-of-date FSVs, design 326M, will be reduced.

In the first half of 2002 completing and commissioning of a site (19 casks of capacity) for temporary (buffer) SNF storage in radiation-protective casks of TK-18 and TUK-108/1 types is planned at SRE "Atomflot". This facility will eliminate the necessity of ensuring simultaneous arrival of the SNF transporting vessel, on the one hand, and of the SNF transporting train, on the other hand, at SRE "Atomflot" site.

Optimum amounts of: -high-priced protective transportation casks to transport SNF and -sites for their temporary storage in SNF accepting areas in order to ensure the planned NS defueling rates were determined.

48 dual-purpose containers (casks), type TUK-108/1, for long-term storage and transportation of NS SNF were made and another 25 casks are under fabrication. These casks (52 units of TK-18 and 73 units of TUK-108/1) allow for defueling reactors of 18-20 NSs a year and, then, realizing SNF transportation for reprocessing with a rate of 18-19 hauls of a special train per year. Additional transport vehicles to remove SNF from regions are also to be constructed (railcars of TK-VG-18A type).

So far, in the Kamchatka region the issues of SNF unloading from decommissioned NSs still remain unresolved. Neither FSVs nor coastal SNF unloading facilities are available in the region. Taking into account both the poor technical status of the FSV based in the Far East Region and its anticipated "shuttle" marches to Kamchatka and back, defueling of the decommissioned NSs based in Kamchatka using this FSV seems to be rather problematical. Therefore the movement of the decommissioned NSs from Kanchatka to the Primorskiy kray followed by their defueling and utilization at the Far East Plant "Zvezda" appears to be most expedient solution from both the technical and economical point of view. However this option also necessitates important funds because, in this case, one needs either to restore the flotation ability for the decommissioned NSs or construct a special transport floating dock. It would be desirable to construct such a dock using funds of the international assistance programs.

Issues of restoring the infrastructure of SNF & RW removal from Gremikha-settlement and Andreeva Bay in the Northwest region also need solution.

Actual NS defueling rates depend on both production capacities of the available infrastructure and the rate of SNF removal for reprocessing. By this we mean that within the actual TTF of SNF management all elements are inseparably linked with each other in such a way that any failure in any TTF element results in faulty operation of the whole flowsheet.

By now thanks to already undertaken measures and work performed, 50% of the decommissioned NSs are defueled. In keeping with our calculations, SNF unloading operations can be completed in the Northwest region by 2006 and in he Far East region by 2007.

Research and practical work to ensure functioning of SNF TTF at facilities of Andreeva Bay and Gremikha-settlement are in progress including activities performed within the international cooperation framework. Among them one should mention the development of a project to construct both a buffer storage facility at PA "Mayak" and an additional SNF transporting train. As mentioned above, their commissioning will allow for realizing a more flexible SNF removal TTF and for stabilizing rates of SNF removal from regions in a case of unexpected events.

However some unresolved problems in the NS SNF management still persist, and the Russian Federation will not be able to manage them alone in the near future. The most important problem issues are summarized below:

1. Lack of coastal centers (sites) for long-term storage of one-compartment units and also of additional room to arrange three-compartment units & multi-compartment units within their actual temporary waterborne storage area in both the Northwest and the Far East regions. Such a circumstance can result in a considerable lowering of the decommissioned reactor defueling rates in the coming years. Most of NSs stored with SNF were in service over 30 years (some NSs over 40 years) because in the prior period only NSs included in international strategic offensive arms reducing agreements were dismantled. Thus, in the coming years the decommissioned NSs housing SNF in their reactors will constitute the most serious environmental hazard.
2. The actual technical status of radiation-dangerous facilities transferred from the Ministry of Defense under the jurisdiction of the RF Minatom gives no way for ensuring safety either for the environment or for the population. The actual infrastructure of RW & SNF storage in Andreeva Bay and Gremikha-settlement is unable to ensure safe handling and removal for processing and reprocessing.
3. At the Far East Plant “Zvezda” (the Far East region) a facility known as “The Infrastructure of SNF Unloading from NPV Reactors” is to be commissioned in 2002. But, to ensure SNF removal, the reconstruction of 38-km piece of railway line up to Bolshoy Kamen’ railway station is necessary.
4. So far in Kamchatka Peninsula over 10 decommissioned NSs containing SNF in their reactors are based. The technical condition of most NSs is unsatisfactory. Neither coastal nor floating SNF unloading facility are available. Actually the facilities of “The Infrastructure of SNF Unloading from NPV Reactors” at the Far East Plant “Zvezda” are in progress. But, one has only one recharge FSV that ensures SNF unloading from NS reactors in the Primorskiy kray, and, what is more, this facility needs a mid-life repair. If arranging the FSV “shuttles” to Kamchatka and back, its long and expensive rebuilding, and also high maintenance costs, should be taken under consideration. To find a proper solution to this problem, an option of hauling the concerned decommissioned NSs from Kamchatka to the Primorskiy kray is under examination.

Thus, reasoning from the above discussion, a consideration of the possibility of rendering aid, within the framework of the international cooperation is advisable.

First of all the following should be considered:

- construction of coastal centers for long-term storage of one-compartment units in the Northwest region and in the Far East region of Russia;
- restoring the infrastructure in Gremikha-settlement to perform SNF & RW removal;

- constructing/granting a sea vehicle to haul decommissioned NSs (with SNF) from Kamchatka to Primorskiy kray;
- reconstruction of a 38-km piece of railway line up to the Bolshoy Kamen' railway station.

STORING AND SHIPPING OF SPENT NUCLEAR FUEL FROM SHIPS: NEW ENGINEERING SOLUTIONS AND PROBABLE RADIATION EFFECTS OF AN ACCIDENT

M. RYLOV AND S. KAMYNOV
RECcenter, St.-Petersburg, Russia

N. ANISIMOV
Acad. A. Krylov TsNII, St.-Petersburg, Russia

M. BARSKOV AND P. SMIMOV
Russian Navy, Moscow

M. BUGREYEV, D. PANKRATOV, AND G. TOSHINSKY
Acad. Leipunsky IPPE, Obninsk, Russia

S. GAVRILOV
PO 6, 123154, Moscow, Russia

1. Introduction

Storage and Shipping of Spent Nuclear Fuel (SNF) from ships is one of the most important tasks created by solving the problem of the Navy's nuclear facilities. The authors have found it reasonable to consider several procedures of safe SNF management starting with general principles for ensuring nuclear safety and reducing risk at all stages of nuclear facilities life cycle.

It is known from the experience of operation of nuclear power plants (NPP), including research, pilot and transport ones, that risk management efficiency is achieved by frequent restoration of material resources and rehabilitation of three characteristics of personnel reliability, namely: cognitive, psychosomatic and motive [1] (also [2]).

Personnel rehabilitation makes it possible to extend nuclear facility service life. But decisions are also required for minimizing human factor influence on the safety (nuclear and radiation included) of facilities being under construction.

It is also necessary to make allowance for "external" terrorism. So only in 2002 after tragic events of September 11, 2001, in New York, there was a report of Chechen terrorists' plan to capture, some years ago, one of the SSBNs of the Pacific Fleet (Russia) to dictate their demands.

Russian experts from some institutions developed a procedure for improving nuclear and radiation safety of SNF stored in the core by draining the light water reactor's vessel (LWR) [3,4]. The logical extension of this specified procedure for nuclear safety of decommissioned Ships up to a complete ruling out of the possibility of nuclear danger is the execution of the following operations:

- filling the dried core up with melted lead or lead-containing materials, e.g., Pb-Bi eutectic with storage vessel with core after metal solidification in the plant of the Submarine or in the light-duty container;

A.A. Sarkisov and L.G. LeSage (eds.),
Remaining Issues in the Decommissioning of Nuclear Powered Vessels, 267–283.

- filling a dried core, including a core with damaged fuel, by radiation-resistant polymers with their solidification, also storing it in RPV with fixed core;
- filling the core (non-drained included) containing SNF or damaged assemblies with solid-phase dispersed material that slows down and absorbs neutrons. The comparison of the safety levels of the proposed procedures with SNF storing and shipping in dual-purpose metal and concrete casks (TUK MBK) [5] has been performed. The site being prepared for SNF interim storage, an area of RTP Atomflot, Murmansk Shipping Co., is chosen as a place for casks storage.

2. Safe SNF storage in the core

Now the possibility of the sinking of the Russian Navy decommissioned Nuclear Powered Submarines and Ships, including ones with SNF in the cores of the light water reactors, is not an abstract issue. It is necessary to find an acceptable decision for the prevention of the threat to the Submarine's crew, and for the prevention of the nuclear and radiation threat to the environment due to the loss of buoyancy and sinking for ships in the holding area (6), and for a reduction in the costs for Ship refloating.

In addition engineering decisions are needed to make SNF removal from Nuclear Powered Ships safe, to reduce costs for this procedure and for SNF on-shore storing, shipping to the site for treatment or long-term storage.

Especially it is necessary to pay special attention to damaged fuel removal from Nuclear Powered Ships and its storage.

2.1. CORE FILLING WITH MELTED HEAVY METAL [7]

First of all it is possible to rule out the possibility of the nuclear danger, to improve personnel, population, environment and radiation safety from any probable emergency situation aboard Nuclear Powered Ships when filling up the dried core and adjacent space of the vessel with melted metal.

This material should have the necessary nuclear and physical, chemical, thermal and hydrodynamic properties as for example liquid -phased heavy chemically inert or weak active heavy metal and have also proper characteristics for its subsequent self-solidification inside the vessel. Such materials are Bi-Pb eutectic, pure lead and some other low-temperature -melting heavy metals and their alloys that are well known, used and recommended for application in the future.

The fixing of the core structure inside reactor vessel prevents any changes in element configuration, and any movement of the elements relative to each other. It makes the occurrence of a self-supporting chain nuclear reaction impossible for any accident scenario.

Certain rules for long-term storage of the reactor compartment even, with nuclear fuel available in the Ship's reactor are not required when the core is filled with the non-active fixing chemical material.

This makes it possible to simplify maintenance of the power plants of the retired Nuclear Powered Ships being stored afloat at the holding areas for long periods of time and to coordinate the rate of SNF unloading with the required number of containers for storing and shipping to the storage facility.

It is essential that not only the possibility of unauthorized removal of enriched nuclear fuel from reactor's vessels of the Russian Nuclear Powered Submarines and Ships [8] is excluded but also that the act of terrorism by both "external" and "internal" terrorists is impossible.

When a fixed core is unloaded into the vessel-container [9] and placed into the container of the simplified design ([10], Fig. 2) the system "reactor vessel with SNF in the solidified metal –container" (Fig. 1) is resistant to any external effects of a technogenic and natural character.

The same considerations apply to the specified system shipping. But in this case it is possible to have a beyond design basis accident with the dumping of the radiation –hazardous vessel from the container.

Remote removal of the facility because of its significant mass and the unavailability of the specific equipment at the area of the accident is a complication.

We note that core filling with metal is not a principle obstacle for SNF removal, if required, because the meeting temperature of the metal is comparatively low. But the operation will require using manipulators and a hot chamber [7].

This problem is not theoretical (see program of Swedish experts [11]). The basic strategic issues of SNF management are discussed including its direct disposal as an extremely hazardous radiation waste and SNF use as a valuable resource.

2.2. CORE FILLING WITH LIQUID POLYMER [7]

Dried core filling with dried polymer makes it possible to achieve results described in the previous subsection and in the first place -to exclude the nuclear danger from a reactor with nuclear fuel.

This material must be mechanically strong, radiation-stable, have minimum shrinkage; chemical, thermal and hydrodynamic properties both for sufficient flow and for complete polymerization of the material inside the vessel. Those materials are unporous epoxy oligomers with double bonds [12]. With the filling of the core of the Ship's plant with fixing chemically non-active thermal plastic it is not necessary to keep a definite schedule of reactor compartment maintenance. This simplifies the maintenance of the power plants of the retired Nuclear Powered Ships stored afloat.

With the fixed core both afloat and during unloading of the solid core from the vessel container into the "outer" container the "reactor vessel with SNF- container" system (Fig. 1) the system is stable with respect to all external effects including terrorism.

There are two differences between filling the core with polymer and filling it with heavy metal.

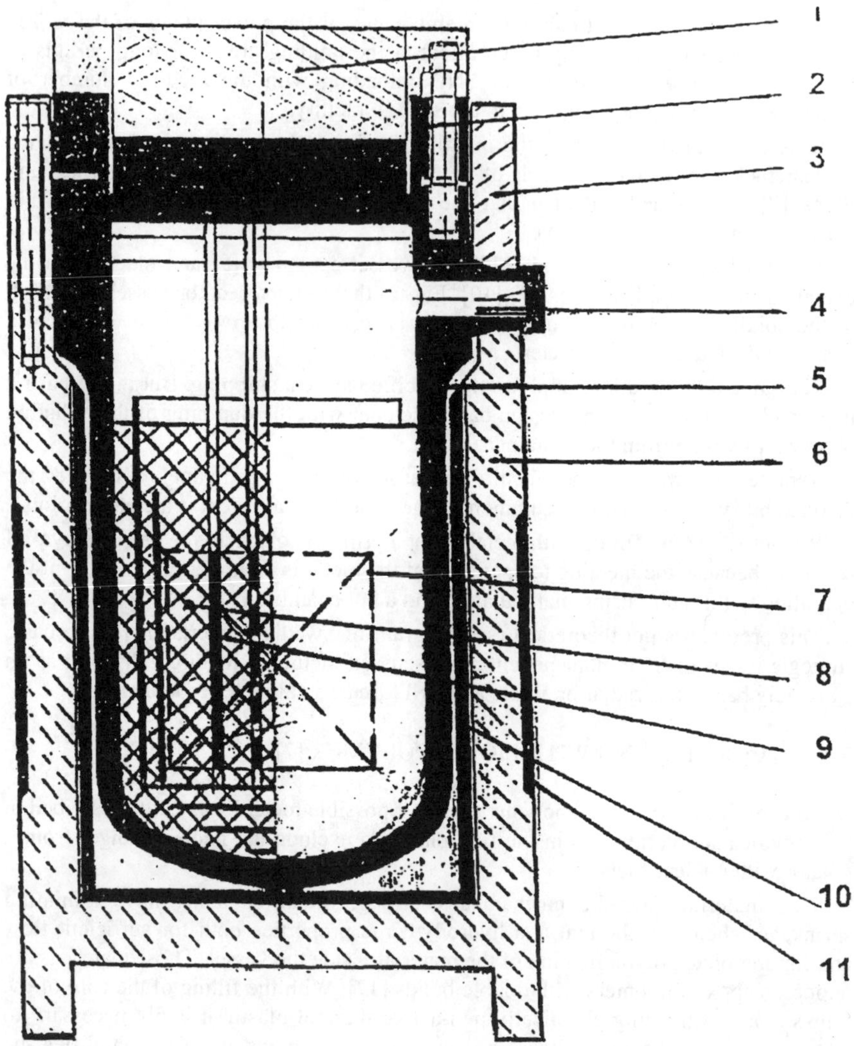

Figure 1. "Reactor Vessel - Container" system for storing and shipping Ship's LWR with SNF in core

1 - RPV head; 2 - Head forcing ring; 3 - Container forcing ring; 4 - RPV connection; 5 - reactor pressure vessel (RPV); 6 - Container case; 7 - Core; 8 - Core fixing matrix;

9 - Spent fuel assembly; 10 - Inner layer of absorbing material; 11 - Outer layer of neutron-absorbing

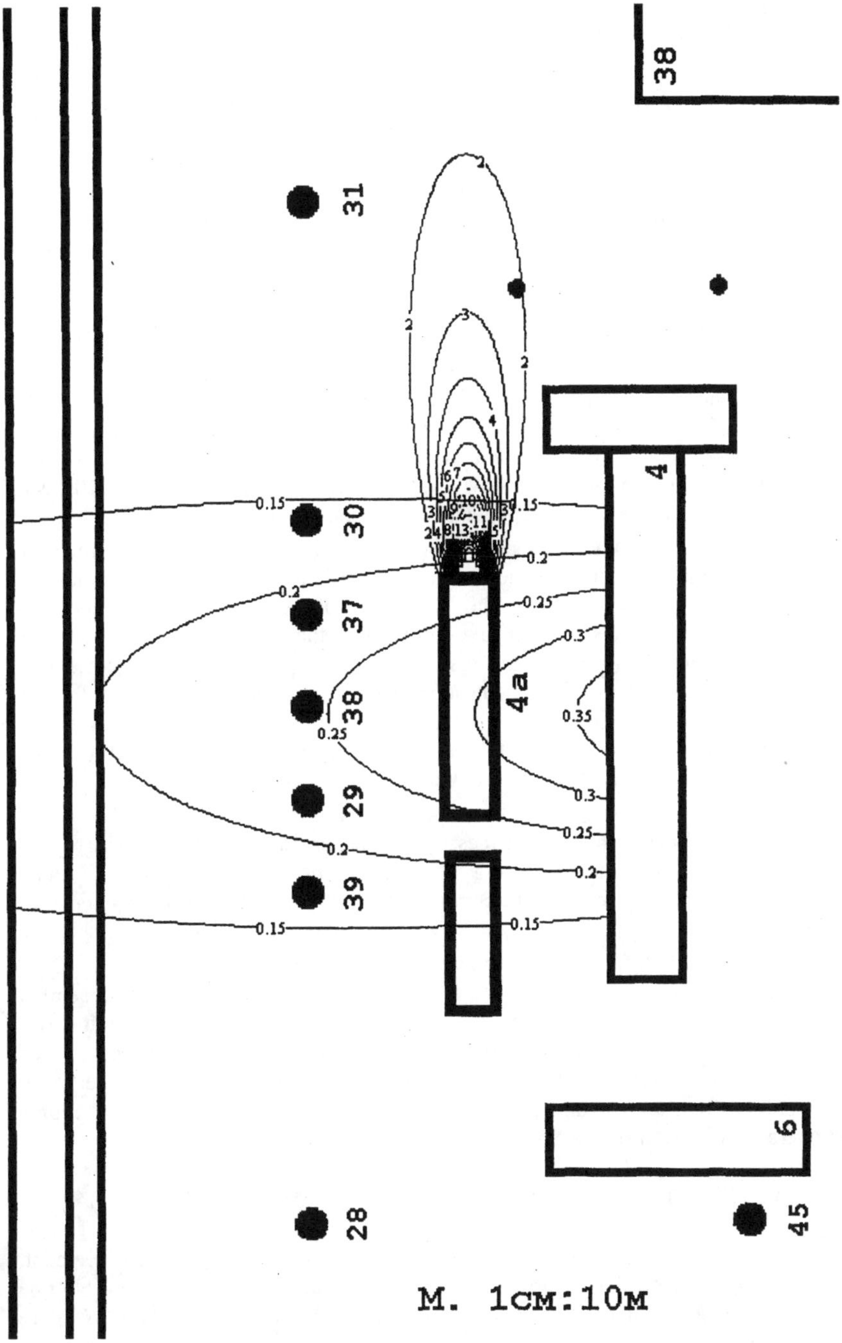

Figure 2. Distribution of effective dose rate, pSv/hr (10^{-12} Sv/hr) inside the aerodynamic shadow for south and east wind directions for "Reactor Vessel -Container Having Core Filled With Melted Heavy Metal" system and accident scenario being under consideration

The unreversible thermal reactive polymer process does not allow removing SNF from the vessel for treatment.

Therefore, for a core with damaged fuel, it is preferable to fill the core and, for quality performance of the operation, it is desirable to dry the space inside the vessel or at least, for a core with damaged fuel, but that is rather difficult.

Shipping of this system for long distances e.g. to the storage facility of "Mayak" is not now likely because the mass of vessel with fixed SNF in this case is lower than when it is filled with heavy metal. It should be mentioned that in some cases, e.g. with a porous heavy metal, to prevent the interaction of the filler with inside space atmosphere [6], the plastic cover could be put on it.

2.3. CORE FILLING WITH DISPERSED SOLID PHASE MATERIAL [13,14]

Another approach to the nuclear safety of the reactor core is filling the core up with dispersed solid -phase neutron absorber bearing material. It improves nuclear and physical characteristics of the core and reduces the neutron radiation from the reactor. This method of preventing a dynamic nuclear accident in an HTGR was developed by German experts independently from specialists in Russia [15].

It is possible, unlike other filling processes, to fill the core with dispersed material absorbing thermal neutrons in the non-drained LWR vessel. Water coolant is partially displaced by filler from the core and the remaining water acts as a moderator of fast neutrons escaping the core. Then the filler and rods of the core control and protect by absorbing the thermal neutrons. To fill up dried cores it is expedient to use a B_4C mix of fast neutron moderators and thermal neutron absorbers such as Hf and B_4C, BeO and Cd that moderate the fast neutrons and absorb the thermal ones. For accident cores with partial deformation the neutron absorbing filler additionally shields the damaged fuel reducing core criticality.

Prevention of "uncovering" of the assembly of the filled core caused by the reactor vessel inclination needs an excess of the dispersed powder layer with regard to the liquid filler level or the filling with solidifying liquid material with small density and high viscosity. Such composition will simultaneously prevent the interaction of the core filled with powder with the atmosphere inside the vessel even with water penetration into it.

Among the positive characteristics of the procedure, the simplicity of assembly removal from the vessel for treatment because of unnecessary filler melting or mechanical destruction of the solidified material should be noted. The proposed procedures ensure nuclear and radiation safety of cores, lowering requirements for their storage afloat and on-shore, direct storage in vessels being the most reliable containers for highly enriched SNF from nuclear ships.

2.4. ESTIMATION OF PROBABLE RADIATION ACCIDENT EFFECTS

The proposed SNF storage procedures exclude the possibility of a nuclear accident, i.e. an emergency with the occurrence of the self-sustaining chain reaction. No one could develop the scenario with such an excursion for both types of LRW core filling.

The "human factor" as it was noted in [1] stipulates the possibility of the accident both at the stage when there are insufficient material resources for safe operation of the reactor and in the period when there are insufficient resources even for the execution of rehabilitation procedures.

Three proposed solutions include the rehabilitation procedures both for SNF and damaged fuel.

The use of these procedures makes it possible to rehabilitate the nuclear safety of the fuel stored in the core during an unlimited period with regard to the time of technogenic civilization development.

That is why the case of the "nuclear terrorism" was examined as the most probable beyond design-basis accident when the system "vessel with SNF in dispersed powder-container" is bombarded by a grenade launcher.

It is conservatively assumed that there is both a destruction of the container and an infringement of the integrity of the reactor. Calculations were carried out in accordance with "Estimation of Economic Damage from Radioactive Materials Used in the National Economy Due to Radiation Accident" approved procedure [16].

In the case when the core is filled with solidified metal (Bi-Pb eutectic and concrete container) the radiation effects of any accident will be minimal, determined by contaminated metal activity (Table 1 and Fig. 2) and by egress of radioactive noble gases from the vessel.

TABLE 1. Effective dose rate (EDR) along an axis of aerodynamic shadow area for two wind directions for the first scenario of accident, pSv/hr (10^{-12} Sv/hr)

Distance from the container (building), m	**EDR (from irradiation in radioactive cloud)**	
	East wind	**South wind**
0	0.36	30
5	0.35	17
10	0.33	11.1
15	0.31	7.9
20	0.3	6.0
25	0.29	4.8
30	0.27	4.0
35	0.26	3.5
40	0.25	3
45	0.25	2.7
50	0.24	2.5
55	0.23	2.3
60	0.22	2.1

The results of the calculation show that the radiation effects of this beyond design-basis accident are negligible.

If the vessel is not drained completely and is filled with finely divided neutron-absorbing material it is possible to have infringement of the vessel integrity and egress of the a part of fission products from the vessel to the environment (Table 2 and Fig. 3). The results of the calculation shows that the effective dose rate for these scenarios is insignificant even near the building and damaged container. They are lower than effective dose limits given in Radiation Safety Standards-99 (NRB-99) for both personnel and population in emergency conditions and they are not dangerous for them.

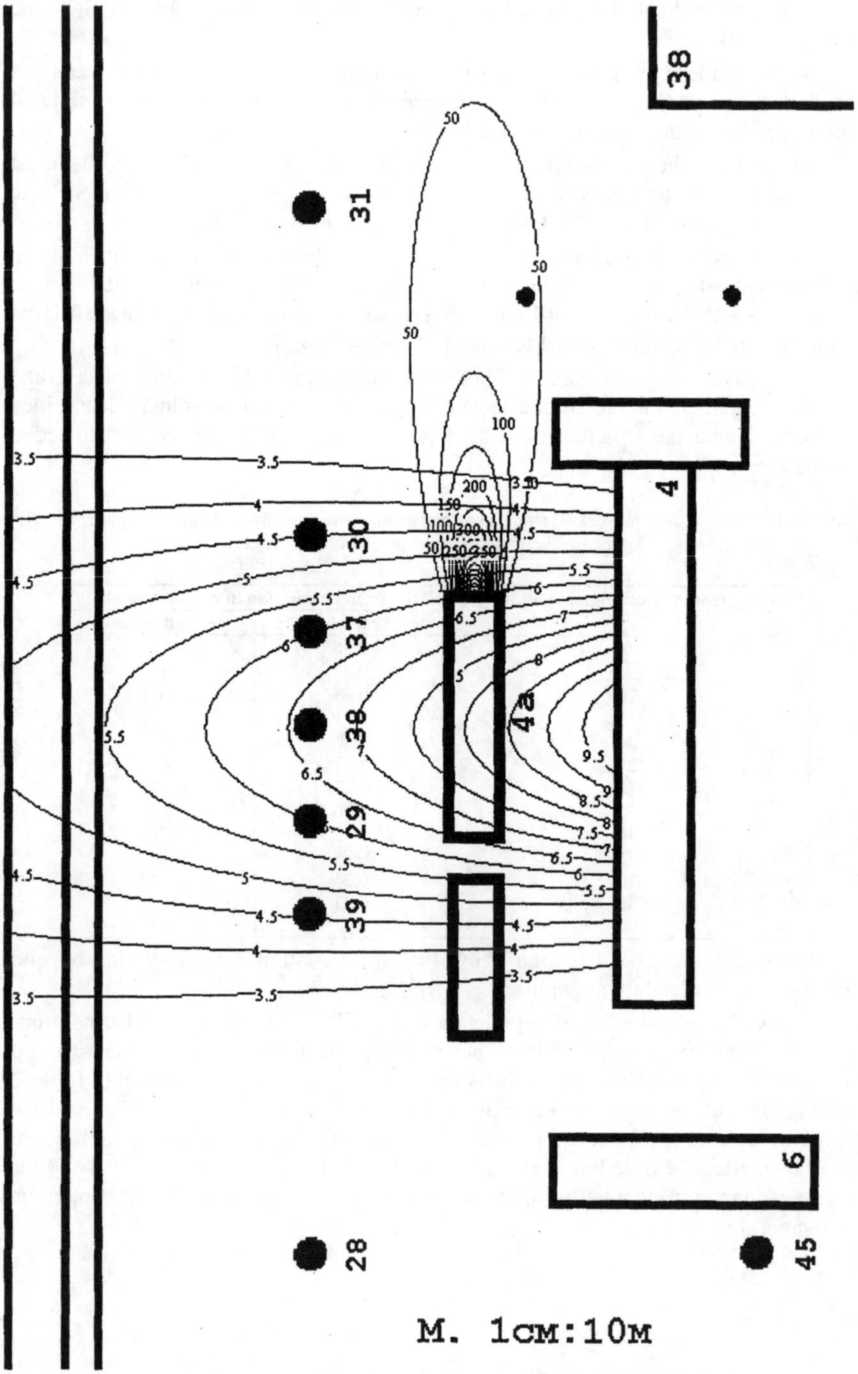

Figure 3. Distribution of effective dose rate, nSv/hr (10^{-9} Sv/hr) inside the aerodynamic shadow for south and east wind directions for the system "Reactor Vessel -Container Having Core Filled Up With Neutron-Absorbing Powder" and the accident scenario under consideration

TABLE 2. Effective dose rate along an axis of the aerodynamic shadow for two wind directions for the second scenario, nSv/hr (10^{-9} Sv/hr)

Distance from container, m	**Dose from external, radiation, nSv/hr**		**Inhalation dose, nSv/hr**		**Cumulative dose, nSv/hr**	
	East-wind	**South-wind**	**East-wind**	**South-wind**	**East-wind**	**South-wind**
0	0.125	10.3	9.9	824	10	834
5	0.118	5.7	9.4	461	9.5	467
10	0.113	3.7	8.9	300	9.1	304
15	0.107	2.7	8.5	214	8.6	217
20	0.103	2	8.2	164	8.3	166
25	0.098	1.66	7.8	132	7.9	134
30	0.094	1.38	7.5	110	7.6	111
35	0.091	1.18	7.2	94	7.3	95
40	0.087	1.04	6.9	83	7	84
45	0.084	0.93	6.7	74	6.8	75
50	0.082	0.85	6.5	67	6.6	68
55	0.079	0.78	6.3	62	6.3	63
60	0.076	0.73	6.1	58	6.1	59

Levels of radionuclides contamination of the equipment's surface and the area of the site near the accident container do not exceed reference contamination levels for personnel.

Therefore only careful inspection of the accident area and, if necessary, "sport" decontamination at the site near the accident container are required.

3. SNF storing in TUK MBK

More than 15 years ago, when it became obvious, that regional SNF storage facilities in most countries would not be built in time, experts from different countries met the problem with a dual-purpose container and a transport-packing set (TUK) construction. Such containers, including metal and concrete ones (MBK), were intended for SNF long-term storing and shipping. This problem is urgent for both SNF from nuclear power plants and irradiated fuel of research and transport reactors. The UK MBK ensures SNF storing and shipping of assemblies. The main part of the TUK MBK is a metal and concrete cask that consists of a case, grid spacer, and inner and outer lids.

The MBK case is a thick-wall cylinder with a bottom and metal sill for lids made from three metals. The space between covers is filled with reinforced concrete. For additional protection against a mechanical load the sill of MBK has a safety-damping lid and its case has a safety-damping belt. The grid spacer is a welded metal construction including a number of membranes of different thickness with holes for SNF covers, and is intended to provide regulated spacing of covers in the inner housing of the cask. The grid spacer and inner cover are made from corrosion-resistant steel. Metal constructions are made from low-carbon steel. Lids are sealed with gaskets. A self-supporting chain reaction inside the tight case was examined as a beyond design-basis accident. The reaction has the character of a short-term neutron flash with great heat liberation that may lead to the cask's seal failure with release of radioactive gases and aerosols into the environment [17].

But the possibility of the accident discussed above for cores with filler has not been taken into account. We are to note that radiation terrorism [19] is dangerous not only from the psychological but also from material point of view, and the scenario under consideration includes both nuclear and radiation effects.

Such terrorism is based on the interaction of "ordinary" explosives and nuclear materials with the occurrence of local self-sustaining chain reaction. It has most probably to do with environment terrorism [20]. As a result it will increase opposition to nuclear use and lead to decline in living standards both in Russia and other countries.

Calculations were carried out in accordance with the available data for TUK MBK and facilities at the storage site using procedure [16]. The highest level of air and land contamination could be at the shortest distance from the accident cask at the site in accordance with the calculations.

The effective dose, resulting from the combined influence of different nuclides for the chosen scenarios, is given in Table 3 and Fig. 4. This influence is caused by both inhalation and irradiation in the radioactive cloud during the accident.

TABLE 3. Equivalent dose rate (EDR) along an axis of aerodynamic shadow for two wind directions for the second scenario, mSv/hr (10^{-6} Sv/hr)

Distance from the cask, m	EDR at south wind, mSv/hr		EER at east wind, mSv/hr	
	Inhalation	From the cloud	Inhalation	From the cloud
0	2.3 10^4	360	270	4.2
5	1.29 10^4	200	250	4.0
10	8.4 10^3	132	240	3.8
15	6.0 10^3	95	230	3.6
20	4.6 10^3	72	220	3.5
25	3.7 10^3	58	210	3.3
30	3.11 0^3	48	200	3.2
35	2.6 10^3	41	200	3.1
40	2.3 10^3	36	192	3,0
45	2.0 10^3	32	186	2.9
50	1.9 10^3	29	180	2.8
55	1.75 10^3	27	174	2.7
60	1.63 10^3	25	169	2.6

It is seen from Table 3 and Fig. 4 that dose exposure from the inhalation of air contaminated with radionuclides and the radioactive cloud effects considerably exceeds dose limits permitted by NRB-99 for both personnel and population (50 mSv/year and 5 mSv/year, correspondingly).

The radioactive cloud spreading may result in a sufficient collective dose. The calculated values depending on the category of atmosphere stability are given in Table 4.

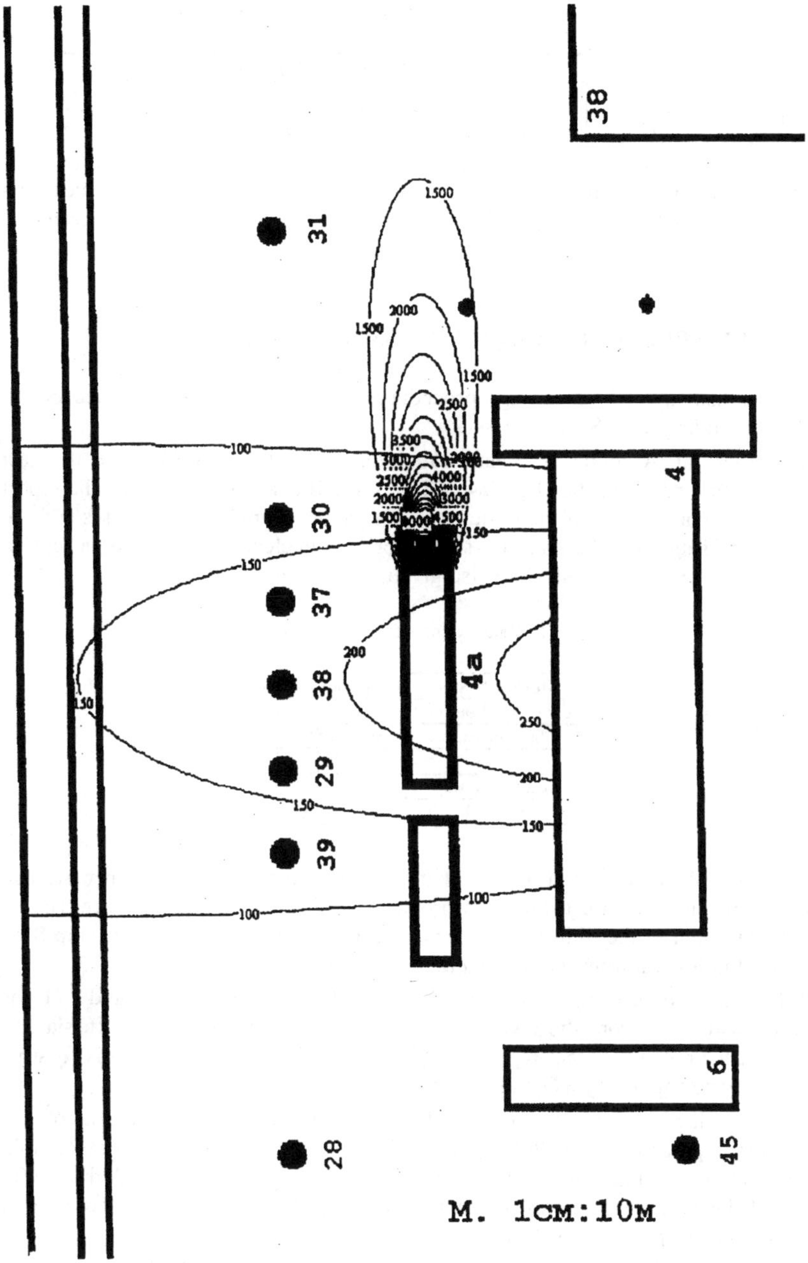

Figure 4. Distribution of equivalent dose rate, mSv/hr (lO⁻⁶Sv/hr), inside the aerodynamic shadow area for south and east wind directions for TUK-MBK accident scenario

TABLE 4. Collective doses received by population while the radioactive cloud is passing

Collective dose(D_3), man-Sv	Category of the atmosphere stability					
	«A»	«B»	«C»	«D»	«E»	«F»
	0,14	0,21	0,18	0,26	0,53	0,96

The surface radiation contamination is a result of radionuclides precipitation from the air. The estimation shows that the contamination exceeds permitted values by a factor of 10^2 or more according to NRB -99 at a distance of up to 100 m from the accident cask.

4. TUK MBK for Ship's SNF. Who is interested in it?

The assessment of SNF storage procedures has frequently been carried out in countries including the USA and Russia.

An expert at Argonne National Laboratory (the USA) [21] gave the last information in a report at the "Russia-NATO" scientific seminar in 2002. The report dealt with the comparison of the basic SNF storage and transport procedures and the relative cost (Table 5). The main causes for private nuclear power plants accepting the use of SNF protective cask were also given in the report.

TABLE 5. Relative cost of the storage options

Storage System	Relative Costs
Cask - dual purpose	1.0
Storage Cask	0.61-0.86
Vault	0.62-0.75
Caisson (silo)	0.5 - 0.6
Drywell	0.29 - 0.6

Calculations of Tables 3 and 5 show that the most expensive SNF interim storage procedure but not the safest indeed is proposed for Russia. It will be necessary to solve the problem of the Ship's emergency SNF management. Why do we ship SNF for disposal in Siberia from the North and the Far East?

At last, the volume of Ship's SNF is insignificant in both ^{235}U and ^{238}U as compared with SNF from the civil nuclear power plants. Experts from Russia and other countries noted that fact repeatedly [22]. But if necessary [7, 13] this feasible material may be applied by using known procedures.

The USA has refused to use the closed nuclear fuel cycle because of the non-economic procedure of SNF treatment, surplus of high-enriched uranium (also in Russia) and the volume of the high-level radioactive waste exceeded the initial volume of the Ship's assemblies in 5 and more [23]. But Russia is now not so rich that it can treat Ship's SNF without respect of expenditures, is it?

Reasons for SNF storage and shipping choices by owners of NPPs in the USA are of interest [21]:

- in the first place, they have already paid SNF shipping to the repository or SNF interim storage until the repository is not in operation; additional expenditures will be returned to them;
- in the second place, storage of SNF from private NPPs at the sites is not allowed because it will lead to civil actions;

- in the third place the US Nuclear Regulatory Commission gave a license for some systems of SNF storage; it makes their use especially expedient for NPPs, whose assemblies have already been held in the storage ponds;
- in the forth place, it is essential that new systems of SNF storage and transportation have been patented, the owners have the unique possibility to profit by sale of the casks or licenses for casks; just as one can not patent the known developments including dry storage in wells.

At present TUK MBK for SNF of LWGR are developed in Russia. The next stage is the development of the metal concrete containers for NPPs with VVER reactors, is not it?

Very likely the introduction of systems for SNF storing and shipping in TUK MBK casks within the RF boundaries will have Minatom of Russia buy licenses (in the USA possibly) for casks for SNF shipping from other countries.

There is another situation with SNF from US public-owned nuclear facilities stored in the dry wells [21] with minimum expenditures. They used this procedure to save money and they were responsible for the decisions.

The procedure of SNF storage in "dry" tanks is used in 10 countries having developed nuclear programs and/or industry. The IAEA has issued some standard documents concerning such kind of storage facilities. The process of licensing these facilities is not difficult in comparison with analogous procedures for SNF protective containers storing and shipping. The long-term storage of damaged assemblies "at site" is more promising than their transportation for thousands of kilometers by land.

It is important that the proposed safer processes of SNF (damaged ones included) storage and transportation allow putting more assemblies into a container and allow a safer process.

What process do we choose? May be our country is so rich that we do not count money for Ship's and damaged SNF storage and transportation as the Americans do.

It is necessary to note that Gosatomnadzor did not certify developed and manufactured metal concrete containers for Ship's SNF storing and shipping.

Based on the given arguments the authors consider that it will be expedient to estimate approaches to Russian Ship's SNF (damaged especially) storage and transportation from the position of our State being the owner of the SNF.

This is noted in some articles concerning Nuclear Powered Submarine decommissioning *(e.g. [6])*.

It is possible that some MBK TUK being manufactured will be shipped to emergency on-shore storage facilities, e.g. Andreev Guba, where the situation is critical [24].

Other container will be shipped to on-shore storage facilities for unloading SNF from Nuclear Powered Submarines and Ships, where the availability of containers became "the weak point" of SNF management.

Currently there are 200 cores in the reactors of NP Submarines located in the holding areas. The number of assemblies at the on-shore facilities and in Tenders is 100 with cores taken into account [25].

It is necessary to have 1500 containers (Navy MBK TUK) to store SNF, moreover some of them are to be up-dated for both damaged SNF and SNF not to be treated. The cost of one container manufactured in Russia is approximately 4,3-mm rubles [26]. The cost of required number of containers without modification and other additional expenditures is approximately 6.5 billion rubles (210 million USD). Can we afford this? Is it necessary for us to store Ship's SNF in dual-purpose containers with limited capacity for a long time?

To solve the problem it was proposed to test and certify advanced Russian systems "Reactor-vessel with core filled with solidified material or filler - Container" for SNF long-term storage and damaged SNF storage until repository construction. The number of such containers will not exceed 200 items. It will handle SNF long-term storage and damaged SNF storage until repository construction. It will make possible reducing the number and/or size of the sites for both types of containers by 2. By locating Navy TUK at the on-shore facilities it will be possible to reduce the size of sites for containers with cores at the Ship Yards in 5-7 and additionally to save expenditures for Ship's SNF storage and shipping.

5. Conclusions

The assessment of the advanced SNF storage and shipping procedures and calculations of the probable radiation accident effects by storing cores of nuclear Ship's LWR reactors in vessels filled with solidified metal or neutron-absorbing powder, as well as in TUK MBK without additional filler, showed:

1. The possibilities of a number critical accident or the theft of assemblies for the purpose of misappropriation of highly enriched uranium are the most dangerous.
2. A procedure has been developed for storing naval SNF in a vessel with a head closure that may be used for LWR reactors of Nuclear Powered Ships being decommissioned. It allows excluding stages of separate assemblies removed from cores.
3. Core filling with melted heavy metal is especially prospective for NPS and Surface Ships being stored in holding areas with cores aboard for long period of time.
4. The same result is achieved when filling cores with polymers for subsequent solidification; it is more interesting for dried damaged nuclear fuel unloaded from the reactors.
5. Core filling with a dispersed neutron moderating and absorbing material is especially advanced for storing cores with damaged assembles, which are extremely hazardous for the personnel, population and the environment.
6. Naval SNF storage using the "Vessel Filled with Solidified Metal or Neutron Moderate and Absorbing Powder - Container" system has some advantages:
   - reduction of neutron radiation level from SNF;
   - non-occurrence of nuclear criticality in long-term SNF storage even in a non-drained core;
   - reduction of γ-emission level from the SNF;
   - facilitation of heat removal from SNF; this allows using the same procedure, e.g., for filling up damaged or decommissioned cores from research LWRs;
   - exclusion of the necessity for dangerous and expensive transfer operations for assemblies, especially damaged ones;

- facilitation of core servicing and control by SNF long-term storage;
- facilitation of removal of SNF assemblies for shipping to reprocessing.

7. The effects of potential radiation when storing LWR core vessels filled with solidified metal or neutron-absorbing filler, are considerably lower than when SNF is stored in TUK MBK. First, this is due to the exclusion of the possibility of a criticality accident in the SNF.
8. Physical protection is improved in this system ("Vessel with Filled Core - Container") due to the impossibility of high-enriched uranium removal from the core without authorized permit.
9. For standard operations of TUK MBK and the design basis accident, when storing or shipping, contamination doesn't exceed the permitted levels of NRB-99 both for personnel and population.
10. For design basis accidents, e.g. the local criticality accident considered in the report, possible contamination from the TUK MBK when storing or shipping exceeds the permitted levels of NRB-99 by a factor of 10^2 or more at a local spot up to 100 m from the accident cask.
11. 20% of the total number of assemblies are to be put into TUK MBR, i.e. it will be necessary to have 5 times more containers than for "Vessel—Container" system."
12. It will take considerable expenditures for manufacturing dual-purpose metal-concrete containers (6.5 billion rubles at the beginning of 2002) without an obvious return in the future.
13. The use of TUK MBK demands large size storage sites, from 2 to 7 times, larger, along with proper costs increasing for guards.
14. Dual-purpose metal-concrete containers are the most expensive system of SNF storing and shipping, and have no potential for military and commercial nuclear facilities that belong to the State.

6. Acknowledgements

The authors would like to thank their colleagues in the participating organizations: Professor Povl Olgaard (Riso National Laboratory, Denmark) for long-term fruitful cooperation concerning Nuclear Powered Ships safety, Mr. O. Keener Earle, P.E. (Argonne National Laboratory, the USA) for the insightful material on deeply the problem of SNF storage and transportation.

7. References

1. Pechenin, N.K. (2000) Risk Management for a Nuclear Plant, *Report for the Interdepartmental Coordination Council on Risk Assessment/or the Nuclear Complex of Minatom of Russia Meeting,* December 22, 22 p. (In Russian).
2. Pechenin, N. (2000) The Concept of Controlled Risk on Historical Examples, *Alternative energy and ecology.* No 1, 116-129 (In Russian).
3. Mitenkov, P.M., Aksionov Ye.I., Vavilkin, V.N., and Sandier, N.G. (1997) Priority Problems of Nuclear Submarines Decommissioning and Utilization, Rus. Atomic energy, 82, No 3, 146-149 (In Russian).

4. Aksionov, E.I., Vavilkin, V.N., Lebedev, V.M., et al. (1999) Enhancement of Safety at Storing Nuclear Power Plants of Warships Removed from Operation Pending their Decommissioning, in A.A. Sarkisov and A. Toutnyol du Clos (eds). *Analysis of Risks Associated with Nuclear Submarines Decommissioning, Dismantling and Disposal,* Advanced Research Workshop, Moscow, Russia, 1997, Kluwer Academic Publishers, Dordrecht, pp. 33-40.

5. Site for Interim Storage of Casks with Naval SNF, *VNIPIET,* Issue No 00-00961, 2000 (In Russian).

6. Barskov, M.K., Birillo N.N., Glushkov, O.A, and Smirnov, P.L. (2002), NFS Out-of Service Nuclear Submarines. Problems of Safety - Status and Forecast *Nuclear Submarine Decommissioning Herald,* No. 1, 38-45 (In Russian).

7. Patent of the Russian Federation, RU 2132098, Int. Cl. 6 G 21 F 9/34, 9/28, 0.04.98, Publ. 27.04.99, *FIPS Bull.,* No 12 (In Russian).

8. Gavrilov, S.D., Kremnev V.A., Smirnov, P.L., et al. (1999) Substantiation of Nuclear Safety Ensuring for Storing LWR Cores, Including Damaged Ones, in NPP of Russian Out of Service Nuclear Powered Submarines, Surface Ships and Vessels, Proc. 7th JSME/SFEN/ASME Int. Conf. On Nuclear Engineering (ICONE-7), Tokyo, Japan, 1999, JSME, ICONE-7080.

9. Patent of the Russian Federation, RU 2129739, Int. Cl. 6 G 21 C 19/00, 19/32, 04.16.97, Publ. 06.20.99, *FIPS Bull.,* No 17 (In Russian).

10. Patent of the Russian Federation, RU 2141138, Int. Cl. 6 G 21 F 5/00, 5/14, 03.24.98, Publ. 11.10.99, *FIPS Bull.,* 31 (In Russian).

11. Methods for Disposal of Nuclear Waste (2002), *Nuct. Engng. Int.,* 47 No 571, 16-18.

12. Mescheriakov, Yu., Matveev, S., and Sorokin, N. (2002) Basic Lines to Minimize Radioactive Waste Amount when Utilizing Nuclear Submarines, Present Workshop.

13. Application for RF Patent No. 99108552/06 of 26.04.99, FIPS *Bull.* (to be published).

14. Barskov, M.K., Birillo, N.N., Gavrilov, S.D., et al. (2000) Ensuring of Long-term Safe Storing of LWR Cores, Including Accident Ones, from Out-of-service Nuclear Submarines of the Pacific Fleet of Russia. Int. Conf. *The Concept and Problems of Nuclear and Environmental Safety of NFS Complex Decommissioning in the Far East of the Russian Federation,* Moscow (In Russian).

15. Kugeler, K.. Deutsches Patentamt, Offenlegungsschrift DE 3812174 Al, Int. Cl. G 21 C 9/02, C 1/22, 16.07.88, Offenlegungstag 18.01.90.

16. Estimation Procedure for Economic Damage from Radioactive Material Used in the National Economy Due to Radiation Accident, *RECcentre,* Issue No P-03/98, St.-Petersburg, 1998 (In Russian).

17. Estimation of the Radiological Effects and Economical Damage from SNF Interim Storage in TUK MBK at the Site of RTP "Atomflot", *RECcentre,* Issue No P-22/01. St.-Petersburg, 2001 (In Russian).

18. Bakin, R., Mitenkova, E., Novilov, N., and Shikm, A. (2002) Estimating the Possibilities of Initiating Spontaneous Chain Reaction in the Navy Transportation Casks, This Seminar.

19. Koldobsky, A. (2002) Radiation Terrorism between Physics and Politics, *Bull. for Atomic Energy,* No. 4, 63-67 (In Russian).

20. Kaurov, G. (2002) Chemobyl Catastrophe and Ecological Terrorism, Ibid., 32-34 (In Russian).

21. Earle, O.K. (2002) Options for the Handling and Storage of Nuclear Vessel Spent Fuel, This Seminar.

22. Barskov M.K., Gavrilov S.D., Smirnov, P.L. et al., (1999) Nuclear Fuel Cycle for Naval Reactors of Russia, Proc. Int. Conf. *Nuclear Fuel for Mankind.* 1998, M., Nuclear Society of Russia, 1999, 585-608 (In Russian).

23. Burritt, J.G. (1999) Risks Posed by Out of Service Nuclear Submarines and Proposed Methods to Minimize the Risks, in A. A. Sarkisov and A. Toutnyol du Clos (eds). *Analysis of Risks Associated with Nuclear Submarines Decommissioning, Dismantling and Disposal,* Advanced Research Workshop, Moscow, Russia, 1997, Kluwer Academic Publishers, Dordrecht, pp. 223-235.

24. Vassiliev, A., Shishkin, V., Mazokin, V., et al., (2002) Issues of Radiological Rehabilitation of Coastal Repositories Storing SNF and RW, This Seminar.

25. Sarkisov, A. A. (2002) Results of the Investigation Carried out of the International Cooperation Research Program Concerning the Radioactive Waste (ICRP RW), *Nuclear Submarine Decommissioning Herald,* No 1, 12-19 (In Russian).

26. Larin, V. (2002) Environmental Problems of the Russian NPS Dismantling or Nuclear Heritage Elimination, *Bull. for Atomic Energy*, No.1, 63-68 (In Russian).

OPTIONS FOR THE HANDLING AND STORAGE OF NUCLEAR VESSEL SPENT FUEL

O. KEENER EARLE, P.E.
Staff Nuclear Engineer
Argonne National Laboratory

Abstract:

There are many options for the handling and storage of spent nuclear fuel from naval vessels. This paper summarizes the options that are available and explores the issues that are involved. In many cases choices have been made, not on the basis of which is the best engineering solution or the most cost-effective, but based on the political realities involved. For example, currently it seems that the most prevalent solution for spent fuel interim storage is in dual-purpose (transport-storage) casks. These casks are robust and, politically, they offer the visible evidence that the fuel is 'road-ready' to be moved from the local area where the fuel is being stored in the interim. However, dual-purpose casks are the most expensive of the storage mediums. Drywell storage (storage in below grade or bermed pipes), on the other hand, the least expensive and most flexible storage option, suffers from an image of permanence (not politically acceptable) and from being improperly implemented in the past. Though these issues are easily resolved from a technical perspective, the option is often not seriously considered because of this past history.

It wasn't too many years ago that spent fuel pools were the storage medium of choice. The pools were never intended for long term storage. As the ultimate disposal path for spent nuclear fuel (processing, repository) became bogged down, however, fuel remained stored in the pools for much longer than intended. Strategies (re-racking, consolidation) were employed to lengthen the storage life of the pools. In some cases, inadequate attention was paid to the wet storage and significant fuel degradation occurred. Pools were then unloaded into dual-purpose or storage only casks as required.

It seems that decisions on spent fuel historically have been short sighted. It is time that the spent fuel situation needs to be evaluated for the long term from a systems perspective. Criteria for the evaluation must consider technical acceptability, safety, flexibility (especially in storage times, fuel condition, and fuel types), active monitoring, costs, security, and, of course, political realities. It is the sense of this author that the political issues may be resolved if a reasoned complete approach is demonstrated.

1. Background

For many years the storage of spent nuclear fuel seemed to be an afterthought. Fuel was typically stored in storage pools that were not designed for long-term storage. In many instances, earlier pools did not maintain adequate water chemistry control and fuel degradation occurred. In other instances fuel was stored in below grade drywells or vaults. These storage mediums also often were inadequately designed and monitored such that fuel degradation occurred. When the time lengthened for the ultimate disposal of fuel

A.A. Sarkisov and L.G. LeSage (eds.),
Remaining Issues in the Decommissioning of Nuclear Powered Vessels, 285–295.

(either through delays in reprocessing or decisions on ultimate disposal in a repository), the reality of the magnitude of high level radioactive waste and spent nuclear fuel issue was highlighted. This realization also seemed to occur at the same time as the 'environmental awakening' --- bringing public scrutiny to the issue as well. Conditions became such that utilities (and later governments) needed to show progress on resolving their spent fuel issues. As a minimum from a technical perspective, the fuel had to be moved to a dry environment. It would also be better if the interim spent fuel storage option chosen could be perceived to move the fuel towards it ultimate disposal path.

1.1 OPTIONS FOR SPENT FUEL HANDLING AND STORAGE

The various types of spent fuel storage that one should consider are summarized below:

1. Pool storage – this type of storage places spent nuclear fuel (SNF) underwater in storage racks. The water provides cooling and shielding, plus allows good visibility for verification. Monitoring the chemistry of the water also provides a good indicator for problems with the fuel. This method is what typically has been used for interim storage at commercial nuclear power plants and the storage of nuclear vessel fuel waiting for processing. Water-cooling is necessary to

Pool Storage in France

allow for the short-term removal of decay heat from the SNF after it is removed from a reactor but pool storage was never intended for long term storage (>10 years) of SNF due to corrosion in the water environment. In fact, corrosion and degradation of SNF in water storage pools has been a major issue in the U.S. (and elsewhere), particularly for non-commercial fuels. Consequently, since reprocessing fuel is no longer permitted in the some places or is uneconomic, a final disposal repository is not yet ready, long term storage of SNF is the only option. Consequently, most commercial entities and governments are moving their SNF to dry storage.

2. Dry storage in casks – this type of storage places SNF in casks designed to serve storage and/or transport requirements. The casks include shielding and passive cooling for the SNF. If designed for transport, exceptional strength and durability to withstand transportation accidents are included in the design. The casks may also include a neutron poison to allow for storage of additional SNF elements. This type of storage has been the predominant storage method chosen by the commercial nuclear power industry in the U.S. Cask vendors have already paid for the initial licensing and incorporated those costs in their cask price. In addition, the USNRC has made it easy to license cask storage as long as the storage is at a licensed nuclear power installation.

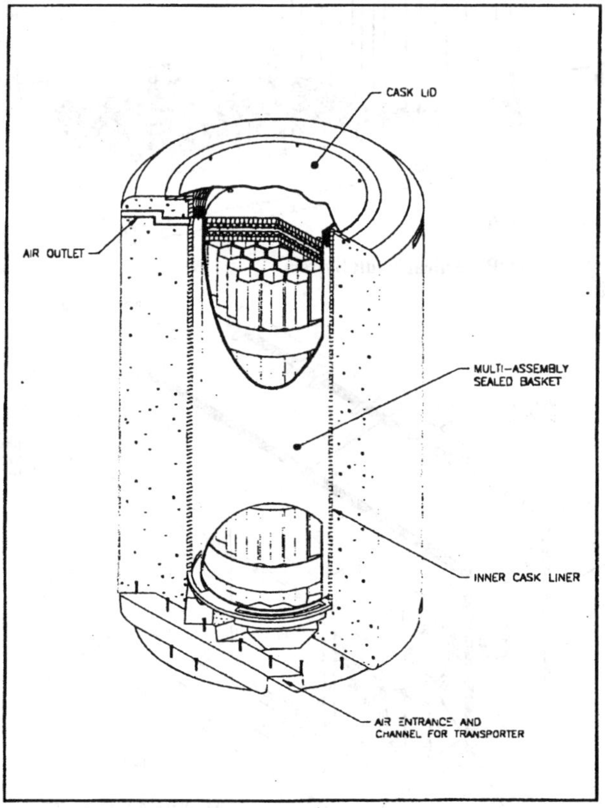

Sirra Pacific Vertical Concrete Cask

3. Dry storage in vaults – this type of storage places SNF in shielded vaults or engineered structures. Typically, individual fuel elements are placed in individual storage pipes. Shielding is typically provided around the entire vault and the movement of fuel elements is done remotely. Passive cooling or active cooling can be provided as part of the design as required. This type of storage was developed in the U.S. particularly to deal with SNF that could not withstand a direct water environment (e.g., graphite fuels). A good example of this type of storage in the U.S. is the Independent Spent Fuel Storage Installation (ISFSI) at the Fort St. Vrain power plant near Ft. Collins, Colorado. Vault storage was also used at DOE's Idaho National Engineering and Environmental Laboratory (INEEL) in Idaho. Internationally a good example is the MACSTOR system in Canada.

It should be noted that the NUHOMS design for horizontal vault storage is a hybrid combination of cask and vault storage.

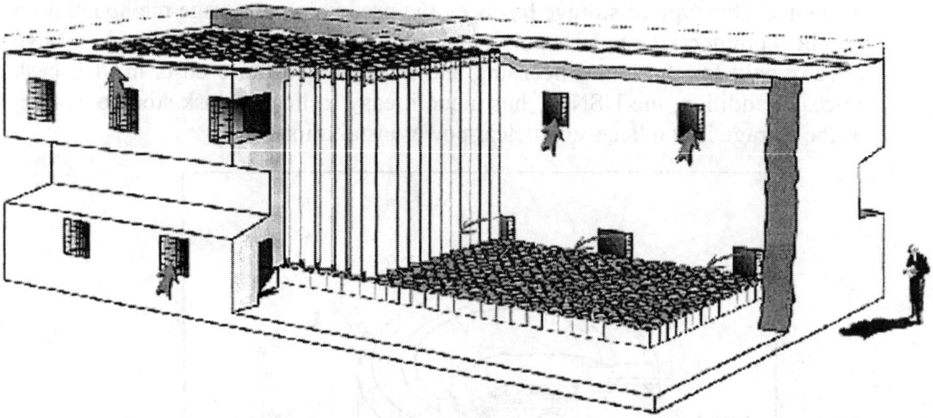

MACSTOR Vertical Vault System

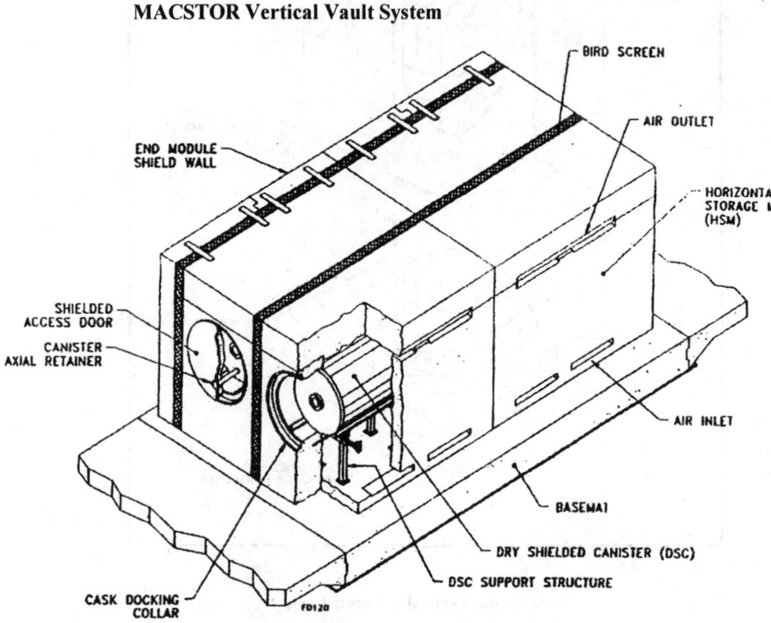

NUHOMS Storage System

4. Dry storage in drywells – this type of storage places SNF in individual canisters which are then placed in pipes ('drywells') that are buried in the ground or in a berm. The ground provides the heat sink as well as the shielding for the SNF. Typically, the drywell pipes have been made of carbon steel and cathodically protected to minimize corrosion. Alternatively, future designs may be able to make use of advanced coating technology to save money. Similar to storage in vaults, this type of storage was originally designed for SNF that could not be directly stored in water. This type of storage has been used successfully in the U.S. at DOE installations such as Argonne National Laboratory-West, Oak Ridge National Laboratory, and the Idaho National Engineering Laboratory. Internationally it has been used successfully at Tokai in Japan and the Lucas Heights Storage Facility in Australia.

Drywell storage at Argonne National Laboratory

Table 1. gives a relative cost of these storage options[1]:

Storage System	Relative Costs
Cask – dual purpose	1
Storage Cask	.61-.86
Vault	.62-.75
Caisson (silo)	.5 - .6
Drywell	.29 - .6

2. Historical Perspectives

Through the early 1980s, the spent fuel industry was fairly balanced in its approach to storage options. In the latter 1980s and on, however, this balance shifted towards storage in casks, particularly in the U.S.:

U.S. Commercial perspective - The U.S. commercial power industry has typically used above ground casks and vaults for dry storage as their spent fuel storage pools have filled up. It is important to understand the driving forces behind these decisions. In a regulated utility environment, costs of spent fuel storage are recovered through regulated rates. Predictability of those costs and stability of the time schedules have a stronger influence on the ability of the utility to recover these costs through the regulatory process than does the magnitude of the cost. The aboveground casks and vaults have a clear advantage because these storage designs have already been issued a Certificate of Compliance (CoC), i.e. they are a tested and proven technology, and local public hearings (with their potential for delays) are not required. Since the above ground storage design has the least interaction with the site, it is less susceptible to court challenges that local site-specific considerations have not been addressed. Storage in dual-purpose (transport-storage) casks also has the advantage of being perceived as 'road-ready' to be moved to the final repository or reprocessing site. This is a naturally positive attribute to the local public.Cask designs also provide a unique economic driver for potential manufacturers or vendors of casks. The motivation for a cask vendor to obtain a NRC license is directly related to the patentability and corporate profit potential of the storage system. For example, a company will develop a dry cask design, patent it, and license it when a market niche is identified for the product. That company is willing to incur the substantial cost of the initial NRC cask licensing process because it holds the patent to ensure that it alone can sell the patented design after it is licensed. In contrast, a simple drywell system

[1] Based on engineering judgement from 'Cost Comparisons for On-Site Spent-Fuel storage Options', EPRI NP-3380 Project 2062-8, May 1984, and "Storage Technology Options Study for BN-350 Spent Fuel Storage, November 28 2000.

that has been described in the scientific literature and used in the AEC/DOE complex for approximately 40 years qualifies as a public domain technology that is not legally patentable. In this scenario there is no profit incentive for a commercial company to pursue the licensing of the drywell storage system, because after they license it, anyone can sell or use the technology.

- U.S. DOE perspective - Many DOE sites have spent fuel storage that was built under DOE regulations. Historically, these regulations have allowed much more flexibility in choice of technologies and options. Suffice it to say, just about every type of spent fuel installation has been built in the DOE complex, including six drywell storage facilities. In the last 10 years, DOE has come under increased scrutiny of its nuclear installations – culminating in detailed reviews and upgrades of many of its installations. Currently and in the future, public scrutiny of new installations will rival that of commercial installations under NRC control. Consequently, DOE is dealing with this process by 'privatizing' the design and construction of nuclear facilities, and requiring the contractor to obtain a NRC license as part of the process. Insofar as future spent fuel storage is concerned, this means contractors bidding on DOE contracts for Independent Spent Fuel Storage Installations (ISFSIs) will also tend to opt for above ground casks or vaults that have CoCs. This makes their costs much more predictable. Though the costs may be relatively high, they are readily defendable and 'pass-through' easily for DOE payment under the contract.

  Another aspect of DOE involvement with spent fuel in the U.S. is its mandated obligation to receive and store commercial spent nuclear fuel. The final repository for such fuel is currently a geologic repository being built at Yucca Mountain, Nevada. Concomitant with this obligation is for DOE to receive commercial spent fuel in the interim and store it in a Monitored Retrievable Storage (MRS) Facility.
- Other Countries' Perspective - Dry storage of spent nuclear reactor fuel began in the United Kingdom in 1972 and in Canada in 1975. At least ten countries have experience with dry storage of spent power reactor fuel. The International Atomic Energy Agency (IAEA) has monitored experience with dry storage in these countries since 1980 and it issues periodic reports on the subject. The most current IAEA guidance for safety criteria associated with spent fuel storage is contained in:
  - "Fuel Handling and Storage Systems in Nuclear Power Plants," Safety Standard 50-SG-D10 Rev. 2 (1997),
  - "Design of Spent Fuel Storage Facilities," Safety Series No. 116 (1994), and
  - "Preparation of the Safety Analysis Report for Spent Fuel Storage Facilities: A Safety Practice," Safety Series No. 118 (1994).

  This guidance is compatible with current NRC licensing procedures although the IAEA does not explicitly recognize the "general license" approach. The guidance is also fairly broad, allowing the full array of storage options, including casks, vaults, and drywells.

  It should also be noted that the spent fuel storage licensing history in other countries has not skewed itself towards 'pre-licensed' casks or vaults as has occurred in the U.S. since their licensing history is different.
- Former Soviet Union Countries' perspective – As in the U.S., the storage of spent fuel in the countries of the former Soviet Union followed a path using virtually all the various storage options. Similar to the U.S., problems with wet storage have led to degraded and damaged fuel in many circumstances. In addition, the final

disposition of spent fuel (typically reprocessing at Mayak) has reached a choke point due to lack of transport, costs, and the lack of a specific need for the reprocessed product. Because of this, it now appears the general storage for naval spent fuel is to have regional storage locations where pads can be constructed for interim storage of spent fuel in either storage or dual use casks. In-country capability to produce casks has long existed and current innovative designs have been developed spurred on by U.S. financial aid (AMEC/CTR). In this regard, the U.S. DOE/DOD preference has been for SNF storage systems that have been previously licensable or are easily licensable in the U.S. As a consequence, the U.S. preference for storage in casks seems to be carrying over to Russia.

3. Technical Issues

- Long Term storage of SNF Greater than 10 years - Most SNF storage concepts developed around the world did not consider the need for long-term storage of SNF greater than 10 years. This is particularly true for pool type storage. The basic premise had been that the pools were for storage for a very limited amount of time until the fuel could be transported for final disposition (either reprocessing or long-term storage at another location). Consequently, when these intended scenarios did not occur, unplanned for consequences occurred - corrosion set in with the fuel, causing it to degrade and release radioactivity to the fuel pools; and fuel pools leaked, contaminating groundwater. Other types of storage including rudimentary drywell and vault storage also experienced problems such as moisture in-leakage and fuel corrosion. The fact was that the storage simply had not been designed considering long-term storage and what could possibly happen to the fuel in the meantime.

As a consequence, in the last 10 years, there has been a large dollar investment in order to:

- characterize SNF that has been in storage to better understand the degradation mechanisms taking place,
- properly condition fuel for dry storage for the long term based on that knowledge gained,
- remove fuel from water pools to dry storage for the long term, and
- upgrade existing dry storage facilities to incorporate features to minimize degradation for the long term.

Basic design requirements for long-term storage developed because of the above include:

1. assuring the SNF is properly dried to inhibit corrosion. For fuel coming out of a water pool, this generally means some type of vacuum drying technique.
2. assuring the SNF is stored in a dry environment – this means that the storage atmosphere is assured to be initially dry and maintained that way through application of leak-tight barriers. If the fuel was degraded or failed, application of an inert cover gas and double canning is necessary.

3. assuring the capability to periodically monitor the performance of the fuel in dry storage by sampling, inspection, gas space monitoring, or by other indirect methods.
4. repackaging degraded or damaged fuel in additional leak tight inerted canisters

- Local vs regional vs remote storage – This is a significant issue. As soon as one transports spent fuel beyond the confines of where it is initially offloaded (such as away from the plant or shipyard) into a public access area, one is usually required to meet very strict transport standards. This requires the use of heavy transport casks designed with impact limiters, etc. which is obviously expensive. If one has the transport casks available and can ship to the final repository area, it is an advantage. If one can stay local and avoid this cost, it is an advantage. If one opts for regional storage, one is usually opting for a cost trade-off between the cost savings of a centralized facility (with the increased transport costs) and that of several smaller local ones with the lesser transport costs.
- flexibility – fuel type variability - life extension – Due to the unknowns in the ultimate disposal path of spent fuel (schedules for reprocessing or a repository), storage designs should be made as flexible as possible. For handling of naval vessel spent fuel, this flexibility is extremely important. For example, if a particular storage design can handle various types of spent fuel merely by changing inserts in the casks, vaults, or drywells, significant savings could be accrued. Likewise, ease of retrieval, robustness of the design (for calculation of any life extension) would also be important.
- ease of fabrication/building – some storage designs are amenable to local fabrication and manufacture. For example, many vault and drywell storage designs can be fabricated using the facilities that would typically be available in a shipyard or large industrial facility. On the other hand, storage or dual purpose casks require very sophisticated manufacturing process that must be tailored to the specific product in order to maintain the desired quality. In addition, since many of the cask designs are proprietary, the details of the design are often not available to the end user.
- licensability – A large portion of the cost of dual purpose casks resides in the licensing certification process. This cost is then passed on to the end user who benefits from knowing that he will not have any licensing or regulatory issues if he uses that design. In many instances, this foreknowledge is worth the extra cost. It should be emphasized, however, that there is no technical reason why the in-place storage options (vaults and drywells) are not easily licensable. It is necessary, however, to be proactive with public involvement to ease the process. Being able to construct the facilities with local labor should be a significant advantage in the communication process.
- Security/MPC&A – With increasing concern about terrorism, security of storage facilities is extremely important. All designs should be amenable to active monitoring and have appropriate layers of security to prevent their compromise.
- Political realities – No one wants nuclear fuel stored in their backyard. However, the most satisfactory solution (shipping the fuel to ultimate disposal or reprocessing) is usually not viable for many years into the future. In the U.S., having fuel stored in dual storage – transport casks has seemed to be an acceptable solution at power

plants since they appear 'road-ready' to leave the state as soon as possible. Though this solution is expensive, utilities do this because they are likely to get reimbursed for their work form ratepayers. Where reimbursement is not practical due to the lack of money, harder political work must be done. Full engagement with the public on the nature of the problems, the costs, and trade-offs involved is necessary for success.

Conclusion

In the last decade, the world seems to have focused on the use of storage or dual use casks for the interim storage of spent nuclear fuel. This is not the most cost-effective approach but seems to have been done mostly for political reasons or because the entity desiring to store the fuel can obtain the money. It is well known from an economic perspective that it is much more cost effective if one can postpone a significant capital expenditure until one actually needs it. A good systems analysis of the spent fuel issue for a region is recommended employing the most cost effective storage technologies (using the criteria discussed above). This should include regional or local storage as much as feasible (to minimize the high costs of 'public' transport and minimizing the number of transport casks required). Proactive public outreach is necessary to be successful under any circumstances. In the end, one would hope to have a flexible spent fuel process sheet as is depicted below.

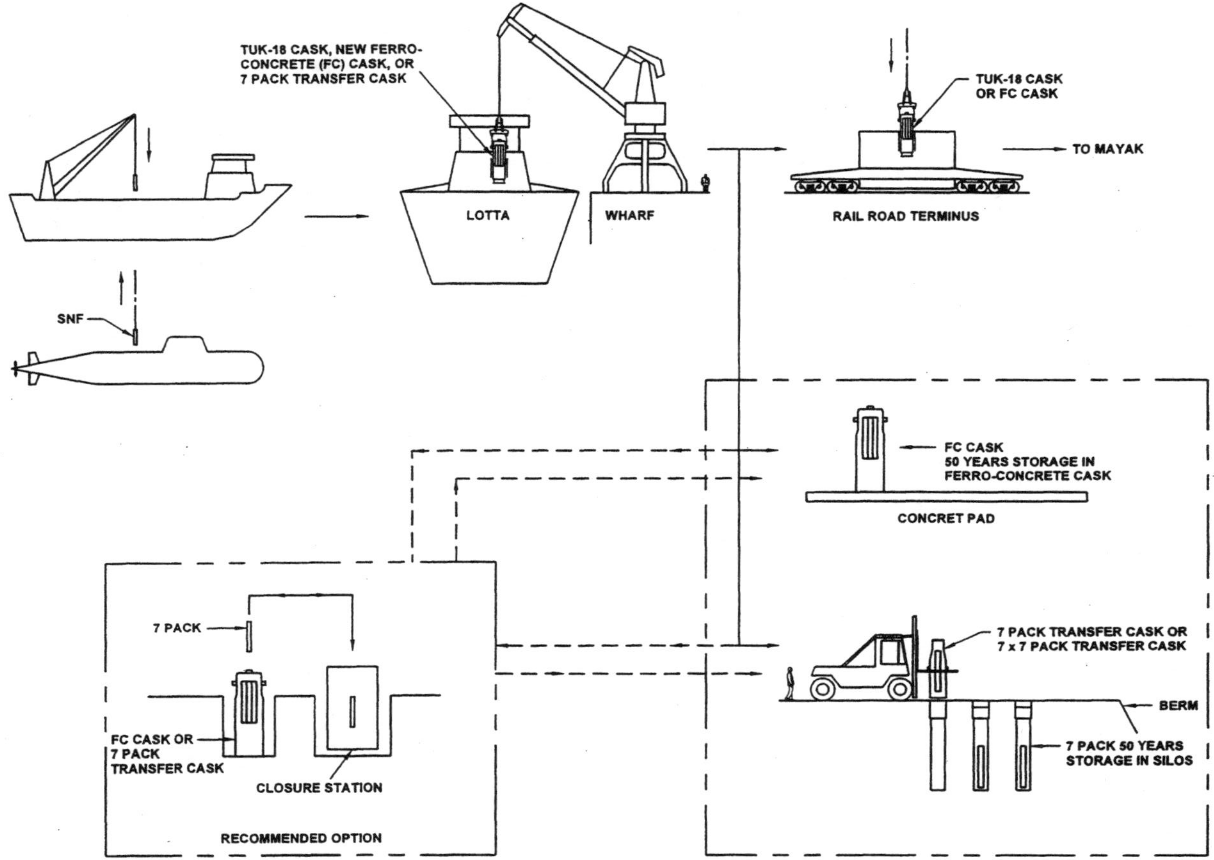

Possible Flexible Process Flow for Interim Storage and Treatment of Russian Naval Spent Fuel

DISPOSAL OF SPENT NUCLEAR FUEL AND RADIOACTIVE WASTE IN THE DECOMMISSIONING OF FRENCH NUCLEAR SUBMARINES

JACQUES CHENAIS
Manager of the Nuclear Propulsion Division

Introduction

The overall strategy for the decommissioning of French nuclear submarines can be brought down to 3 phases:

1. Level 1 dismantling which essentially consists in unloading the spent fuel and storing it in a pool.
2. Level 2 dismantling consists in isolating the nuclear boiler compartment from the rest of the submarine and conditioning it for interim storage on the construction site in Cherbourg, and full dismantling (level 3) for the rest of the submarine (excluding the nuclear boiler compartment)
3. Level 3 dismantling of the nuclear boiler compartment after a storage period

The SSBN Le Redoutable, first French nuclear submarine which was removed from active service in 1991, underwent the first two phases. Three other SSBNs removed from active service have undergone level-1 dismantling and are waiting for phase 2.

What kind of waste is produced and in what quantities? What means are used to condition and treat this waste?

I propose giving some answers to these questions, by discussing firstly the spent fuels (the only high-activity waste), and secondly the solid and liquid waste of low and medium activity.

1. Spent fuels (high activity waste)

The dismantling of submarines does not pose any specific problems as far as the evacuation of high activity waste - that is to say the irradiated fuel elements - is concerned.

In fact, these complete unloading operations may have been performed several times during the service life of the submarine. Consequently, the operations are well known and the corresponding tools and infrastructures exist. (see figures)

The following principles are applied:

- The fuel elements are unloaded under water and placed in a container above the reactor vessel.
- The fuel elements are transported in this container, which can accommodate two elements and is equipped with systems for evacuating the residual power.
  A 3rd barrier is created around the container, in the form of either containment around the container or a transfer tunnel between the reactor and the pool.
- The irradiated fuel elements are then stored in a pool.

The control fuel absorbers and the neutron startup sources will be transferred at the same time as the fuel elements that contain them.

Beyond these operations that lead to the storage of the irradiated fuel elements in the pools of the Ports of Brest, Toulon, Cherbourg, the question that remains concerns the final disposal of these fuels.

A.A. Sarkisov and L.G. LeSage (eds.),
Remaining Issues in the Decommissioning of Nuclear Powered Vessels, 297–299.

Nuclear propulsion fuels can be divided into two large families:

- The metallic fuels are based on a UZr metallic alloy with highly enriched Uranium, and were used in the first cores of SSBNs such as Le Redoutable, and for the first on-shore reactor at Cadarache (PAT).
- The oxide fuels use slightly enriched Uranium. Although the technology is different, it uses components similar to those of the cores of nuclear power plants. This fissile material is sintered UO_2 in the form of plates covered with zircaloy. All the fuel elements to be treated in the foreseeable future will be of this type.

The strategy regarding the disposal of the spent fuels differs depending on the family they belong to.

No reference process for the reprocessing of metallic fuels exists at present in the French industrial facilities. Moreover, the fact that the quantities are limited has led to the adoption of the solution of dry storage in the CASCAD facility on the CEA site of Cadarache, which is designed for a lifetime of 50 years.

As can be seen in the figure, the fuel elements are grouped in baskets and stored inside sealed metal pits in a storage bunker. This bunker communicates with the atmosphere through an air intake and a stack. The heat produced by the decay of the radioactive matter in the fuel elements, heats up the metal wall of the pits by internal convection, conduction and radiation.

The air around the pits warms up through contact with them, becomes lighter and rises. The hot air accumulates under the roof of the bunker and is evacuated by thermosiphon through the stack.

The irradiated fuel elements are placed dry in containers and sealed before they are transferred to the CASCAD facility. These operations are carried out in a hot cell on the Cadarache site.

All transfers from the port pools to Cadarache are made in dedicated large-capacity containers, and pose no particular problems.

The feasibility of reprocessing oxide fuels is established due to the fact that the technology is similar to that of nuclear power plant fuels. Technically speaking, oxide fuels can be reprocessed in the industrial facilities of the COGEMA, on condition that specific means are set up for the cutting, handling and transportation of the fuel.

The other methods explored for fuels from civilian sources have also been examined for military fuels. Among these we can mention:

- long-term interim storage prior to geological storage in triple-function "transport-interim storage – final storage" containers, followed by deep storage in about 2030
- The maintaining of interim storage in either a pool, a CASCAD-type facility, or the containers mentioned above.

At the present time, the oxide fuels are stored in pools. The existing facilities and those under construction (a new pool at Cadarache) allow the decisions to be postponed beyond 2015, and to integrate the strategy for the disposal of military fuels into that for civilian fuels, which is to be debated in the French Parliament in 2006. The economic aspects will obviously be taken into consideration.

2. Liquid and solid waste of low to moderate activity

The liquid effluents result from phases 1 and 2 and come from the complete emptying of the different circuits and the pool. The quantities and activities in question are low for a nuclear boiler of about 65 cubic metres with a mean activity of some MBq/m3 for the primary circuits. The liquid effluents are treated by evaporation, and the concentrates are conditioned and stored by ANDRA.

The solid waste from phases 1 and 2 (down to level-2 dismantling of the nuclear boiler compartment, and down to level 3 for the rest of the vessel) from one nuclear boiler can be held in 3 containers of 10 cubic metres and about two hundred 200-litre drums with a mean activity of some GBq/m3. The containers are transported to the ANDRA centre for conditioning and storage. The drums are transported to the CENTRACO incineration centre, and the residues from this are conditioned and transported to the ANDRA centre.

The dismantling of the nuclear boiler compartment down to level 3 (phase 3) is envisaged in a specific building to be created on the Cherbourg site. The procedure envisaged today is:

- To produce type-A packages of the size of the individual components of the nuclear boilers: vessel, pressurizer, steam generator, etc.
- To transport them to the ANDRA storage centre.
- Conditioning and final storage by ANDRA.
- The hull section and its bulkheads and interior seatings are considered to be very low activity waste. It is planned to cut them up into pieces of a size compatible with the accommodation capacities of the future storage centre for this type of waste.

3. Conclusion

From the moment a single core does not last the service life of a submarine, the unloading of the spent fuel is not specific to the submarine dismantling operation. It is a well-mastered operation that poses no problems.

The disposal of spent fuel from military sources will, insofar as possible, fit into the strategy adopted for fuel from civilian sources. At present, the only methods used are underwater storage in pools or dry storage.

The quantities and activities of the effluents and waste resulting from level-2 dismantling are limited, apart from the nuclear boiler compartment which is stored on the site.

A scenario for the level-3 dismantling of this compartment after 20 to 40 year's storage has been studied. The products from this final dismantling phase will be stored in the French storage centres.

ASSESSMENT OF THE PROBABILITY OF INITIATING A SPONTANEOUS CHAIN REACTION WITHIN METALLIC-CONCRETE TRANSPORTATION CASKS: CASE OF TEMPORARY STORAGE OF VESSEL SPENT NUCLEAR FUEL OF THE RUSSIAN NAVY

R. BAKIN, E. MITENKOVA, N. NOVIKOV, A. SHIKIN
Nuclear Safety Institute
Moscow, Russia

When considering beyond design-basis accidents in Transportation Casks (TC) for temporary storage of Spent Fuel Assemblies (SFA) of Nuclear Submarines (NS), the most hazardous radiological consequences occur if a Spontaneous Chain Reaction (SCR) is initiated. Despite a considerable safety margin for TCs against mechanical and other external impacts, an investigation to assess the type of destruction and the scale of external impact under which a SCR can be initiated is advisable.

Detailed analysis of SCR initiation conditions is rather difficult because very little or no data on fuel composition and the operational history NS reactors is available in open sources. This thesis especially concerns Russian NSs of the second- and the third generations. On the contrary, information on "Lenin" nuclear icebreaker steam producing facility is almost fully declassified.

Radionuclide concentrations in TCs were assessed with respect to a water-moderated reactor of the 2nd-generation NS based on the following assumptions:

1. The Reactor Core (RC) consists of 250 Fuel Assemblies (FA) and comprises 350 kg of uranium at 20%-235U enrichment.
2. One TC stores 49 SFA that is almost 1/5 of a RC.
3. Integral energy-producing capacity is 17 GW*day.
4. Time of RC storage prior to its reloading into the TC equals 5 years.

To estimate radionuclide concentrations within the TC, Spent Nuclear Fuel (SNF) radiation characteristics of different NSs and "Lenin" icebreaker dumped near Novaya Zemlya (New Land) Archipelago we used (see [1]).

Data analysis performed in [2] has demonstrated that long-lived fission product activities can be calculated to an acceptable accuracy even in the case that the real history of reactor operation at full power is unknown. Under long-term storage the activity of fission products is determined by integral energy production and depends almost linearly on this parameter.

When assessing actinide concentrations, one has to take into account not only real operational modes of a nuclear facility but also different fuel enrichment in the icebreaker and NSs. Because the first-type data are inaccessible, as a first approximation an assumption was made that the rate of actinide nuclei generation (excepting 238Pu) was inversely proportional to power density in the RC. For 238Pu a coefficient normalized to unit of energy production was used (the icebreaker case).

A.A. Sarkisov and L.G. LeSage (eds.),
Remaining Issues in the Decommissioning of Nuclear Powered Vessels, 301–305.

238U activity was calculated based on its integral concentration in the TC. 235U concentration was obtained based for its burnup.

In Table 1 the results of assessing radionuclide activities in a TC based on the above assumptions are presented.

TABLE 1. Activity of radionuclides in TC-108/1

Nuclide	Activity in TC, Bq		Nuclide concentration in one fuel element, g
	By the reactor shutdown time of	SNF storage after 5 years	
Long-lived fission products			
^{3}H	2.1+11	1.6+12	1.3-6
^{85}Kr	5.0+13	3.6+13	7.3-4
^{90}Sr	3.1+14	2.7+14	1.6-2
^{99}Tc	4.8+10	4.8+10	2.3-2
^{129}I	7.8+7	7.8+7	3.5-3
^{134}Cs	7.6+13	1.4+13	8.7-5
^{135}Cs	2.4+9	2.4+9	1.7-2
^{137}Cs	3.4+14	3.0+14	2.8-2
^{147}Pm	1.0+15	2.7+14	2.3-3
^{151}Sm	7.3+12	7.0+12	2.1-3
^{152}Eu	6.4+10	5.0+10	2.3-6
^{154}Eu	1.2+12	8.2+11	2.5-5
^{155}Eu	8.4+12	4.0+12	6.9-5
Actinides			
^{235}U		8.4+8	3.9
^{238}U		7.0+8	20.7
^{238}Pu		8.7+10	4.0-5
^{239}Pu		4.4+11	5.7-2
^{240}Pu		2.0+11	7.1-3
^{241}Pu		2.3+13	1.8-3

Estimates of neutron-physical characteristics of the SFA loaded into TC-108/1 under different hypothetical accident scenarios were performed using the MCNP Code [3] using the Monte-Carlo method in 3D-geometry.

Data on construction of both TC-108/1 and packages used to arrange SFA were obtained from the relevant engineering documentation. A cylinder 450-cm in height and 160 cm in diameter simulated the TC carcass. The cylinder possessed an internal steel cover and an external steel cover, the void between them being filled with reinforced concrete. The height of the TC internal housing made up 350 cm and the diameter 80 cm. Special spacer grids within the TC designed to arrange packages were not taken into account in calculations. At full loading one TC comprises 7 packages including 7 SFA each.

A package represents a welded structure made of corrosion-resistant steel, which consists of seven tubes fixed by the TC base from below and by a removable lid from above.

It was anticipated that Fuel Elements (FE) were cylinder in shape and had equilateral hexagonal packaging within the TC. Geometric parameters of both FA and FE were chosen with consideration for full RC loading and the package dimensions. The composition geometry used in calculations with the application of the MCNP Code is depicted in Figure 1.

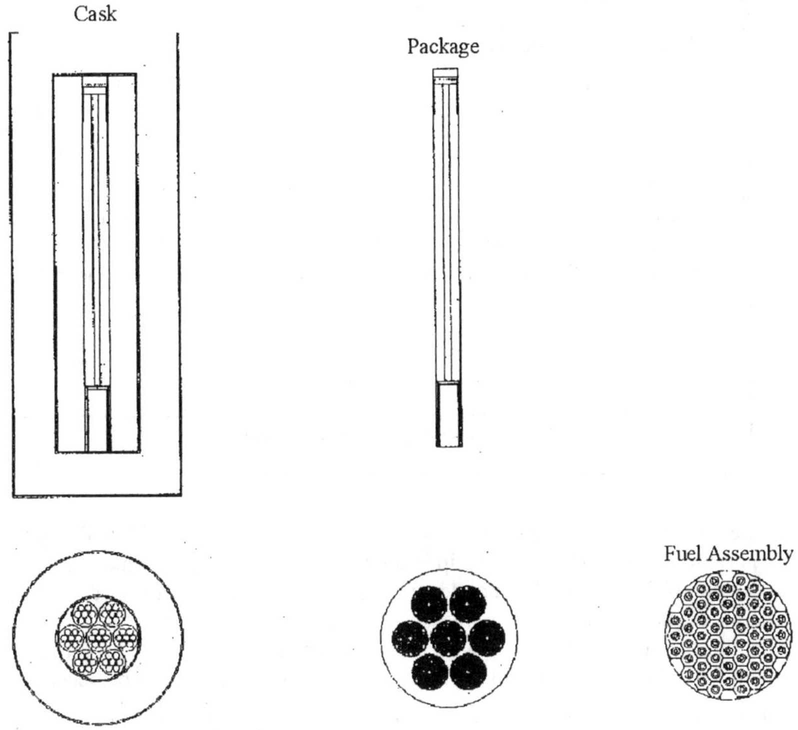

Figure 1. Geometry of the Calculated Composition

Continuous energy nuclear data libraries were employed in the calculations. Pointwise cross-sections of neutron interactions for given temperatures were prepared using the NJOY System [4] based on standard files of nuclear data in the ENDF/B format [5].

In our calculations the effects of changes in the system construction on Keff multiplication factor were assessed. Stable nuclides produced during the reactor operating period (a part of which had a large neutron absorption section) were not taken into account in fuel composition. Thus, the calculation results demonstrated in Table 2 can be considered as conservative.

TABLE 2. Keff of the System at Different TC-108/1 Hypothetical Accident Scenarios

№	Scenario type	Keff $\pm \sigma$
1	Normal TC maintenance, fuel temperature T=600K, other material temperature T=300K	0.1325 ± 0.0010
2	Emergency situation: lid broken, TC filled with water, temperature T=300K Similar emergency scenario with an increased water density: ρ=1,12 g/cm3	0.7182 ± 0.0010 0.7220 ± 0.0020
3	Homogeneous option: both fuel and water are placed evenly over the inner TC housing, fuel mass being determined by the initial loading, temperature T=300K Similar emergency scenario but uranium + plutonium fraction is increased by a factor of 1,8	0.5635 ± 0.0010 0.8500 ± 0.0020
4	Option 2 with 36,8% 235U enrichment	0.9100 ± 0.0030

The data of Table 2 demonstrates that the presence of water within TC contributes to an increase of the effective multiplication factor in the system and, thus, has to be strictly controlled during TC maintenance. Despite this, the FA under consideration still remains subcritical. Even an increase of fuel enrichment up to 40% (36,8% following unloading) is unable to produce Keff values at which more detailed investigations of the system from the viewpoint of analyzing SCR initiation risks would be necessary.

It is worthy of noting that the investigation was methodical in nature. By this it is meant that Keff temperature dependencies, the effects of extraneous agents, etc. have not been studied.

The application of the above approach will allow for improving the TC construction from the viewpoint of increasing SNF loading capacity. As a consequence, expenses related to NS SNF transportation and storage can be appreciably reduced.

References

1. Rubtsov, P. and Ruzhanskiy, P. (1996) Estimates of Spent Nuclear Fuel Radiation Characteristics of Nuclear Submarine & "Lenin" Icebreaker Reactors Dumped in Novaya Zemlia Area, *Atomnaya Energija (Atomic Energy)* **81(3)**, 212-219 (in Russian).

2. Bogatov, S., Gavrilov, S., Danilyan, V., Kisselev, V. (2001) *Estimates of Radionuclide Release for Different-type Hypothetical Accidents at the Russian Navy Facilities.* IBRAE's Preprint № IBRAE-2001-07, Moscow (in Russian).

3. Brismeister, J. (ed) (1997) *MCNP – A general Monte-Carlo N-Particle transport code, Version 4B,* Report LA-12625-M, Los Alamos Laboratory.

4. MacFarlane, R. *The NJOY Nuclear Data Processing System, Ver. 91.0,* Prepr. LA-9184-M.

5. Dunford, C. (1992) *ENDF Utility Codes Version 6,7,* Prepr. IAEA-NDS-29, Vienna.

RADIOACTIVE WASTE MANAGEMENT

RADIATION-EQUIVALENT APPROACH TO RADIOACTIVE WASTE MANAGEMENT

E.ADAMOV, I. GANEV, A. LOPATKIN, V. ORLOV
Russian Design Institute of Power Engineering
Moscow, Russia

Today the most used mode of Radioactive Waste (RW) ultimate disposal, with respect to high-level RWs, consists of placing it into a deep geological repository after a few decade storage of the Spent Nuclear Fuel (SNF) in open-cycle conditions and of vitrificated of RWs under closed-cycle conditions. During the initial disposal stage the integral RW activity must not lead to an inadmissible heating of surrounding rocks; however there is not a limiting value, and the activity will persist at a rather appreciable level up to tens and hundreds of millions of years. With due regard for such protracted periods of time, the ecologists require proof of the concerned repository safety with consideration for the risk of potential natural cataclysms. This is a very difficult task.

From the environmental safety standpoint an alternative way is worthy of consideration, which consists in an investigation of the radiation equivalency between the hazard due to long-lived high-level RWs and that of natural uranium. Such an approach makes it possible to establish an initial hazard level below which the activity of disposed RWs becomes equivalent to that of extracted natural uranium. In this case no safety proof is needed just as they do not need if considering natural uranium deposits. For comparison one can consider the Potential Biological Hazard (PBH) of both RWs and natural raw materials (Sv) calculated by multiplying a dose coefficient (Sv/Bq) taken from the Russian Radiation Safety Standards (RSS) into the activity of a given nuclide mass at a given time (Bq). To reach the radiation equivalency, when envisaging further development of large-scale Nuclear Power Industry (NPI), one needs:

- modify the uranium mining process by eliminating the procedure of extracted uranium separation from the accompanying components of the same family, i.e. radium and thorium, that will simultaneously reduce the long-lived activity in waste piles;
- realize a transmutation closed fuel cycle;
- apply long-term controlled RW storage within special repositories and, if necessary, improve packaging of the most hazardous elements prior to their ultimate disposal;
- accompany and complete all uranium-extraction operations by reclamation activities in order to replace the extracted natural uranium by nuclear fuel cycle wastes at a later time; thus, in some cases RWs will be housed within natural radioactivity areas;

A.A. Sarkisov and L.G. LeSage (eds.),
Remaining Issues in the Decommissioning of Nuclear Powered Vessels, 309–317.

- PBH of RWs at the disposal instant should approximate that of the extracted natural uranium; thus, in a way NPI will become a radiation wasteless industry.

On selecting a RW management strategy one should make efforts in reaching, at least, a nuclear power radiation balance (note that uranium export should also be considered in this case). If, when performing ultimate RW disposal, the principles of:- reaching a local radiation balance, and -ensuring better RW packaging (using mineral-like matrixes), as compared to natural conditions, are observed, there will be no need to prove the repository safe against potential natural cataclysms (meteorites, volcanoes, glaciers, etc.) for a period of tens and hundreds of years.

The radiation balance between the used raw materials and produced RWs can be considered and presented in the reactor design documentation to indicate both the unbalance causes and values. Ultimate RW disposal within a repository should be performed when the radiation equivalency is reached; however an earlier disposal is also admissible if, in keeping with the performed calculated estimations, the radiation balance can be reached after a relatively short period of time. To reach the radiation equivalency, the most radical solution consists in a prolongation of RW storage within storage facilities for a long period of time (e.g. up to completing the long range, large-scale NPI development with burning in Fast Reactors (FR) of the available depleted uranium stocks).

Transmutation Nuclear Fuel Cycle (NFC) represents further improvement of the closed fuel cycle needed to ensure long-term development of the large-scale NPI. Thus, the so-called "simple closed-fuel cycle" (wherein only U and Pu are extracted from irradiated fuel) should be developed up to the transmutation cycle, wherein:

- such elements as: Pu, Am and Cm are transmutated by fuel recycling via reactors (Cm after 50-100-year storage and transition to Pu);
- Sr, Cs and Np are extracted for beneficial use or storage whereas I, Tc and, possibly, Np for transmutation;
- prior to ultimate disposal, RWs are stored under control for 100 to 200 years and, next, are packed into stable matrixes;
- long-lived ^{226}Ra (dissolved and extracted with uranium up to 90%) and ^{230}Th (dissolved and extracted up to 85%) are extracted together with natural uranium and disintegrated (transmutated in neutron flux or using proton accelerator); simultaneously the problem of long-term tailing dump activity is resolved. When dealing with this multi-purpose task, one should take into account a triple increase of the uranium activity if adding Ra and Th.

The construction of storage facilities to ensure controlled RW storage during 200 (or even more) years facilitates the task of substantiating RW ultimate disposal.

The radiation equivalency concept taken together with that of the natural safety of reactors [2] introduces into the nuclear industry:

- a higher safety level (elimination of accidents requiring evacuation of the population);
- a new technology of nuclear & radioactive material management to solve the environment protection issues (reaching a radiation equivalency between RWs and raw materials);

- acceptable economy level (the cost of new-generation reactors does not exceed that of the present-day LWR).

As a whole, the realization of the above tasks, with consideration for unlimited fuel resources within the closed NFC, allows for solving the problem of long-term and safe energy-supply thanks to nuclear energy.

Let us analyze both the possibility of reaching the radiation equivalency and the NFC parameters, obtained, when considering the NPI development scenario worked up on basis of "The Strategy…"[1]:

- naturally-safe BREST-1200 facility is chosen as the fast breeder operating at a 1,05 Core Breeding Ratio (CBR) that is due to an increase in safety due to reducing the reactivity changes when burning [2];
- the Excess Core Breeding Ratio (ECBR) of 0,05 allows for producing a surplus of ^{239,241}Pu and using this material to start fuel charging for the BREST facility during the initial stage or, when developing the reactor system at a rate of 0,6-0,7%/year after the use of SNF from Thermal Reactors (TRs);
- the second NPI component comprises VVER-1000, VVER-440 and RBMK-1000 reactors, which use enriched uranium; VVER-440 and RBMK reactors are to be decommissioned as their service life is over; the VVER-1000 system is, first, to be developed, next, also decommissioned when its design life time is reached;
- there is a possible issue of introducing TRs into Th-U NFC in order to solve other tasks in addition to power generation (this issue, however, is not considered below).

Reactor Stock Initial Data. Let us consider the actually-operating Nuclear Power Plants (NPP) and also those NPPs, whose commissioning is planned up to 2020 in keeping with "The Strategy…". Thus, according to the TR scenario, the development of NPI during the concerned operation period will result in the consumption of 350 or 490 thousand tons of natural uranium (at 0,1 or 0,25% ^{235}U concentration in dumps, respectively) and the production of 62,100 t of SNF. According to "The Strategy", TR SNF reprocessing will begin in 2030, and the first serial BREST-1200 facility commissioning will take place the same year. The pace of Fast Reactor (FR) introduction is estimated at 3 reactors per 2 years until SNF from TRs is available. The volume of reprocessed TR SNF corresponds to needs in Pu. Minor Actinides (MAs) of the reprocessed SNF of TRs are to be used in the initial FR fuel charging. Any fuel surplus is to be used for new FR commissioning. When a FR service life is over, the remaining fuel of the shut down FR is to be used for putting new FRs into operation. TR SNF storage prior to the reprocessing onset is to be equal to at least 10 years. SNF reprocessing is to begin from SNF of VVER-1000. If this-type SNF is lacking, RBMK SNF is to be reprocessed. In Figures 1 through 5 the results of calculations [3] are demonstrated, as applied to the NPI development scenario of putting 1,5 BREST/year into operation on average.

In Figure 1 demonstrates the possibility of reaching the radiation equivalency in approximately the year 2100 in a case that the following conditions are observed:

- the NPI development progresses in keeping with "The Strategy…" [1],
- the first FR is put into operation in 2030 and

- natural uranium resources (490, 000 tons at 0,25% of ^{235}U in dumps) are used.

The uranium balance is considered for SNF of the BREST-1200 reactor (34,000 tons of heavy atoms), SNF of the RBMK (23,000 tons of heavy atoms), SNF of the VVER (39,000 tons of heavy atoms) and for Long-Lived High-Level Wastes (LLHLW), wherein, when performing chemical fuel reprocessing, the following elements are added:

- -Sr, Cs, Tc, I (5% each),
- -other fission products (100 %),
- -U, Pu, Am, Cm (0,1% each) and
- -Th, Pa, Np, Bk, Cf (100%).

The used uranium–SNF balance is reached in the open cycle at 10^5-10^6-year storage, whereas the uranium-LLHLW balance in the closed cycle at 100 to 200-year storage.

In Figure 2 the contribution of individual elements into the potential biological hazard of the BREST-facility SNF is demonstrated. Thus, the largest contribution is due to plutonium and americium. The contribution of uranium and neptunium becomes appreciable after 10^6 years. The pattern for Sr, Cs and Cm is similar up to 200 years; thereafter Cm is noticeable up to 10^4 years, whereas Sr and Cs are disintegrated rather rapidly. The contribution of ^{14}C is appreciable up to 10^4 years.

Depicted in Figure 3 are: -the accumulation of Pu and MAs up to 2030 (the year of the very first FR planned initial operation) in TR irradiated fuel and -the decay of Pu and MAs used in the initial FR fuel charging. Figure 4 illustrates the TR power increase/decrease and an increase of FR installed power. In Figure 5 the MA mass–fissionable Pu mass relationship ($\delta_{MA/Pu}$) within the initial FR fuel charging is given. In order to ensure nuclear safety in a case of loss-of-coolant in FRs, this relationship must not exceed 0,3-0,35. After several fuel cycles in the reactor the relationship tends to an equilibrium value of 0,08.

MA Transmutation Results. All MAs (Np, Am and Cm) produced by TRs and extracted from TR SNF, when reprocessing, can be introduced into the BREST reactor initial fuel charging. In this case the mass fraction of the above MAs in the BREST fuel will not exceed ~ 3,3%. Such MA amount will not make appreciably worse the BREST safety characteristics, when removing lead coolant, as this is expected today. 40% of loaded MAs will burn during the first Fuel Operation Period (FOP), whereas excessive MAs will burn almost in full in the course of five FOPs. However in this case a substantiation of the technology of both nitride fabrication and fuel assembly fabrication with consideration for a high MA concentration in fuel is necessary, since the equilibrium MA concentration in the BREST fuel makes up 0,7%.

In Figures 6 and 7 are demonstrated the effects of an increase of both the burnup and TR SNF storage period on a relative level of MA concentration in BREST-1200 fuel. An increase of the burnup makes appreciably worse the MA mass–fissionable Pu mass relationship since it contributes to an increase of MA mass with respect to plutonium. The value of this relationship increases further as SNF is stored (and accompanied with ^{241}Pu disintegration into ^{241}Am) and is able to reach a critical value for realizing transmutation of the concerned MAs within the FR.

Taking into account potential long-term SNF storage and excessive MA accumulation in FR initial fuel charging, other MA transmutation scenarios can be proposed including MA storage with partial transmutation within the FR system and, possibly, the creation of specialized reactors performing transmutation of the MA surplus.

The foregoing showed a possibility of reaching the RW - natural uranium PBH balance, i.e. their radiation equivalency if putting into operation thermal reactors in keeping with "The Strategy...". During further NPI development (after 2100) the time of reaching the radiation balance will increase due to an augmentation of RW mass at a constant mass of natural uranium consumed by the system. As a potential reserve, a step-by-step upgrading of the irradiated fuel reprocessing technology including the extraction of most of the hazardous elements out of SNF, to perform recycling, and a consideration for radionuclide migration peculiarities, when analyzing the radiation balance, is recommended.

It is possible that the problem of RW final disposal would be worthy of consideration only when the stocks of natural depleted uranium (used to make up FRs) were exhausted. Until then, long-lived high-level RWs can be kept within storage facilities for controlled long-term storage. In keeping with our estimates, RW amounts will be minor due to high energy content of nuclear fuel and, respectively, to small mass of irradiated fuel. Thus, for NPI of 300 GW power during 1000-year period of operation the volume of high-activity RW produced is estimated at 80,000 m^3 at most, if about 10% of BREST-1200 SNF are introduced into RWs of 5 t/m^3 density (note that at 8% average burnup SNF contains 12 t of heavy atoms /GW·year including about 1 t of fission products).

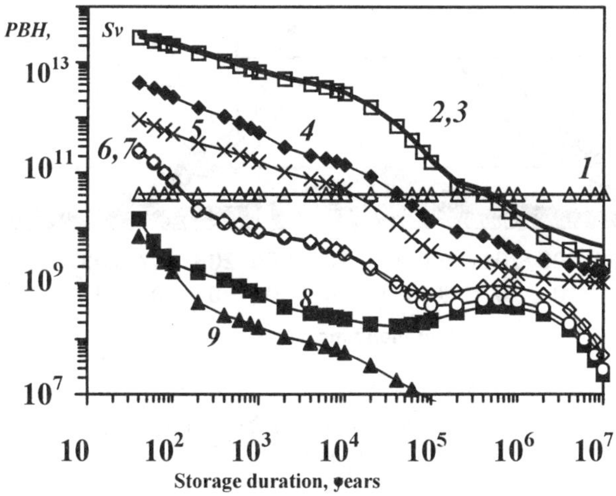

Figure 1. Reaching a balance between PBH (ingestion with food) of the consumed natural uranium (1) *and PBH of SNF* (2-5) *or of Long-Lived High-Level Wastes (LLHLW)* (6-9) *of VVER-1000* (4, 8), *RBMK-1000* (5, 9) *and BREST-1200* (3, 7) *by 2100:* 2 – *SNF sum,* 6 – *LLHLW sum*

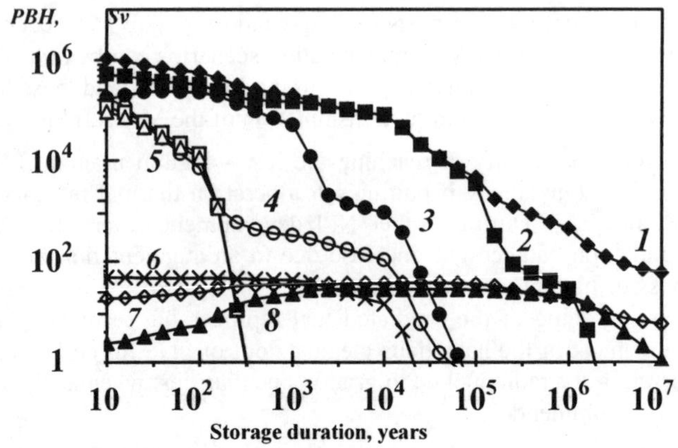

Figure 2. Contribution of individual elements: 1- SNF, 2 – Pu, 3 – Am, 4 – Cm, 5 – Sr(Y)+Cs(Ba), 6 – ^{14}C, 7 – U and 8 – Np into PBH (ingestion with food) of BREST-1200 SNF (normalized to 1 kg of irradiated actinides or to 1,06 kg of nitride fuel)

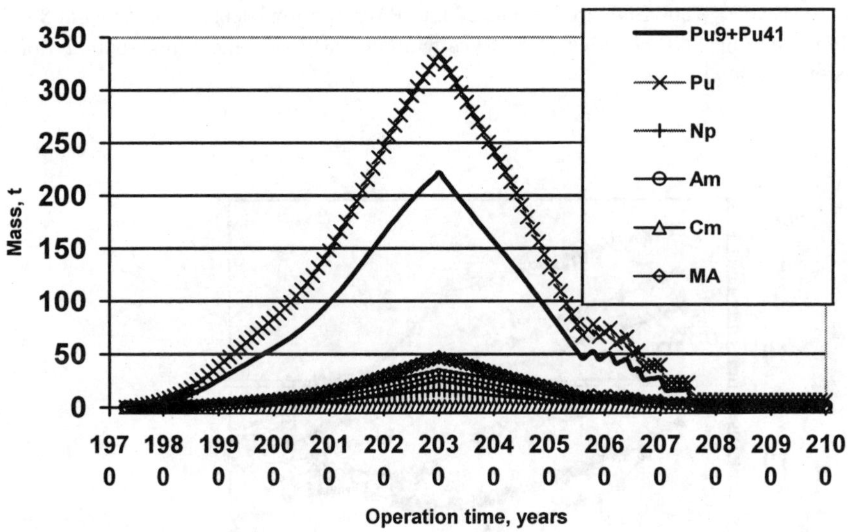

Figure 3. Pu – minor actinide (MA=Np+Am+Cm) balance in SNF of TRs (case of the NPI development scenario presuming 1,5 BREST-1200/year putting into operation)

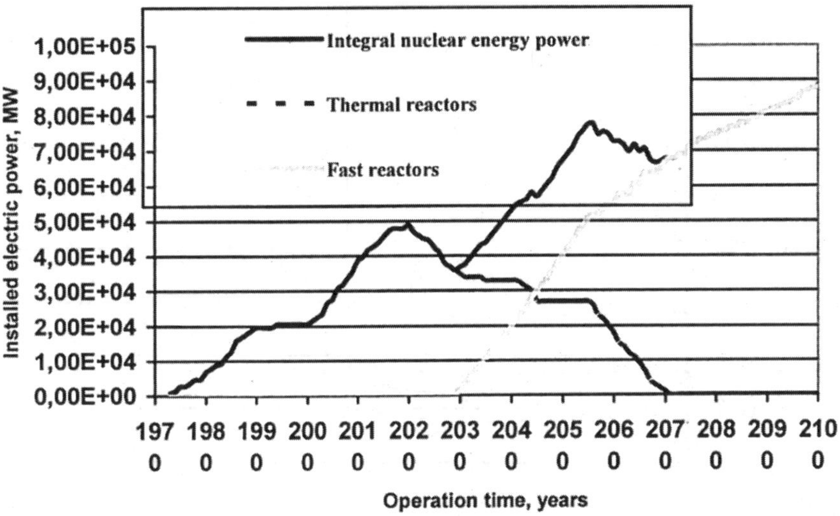

Figure 4. Time variations in the installed power of NPPs with TRs and FRs: case of the NPI development in keeping with «The Strategy.

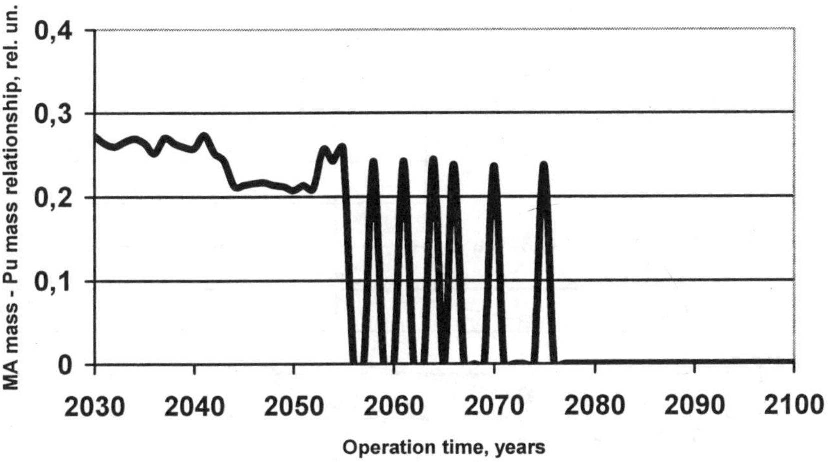

Figure 5. MA mass - fissionable Pu mass relationship in the first charging of BREST-1200 fuel

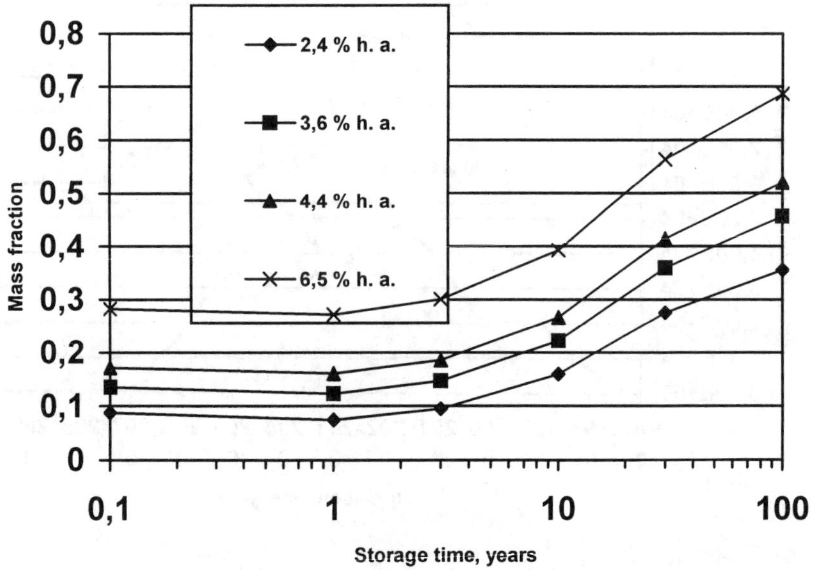

Figure 6. Dependence on the duration of storage of MA mass - fissionable Pu mass relationship in irradiated fuel of RBMK-1000 for different burnups

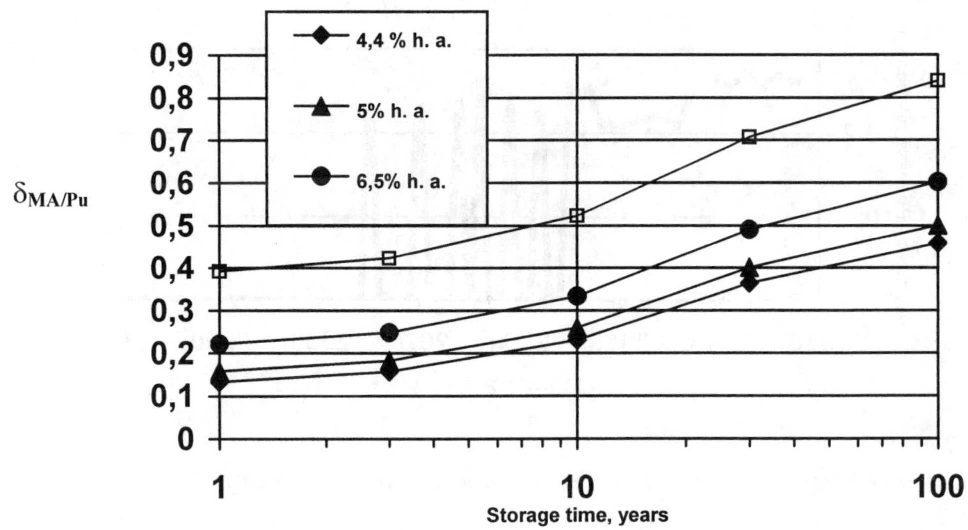

Figure 7. Dependence on the duration of storage of MA mass - fissionable Pu mass relationship in irradiated fuel of VVER-1000 for different burnups

References

1. *The Strategy of Nuclear Industry Development in Russia during the First Half of the XXI Century. Underlying Positions. Approved by the Russian Federation Government on May 25, 2000. Proceedings 17.* (2000) Minatom, Moscow (in Russian).

2. Orlov, V., Leonov, V., et al. (2000). BREST-Reactor Design. Experiments to Substantiate the BREST-Reactor Concept: Results & Prospects. *Proceedings of the International Conference «Fast Reactor and the Fuel Cycle of Natural Safety in the Large-Scale Power Industry. Fuel Balance, Economics, Wastes, Non-Proliferation»*. Minatom, Moscow (in Russian).

3. Ganev, I., Lopatkin, A, Muratov, V., et al. (2002) *Parameters of Nuclear Fuel Cycle and Fast Reactor Fuel when Developing Nuclear Industry in Russia in Keeping with «The Strategy...»*, FSUE RDIPE Report, Moscow (in Russian).

4. Adamov, E., Ganev, I., Lopatkin, A., et al. (1999) *Transmutation Fuel Cycle in the Large-Scale Russian Nuclear Industry: Monograph.* FSUE RDIPE, Moscow (in Russian).

NUCLEAR VESSEL UTILIZATION: BASIC PATHWAYS OF REALIZING ENGINEERING SOLUTIONS OF THE CONCEPTUAL LRW & SRW MANAGEMENT FLOWSHEET IN THE NORTHWEST REGION OF RUSSIA

V. SAFUTIN, A. KIRSANOV, A. SHVEDOV, V. SOROKIN,
A. DEMIN, L. ZVIAGINA
All-Russian Research Institute of Applied Energy Technologies
St. Petersburg, Russia

The paper considers basic pathways for realizing technical solutions related to management of Liquid Radioactive Wastes (LRWs) and Solid Radioactive Wastes (SRWs) produced during Nuclear Submarine (NS) operation, repair and utilization operations. These pathways were developed by the All-Russian Research Institute of Applied Energy Technologies (ARRI AET) and are aimed at appreciably reducing Radioactive Waste (RW) amounts and, thus, improving the environmental situation in the Northwest region of Russia.

1. Actual status of RW management problem in the Northwest region

RWs produced by at least 200 power reactor facilities of the Arctic Navy and NS servicing enterprises have been accumulated at special sites located along the Kola Peninsula coast and in the vicinity of Severodvinsk (Arkhangelsk region).

In keeping with the Russian Federation (RF) law "On the Environment Protection" (1992), marine disposal of RWs is prohibited.

At present SRWs (stored in containers) are being accumulated at open-air concrete sites at the Arctic naval bases and constitute an environmental hazard. In fact, to process LRWs & SRWs, no technology is actually available. RW storage facilities at naval bases are full; many of them are in emergency state and are to be dismantled.

Proper solution of RW management issues that can ultimately ensure stabilization and improvement of the environmental situation at every concerned site/enterprise and throughout the Northwest region is possible only if a multi-purpose approach is applied.

2. Conceptual Flowsheet of RW Management

The below-considered flowsheets of RW management are aimed at: -first, RW processing and conditioning, i.e. considerable reduction of their volumes and transformation into an environmentally safe state, and –next, RW removal from the technogenic sphere (RW disposal).

A.A. Sarkisov and L.G. LeSage (eds.),
Remaining Issues in the Decommissioning of Nuclear Powered Vessels, 319–328.

2.1. BASIC PRINCIPLES FOR REALIZING RW MANAGEMENT FLOWSHEET

- Use all available, under construction and projected facilities & technologies for RW management, as much as possible,
- Construct two coastal centralized sites for deep processing of the largest part of the RWs being produced in the Northwest region;
- Perform treatment of most of the process LRWs produced by the nuclear icebreaker fleet and the facilities of both the Russian Ministry of Defense & the Ministry of Economics at "Atomflot" Service & Repair Enterprise (SRE), ensure their containerization and the temporary storage of LRW packages;
- Process most of the SRWs produced by facilities of Ministry of Defense, Ministry of Economics and "Atomflot" SRE at "Nerpa" Ship-Repair Yard (SRY) or SRY-10, and next, ensure their containerization and temporary storage of the SRW packages;
- Construct a facility at FSUE "Zvezdochka" to process RWs produced in the course of NS decommissioning operations;
- Organize large-scale deactivation in situ of low-level LRWs coming from special-purpose laundries and decontamination centers on the basis of the type modular installations, which use a reduced flowsheet (save for saline concentrate hardening).
- Construct a mobile modular installation to deactivate both low-salt waters and "accidental" LRWs for subsequent use at individual facilities.
- Organize centralized "sending" sites to ensure sending of conditioned RWs for disposal.
- Construct a repository for disposal of conditioned RW within the mainland or in the Novaya Zemlia (the New Land) archipelago.

2.2. LRW MANAGEMENT FLOWSHEET

The basic lines of LRW management are depicted in Figure 1.

LRW collection is to be organized in situ in temporary-storage tanks (both floating and coastal ones).

LRW transportation to a central treatment center at "Atomflot" SRE is to be performed using the "Serebrianka" special vehicle.

LRW processing

- Most of LRWs are to be processed at "Atomflot" SRE using a modernized processing facility installed within the available buildings of the Repair & Process Facility and the Decontamination Center as well as in the Building for Special Water Deactivation (under construction);
- Processing of medium-level and "accidental"-origin LRWs is to be performed at a few facilities using mobile modular installations;
- Treatment of low-level LRWs issued from decontamination centers and special laundries is to be performed in situ using a type of modular facilities,

which employ a reduced flowsheet (save for saline concentrate hardening). Saline concentrates are to be transported to "Artomflot" SRE for processing;

- Treatment of LRWs produced during NS decommissioning at FSUE "Zvezdochka" is to be realized in situ (full-treatment cycle);
- Both interim and temporary storage of containers with already treated LRWs is to be carried out at new storage facilities of both "Atomflot" SRE and FSUE "Zvezdochka";
- Conditioned RWs are to be packed into reinforced concrete non-return containers of a unified type;
- Conditioned LRWs are to be removed to a special-purpose underground repository in the mainland or in the Novaya Zemlia archipelago.

2.3. SRW MANAGEMENT FLOWSHEET

Basic lines of LRW management are depicted in Figure 2.

SRW collection, grading and containerization as well as interim storage of containers with SRWs is to be carried out in situ within the available and newly constructed facilities (at centralized sites of SRW collection, grading and containerization and in SRW temporary storage facilities).

It is planned that SRW unloading from storage facilities to be reconstructed (such as: CSEs of the RF Navy, "Radon" Special-Purpose Combine in Murmansk and "Mironova Hill" facility) will be performed using temporary complexes of SRW unloading and packaging into special containers.

Transportation of SRW containers for processing is to be realized by the "Amur"-vehicle after dismantling of the available radioactive water treatment facility and the ship conversion into a container vessel.

Temporary storage of SRW containers prior to their processing will be within a site of the centralized SRW treatment complex.

SRW processing:

- Most of SRWs are to be processed within "Nerpa" SRY or SRY-10 at a new centralized SRW processing complex;
- Processing of burnable low-level SRW produced by "Atomflot" SRE is to be performed in situ using the available incineration facility after its reconstruction;
- Processing of SRWs resulting from NS decommissioning at FSUE "Zvezdochka" is to be organized at new processing facilities installed at the same enterprise;
- Containerization of already treated SRWs is to be performed using reinforced concrete non-return containers of a unified type;
- Containerization of nonprocessable SRWs is to be carried out using metal containers with a special cladding. For transportation, protective steel containers of return-type are to be used;
- Both interim & temporary storage of the containers housing already treated SRWs is to be realized within the available and new storage facilities at "Nerpa" SRY and FSUE "Zvezdochka";

- The containers containing conditioned and nonprocessable SRWs are to be removed to an underground repository in the mainland or the Novaya Zemlia archipelago.

3. Main Solutions to Realize the Conceptual Scheme of LRW Management

3.1. DESIGN CHOICE TO PROCESS LOW-LEVEL LRW DUE TO NUCLEAR SUBMARINE UTILIZATION AT THE "ZVEZDOCHKA" PROCESSING COMPLEX

Multi-purpose procedures & techniques of different-type LRW processing used at FSUE "Zvezdochka" (Building 160/161) are based on the relevant latest Russian and world achievements and include:

- Use of the principle of radionuclide selective sorption on non-organic sorbents of manufactured in Russia.
- Use of a reverse-osmosis facility to decontaminate LRWs produced by the Russian atomic fleet;
- Application of an advanced American concentrating system (manufactured by the "Ohmptemp" Company) allowing for the maximum possible LRW concentration up to dry salts (without recourse to evaporators) that is satisfactory for ultimate disposal;
- Use of: -protective containers to dispose of spent sorbents and -double-protection containers (which include temporarily sealed inner coffins ensuring double protection when storing conditioned RWs) to dispose of dry salts.

The above system is able to process: -800 m^3/year of low-salt primary circuit solutions, of deactivation and mixed solutions; -700 m^3/year of solutions issued from the reactor biological protection tanks and -2500 m^3/year of solutions from the special laundry.

A distinctive feature of the concerned LRW treatment system consists in the fact that most hazardous radionuclides, such as ^{137}Cs, ^{60}Co and ^{90}Sr are extracted almost in full at a stage of pre-decontamination of tbe initial solutions using selective non-organic sorbents.

The above design choice related to processing of low-level LRWs due to NS utilization operations is to be realized at FSUE "Zvezdochka" following the commissioning of the LRW processing complex in November 2000.

3.2. DESIGN CHOICE TO PROCESS LRW AT "ATOMFLOT" SRE

All LRWs produced in the Northwest region of Russia are to be processed at "Atomflot" SRE.

A modernized installation is to be installed within the available buildings of the repair & process facilities and special laundry. Its capacity is estimated at up to 5000 m^3/year.

The following types of LRWs are to be reprocessed: -low-salt process (circuit) and low-level non-process waters; -saline decontamination waters and -high-salt decontamination waters. To enlarge the capacity of LRW-processing installation regarding LRW acceptance and treatment, tank facilities of the Special-Water-Deactivation Building are to be used.

The modernized LRW processing installation will comprise the assembly units of: -LRW acceptance; -precipitating deactivation of saline LRWs; -LRW deactivation at selective sorbents; -filtering & microfiltering; -electro-membrane deactivation of salts and their concentration; -LRW afterpurification at selective sorbents; -LRW afterpurification at ion-exchange resins; -salt concentrate & residuum cementation; -cementation of ashes of SRW burning installation and -containerization of both cement compound and spent sorbents.

All the above installations have already been constructed at "Atomflot" SRE and their pilot testing has just started.

3.3. DESIGN CHOICE TO PROCESS LRW USING MOBILE MODULAR FACILITIES

LRW processing at modular facilities allows for reducing the integral LRW volume by more than 100 times (from 1200 m^3 down to 11 m^3).

Thus, 6400 m^3/year of collected and 1300 m^3/year of annually produced LRWs are to be processed using the proposed flowsheet.

To treat all-type LRWs, the following design choice is suggested: -selective sorption of ^{137}Cs and ^{60}Co using non-organic sorbents at a stage of pre-deactivating all-type solutions; -use of a membrane-centrifuge unit to perform separation of suspensions and oil products (if needed) of LRWs at the pre-deactivation stage; -use of a membrane-reverse-osmosis unit of basic deactivation from radionuclides and chemical impurities; -application of a sorption after-deactivation unit and of -a rotor-solution concentrator that makes it possible to obtain ultimately a cement compound packed into standard 200-litter barrels.

To realize operation of the above facility, a preliminary design has already been prepared and is actually under consideration and approval from supervisory bodies.

4. Main Solutions Aimed at Realizing the Conceptual Scheme of SRW Management

4.1. DESIGN CHOICE TO PROCESS SRW AT FSUE "ZVEZDOCHKA"

If realized in full, the proposed SRW treatment methods will provide for reducing the volume of SRWs to be stored after treatment from 200 m^3/year (initial SRW amount) to 36 m^3/year (treated SRW amount).

Non-metal SRW processing. STWs to be pressed include: combustible materials (rugs, paper, planks, old clothes, etc.) and plastics (polyethylene, elastron, PCV, plastics, rubber, coatings of "dry" deactivation, etc.).

The volume of SRWs to be pressed is estimated at 140 m^3/year.

The volume-reducing coefficient with respect to mixed SRWs processed via pressing method is estimated at 6.

After pressing the volume of processed SRWs will make up 23,5 m^3/year or 117,5 unit/year if considering packages (metal barrels) of V=55 gal.

Metal SRW processing. Within a special SRW-grading building SRWs are reduced to fragments and categorized, metal SRWs being separated.

After fragmentation, SRWs to be decontaminated are put into special barrels with shells and transported to a decontamination site within the temporary storage zone.

Metals are decontaminated at a transportable installation, which uses the principle of electrochemical impact on surface contaminants.

It is planned that both fragmentation and electrochemical decontamination will allow for reaching 40 m^3/year of metal SRWs decontaminated to permissible limits. As the result, their "normal" storage at scrap metal storage sites will be possible.

If optimizing the barrel loading procedure, SRW volume can be reduced by a factor of 2. Thus, of 15 m^3/year metal SRWs to be loaded into barrels one can obtain 7,5 m^3/year of packed wastes or 37,5 barrel/year.

Processing of bulky metal SRWs. The fraction of bulky hull structures (dimensions up to 2 x 1,5 x 1,5 m and up to 5-t weight) in metal SRWs is 5 m^3/year.

Bulky SRWs are transported to a fragmentation site wherein they are dismantled and fragmented and, next, loaded into 200-l barrels.

Cutting is performed using an air-plasma cutting installation and other portable electric tools.

Dismantling operations will make it possible to: -decontaminate some metals down to permissible storage limits allowing for their "normal" storage within scrap metal storage sites; -obtain minor amounts of non-decontaminable metal put into 200-l barrels and -an insignificant amount of pressed nonmetal fragments.

The above design choice to process SRWs resulting from NS decommissioning was realized in November 2000 following the commissioning of FSUE "Zvezdochka" processing complex.

4.2. DESIGN CHOICE TO PROCESS SRW USING MOBILE MODULAR FACILITIES

The SRW Mobile Modular Facility (MMF) is designed to perform primary processing of SRWs collected and produced in the Northwest region at the Russian Navy facilities in the course of NS utilization and also during rehabilitation operations at both radiation-dangerous installations and within the contaminated territories.

MMF is to prepare SRWs for pressing (size reduction, drying, loading into special barrels) and to carry out their pressing. SRWs that cannot be processed at MMF are to be graded and, next, loaded closely into packages for subsequent storage and treatment at a regional processing center.

MMF consists of the following workcells: -SRW acceptance installation; - system for bulky SRW fragmentation; -SRW grading plant; -nonprocessable SRW packaging installation; -SRW disintegration plant; - SRW drying system; -SRW pressing plant; -a sanitation section (sanitary lock-chamber, radiation control station);

-a management & control section; -an express laboratory; -a section of deactivating solution preparation; -engineering & technical support section (power supply, ventilation, water, steam, gas, etc.).

To realize the above proposals, a statement of work has been developed and a detailed design is being elaborated.

4.3. DESIGN CHOICE CONCERNING SRW PROCESSING COMPLEX AT "NERPA" SRY OR AT SRY-10

At SRY "Nerpa" or at SRY-10 the construction of a SRW Processing Center is planned. The Center will include the following facilities:

- an installation ensuring SRW cold superpressing (up to 1 m^3/h capacity with respect to initial SRWs of the 1st-activity group, SRW volume-reducing coefficient is equal to 5);
- a combustible SRW burning plant (up to 50 kg/h capacity for initial SRWs of the 1st-activity group, SRW volume-reducing coefficient is equal to 80);
- a hot pressing installation for combustible incompressible SRWs (up to 100 kg/h capacity with respect to initial SRWs of the 1st-activity group, SRW volume-reducing coefficient is up 4 to 5);
- a workcell of SRW grading and fragmentation (the workcell capacity up to 2500 t/year for initial SRWs, SRW volume-reducing coefficient after well-ordered placing of metal into primary packages is 2 to 4);
- a workcell to perform deep decontamination of metal SRWs (the workcell capacity up to 2000 t/year for initial SRWs of the 1st-activity group; return of up to 70 % of the metal for reuse);
- a section of nonprocessable SRW containerization.

A technical and economic assessment (feasibility study) to construct a two-site SRW Processing Center (at "Nerpa" SRY and SRY –10) has already been performed.

4.4. DESIGN CHOICE RELATED TO SRW PACKAGING AND STORAGE ("NERPA" SRY OR SRY-10)

Packages containing bitumized RWs are to be stored within special modules made of monolithic reinforced concrete within the hard-wall part of buildings.

Packages with nonprocessable metal SRWs with exposure dose rates up to 1000 mrem/h are to be stored in an individual module within special reinforced concrete cells in the hard-wall building section. After fragmentation nonprocessable bulky RWs are to be also stored within the hard-wall section of the building.

Cemented ashes of the incineration facility, polymeric hot-pressing blocks, pressed SRWs of the superpressing facility, nonprocessable metal RWs of up to 300 mrem/h exposure dose rates, solidified LRWs, cured ion-exchange resins and accumulated ashes, cemented sludges of etchants are stored within a common room (light structural units).

A technical and economic assessment (feasibility study) to construct a storage facility for conditioned SRWs at "Nerpa" SRY and SRY -10 has been also prepared.

5. Design choice related to RW disposal

The technical approach used to solve this problem is based on a proposed method of underground sealing of both RWs and nonprocessable Spent Nuclear Fuel (SNF).

At present the 2 following options of RW removal from the technogenic sphere are being developed:

Option 1: Ultimate container disposal of RWs within repositories created in rocks (granites) of the Northwest region following their long-term interim storage. The project contains a technical & economic assessment of seven sites proposed as possible alternatives to the future RW repository in Murmansk region.

Within the framework of the project a conceptual design of an underground repository has been developed using funds of the TACIS Program. The conceptual design presumes the construction of a buffer interim storage facility for conditioned RWs coming from the above-cited enterprises of both the Murmansk region and the Arkhangelsk region following their processing.

Option 2: RW container disposal (after processing and long-term interim storage at enterprises) in permafrost rocks and the project implementation at an Experimental-Industrial Facility (EIF) in the Novaya Zemlia archipelago.

RWs and long-term storage SNFs are to be stored in underground conditions within permafrost rocks in deep trenches, different-type large-diameter boreholes, pit-shafts and galleries.

A technical and economic assessment (feasibility study) to substantiate Option 2 of RW & SNF disposal has been performed and presently is undergoing an international expert examination.

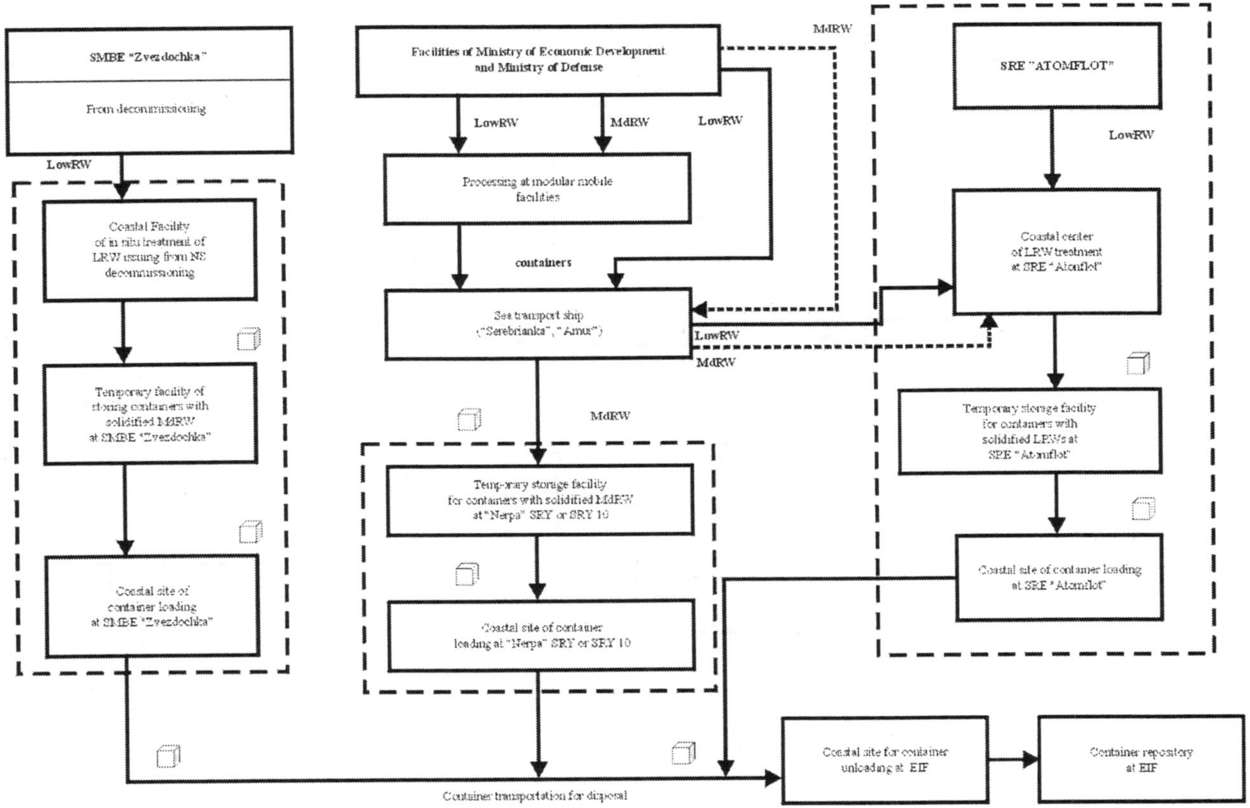

Figure 1. Basic Principal Solutions on Creating the Infrastructure of LRW Management in the Northwest Region of Russia

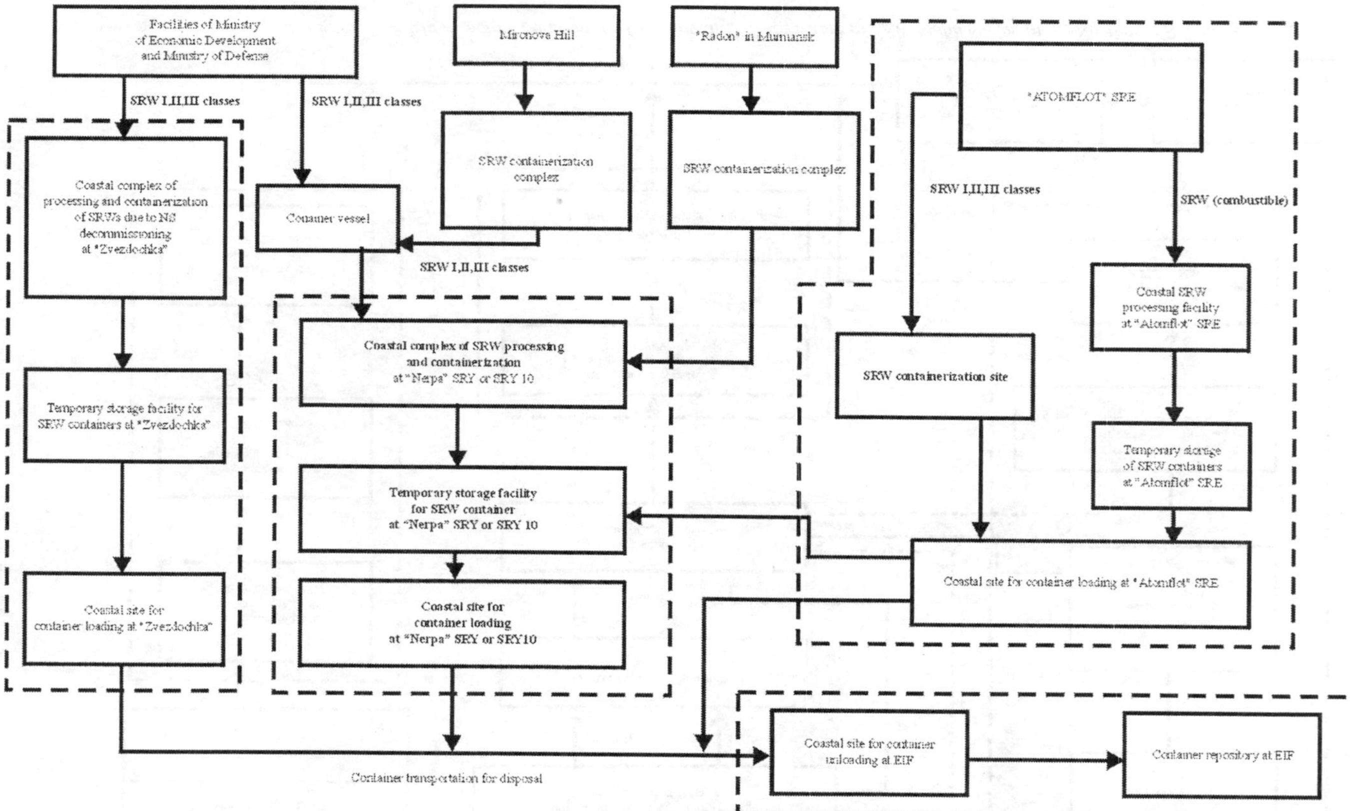

Figure 2. Basic Principal Solutions on Creating the Infrastructure of SRW Management in the Northwest Region of Russia

WAYS OF MINIMIZING AMOUNTS OF RADIOACTIVE WASTES CREATED DURING NUCLEAR SUBMARINE UTILIZATION

Y. J. MESCHERIAKOV
"Electropribor" Close Corporation
Moscow, Russia

N. M. SOROKIN
Minatom Mounting Technology Institute
Moscow, Russia

S. A. MATVEEV
Defense Ministry Central Research Institute
Saint-Petersburg, Russia

ABSTRACT

The paper considers scientific backgrounds for solving tasks of ensuring the radiation safety and minimizing radioactive waste amounts during radiation-dangerous work performed under normal operation of both commercial and defense nuclear power installations, when eliminating the consequences of accidents and collecting or utilizing radioactive wastes. For these purposes the use of aromatically conjugated oligomers and polymeric compounds made thereof is proposed.

1. INTRODUCTION

Created more than 60 years ago, nuclear power industry predetermined the invention and fabrication of many new materials including those based on polymeric components. However neither notion nor parameter is still available for developing a concept of radiation resistance – polymer structure relationship. The existing polymers and polymeric compounds made thereof are unable to ensure the needed radiation resistance in all situations related to radiation-dangerous work under normal operation of nuclear installations, when eliminating accidents, collecting and utilizing Solid Radioactive wastes (SRW) & Liquid Radioactive Wastes (LRW), preserving Fuel Assemblies (FA) & reactors of damaged nuclear submarines, etc.

A.A. Sarkisov and L.G. LeSage (eds.),
Remaining Issues in the Decommissioning of Nuclear Powered Vessels, 329–337.

2. BACKGROUNDS FOR INVENTION AND FABRICATION OF POLYMERIC COMPOUNDS TO ENSURE SAFETY OF RADIATION-DANGEROUS OPERATIONS

The existing polymeric compounds based on furane monomers & oligomers, polyester-maleated styrene resins, oligo-ether-acrylate resins, epoxides and phenol resins have many imperfections, namely, the toxicity and complexity when reprocessing (component dosing and curing) as well as a deficient structure of the final products. Thus, they are unable to ensure full solution of the tasks related to both radiation safety and minimization of Radioactive Waste (RW) amounts.

In the context of the subjects considered by "Unresolved Issues Remaining in the Decommissioning of Nuclear Powered Vessels and in the Environmental Remediation of their Supporting Infrastructure" NATO-Russia Advanced Research Workshop (Moscow, 2002) an attempt is made to formulate principal ways of research and examine the theoretical basis of creating polymeric compounds as components of materials designed to ensure the above tasks.

2.1. BACKGROUNDS FOR GENERATING THERMOSETTING OLIGOMERS AS A BASIS FOR POLYMERIC COMPOUNDS

Research is focused on generating products with an oligomeric-unit structure that resists the radiation impact through redistribution of the obtained energy via both oligomeric chain and 3D polymer lattice. Such polymers have a minimum contraction and, as components of polymer compounds, can be reprocessed into coatings, filling compounds, plastic laminates, press pulps, hermetics and adhesives providing for proper solution of the task under specific operating conditions.

It is anticipated that such oligomers exert precipitating and curing action on LRW, whereas the coatings made thereof resist corrosion under a radiation impact and, in addition, are easily decontaminated. Compounds of special composition generated of such oligomers allow for making coatings for casks and containers designed to store, transport and dispose Spent Nuclear Fuel (SNF), RW and fuel for a long period of time, as well as to ensure preservation of FA, reactors and their elements.

Let us consider oligomeric components of reactive binding agents ranking first when producing coatings, compounds and binders.

In keeping with Professor A. Berlin's classification [1], the above substances are subdivided into the following three groups:

- Polymerizable oligomers;
- Oligomers with functional groups;
- Oligomers with conjugated bonds.

However, when using the above subdivision, such widely employed substances as polyesters, phenol-formaldehydes and epoxies should be considered as mixed oligomers since they possess features typical for all of the three groups.

In deciding on a particular oligomer to solve a specific task, one should start from the fact that variations in volume and dimensions, when binder curing, result in concentrations of internal stresses in materials and articles, which lead to deformations under operational impacts and are even able to produce local failures. E.g., polymerizable oligomers (polyesters, oligo-ether-acrylates and acrilates) possessing double bonds, i.e. nonsaturation, are cured by radical mechanism under the impacts of peroxidates, heat or power and produce the largest (up to 10%) variations in volume due to approaching of carbon-carbonic segments when double bond opening.

In the case of phenol-formaldehyde oligomers or furane oligomers such variations make up 3 to 5 % when interacting methylol groups under the impact of catalysts, heat and pressure.

Epoxy oligomers are cured by a mixed mechanism under the action of catalysts and curing agents carrying a mobile proton. In this case variations in volume do not exceed 0,5 – 3% at epoxy ring opening that offers some advantages of epoxy oligomers over other types.

In any case, in addition to oligomer & polymer shortcomings predetermined by their structure, a number of phenomena resulting from incompatibility of components and filling agents, unregulated exothermicity or decreased functionality, steric problems when interacting functional groups, et al. take place in cured systems. As a consequence, polymers with imperfect structure (defect areas, incomplete polymerization, etc.) are produced, and this circumstance gives no way of obtaining a complex of valuable mechanical, thermal and electrophysical properties in materials.

Let us consider technological potentialities of oligomeric product reprocessing in plastic production on the condition that the end product possesses properties predetermined by the material nature rather than has "some preset properties" and using the established (optimal) operational modes needed to proceed physical and chemical reactions. Any attempt to reach "a complex of preset properties" can be compared to the futile endeavor of Swan, Crayfish and Pike to pull the loaded cart of a well-known fable, see *Figure 1* [2].

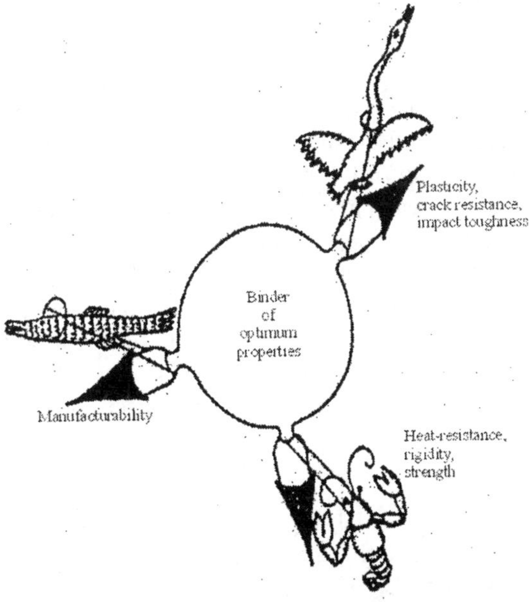

Figure 1. Comic illustration of the inconsistency of binder properties

With consideration for the foregoing one believes that to generate up-to-date compounds when solving tasks of utilization and rehabilitation of radiation-dangerous units & installations in the Russian Navy, the use of epoxy oligomers with conjugated bonds may be of interest.

The backgrounds are substantiated as follows:

- Ar – Ar is the strongest chemical bond;
- Ar – Ar conjugation possesses the largest heat-resistance and thermal stability;
- Ar – Ar bond is able to resist high-power impacts (dynamic conditions) due to redistribution (de-localization) of electronic density along molecules as well as within the polymer structure. This means that in addition to the induction (I) effect the bond can possess the conjugation (+/- C) effect depending on the location within aromatic nucleus.

The +C effect is most important. In the case that ether oxygen or nitrogen of amides are substituents, these elements, being involved into a conjugated-bond system, play a part of electron-donor substituents influencing both oligomer polarization and reactivity (note that the more right-hand position takes an element in the periodic table row, the less is +C effect). [3] It is obvious that oligomeric-unit structure oligomers are polyfunctional compounds.

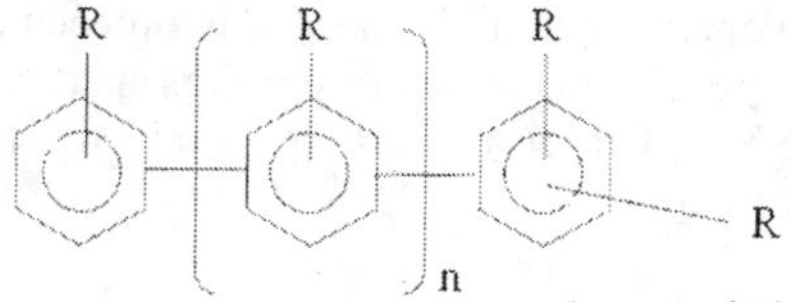

, where $n = 0 \div 2$

R-OH, NH_2

Evidently, such structures will possess the closest packing and the least contraction when structuring. Under intense thermal/power impact and in the case of strong oxidant action additional structuring of the system may occur resulting in the generation of a high-strength matrix with up to 70%-mass carbon residue.

In already cured blends based on ARGOF compound (created during our investigations) low porosity and stability in non-polar hydrocarbons (resistance to aviation fuel impacts) are anticipated.

The availability of aromatic conjugation in oligomers is confirmed indirectly by color variations (from straw to orange, next, to red-brown and, finally, to black) as their length, i.e. the number of fragments, increases.

Using models we managed to calculate and obtain the following compounds:

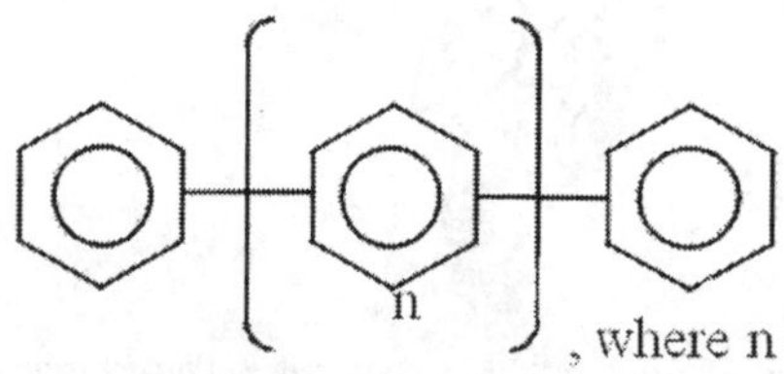

, where $n = 0 \div 2$

and, next, functional groups which received the name ARGOF:

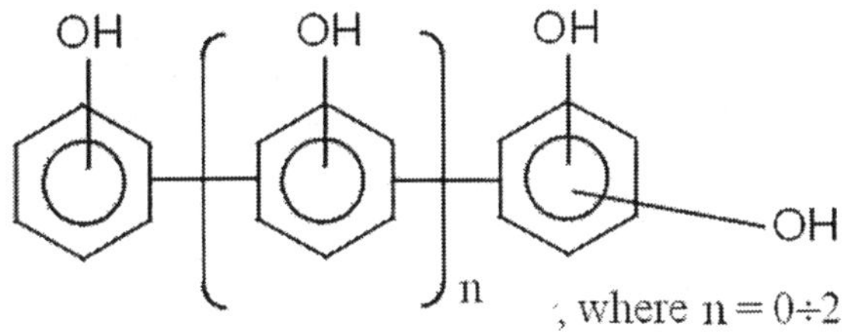

Such oligomeric models are water-soluble when n=0 (than their solubility decreases) and are insoluble in benzene, gasoline and toluene. These are high-reactivity oligometric models.

In hydroxyl groups indirect epoxidation is possible resulting in the generation of the following structure:

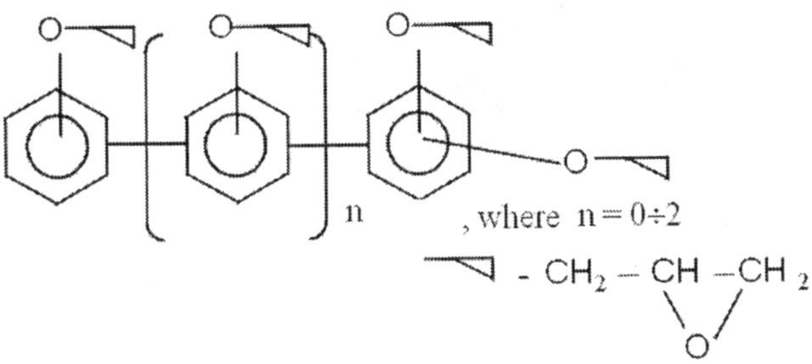

3. IMPROVING TECHNOLOGIES OF POLYMER COMPOUND RE PROCESSING

As mentioned above, reprocessing of oligomers and compounds made thereof necessitates further improvement in order to optimize contradictory requirements and stabilize parameters responsible for the process reproduction during operation.

Any technology of coating/material fabrication consists of the following three or four operations (*Figure 2*) [4]:

I.- oligomer synthesis or component dosing (15 – 250 °C);

II.- dosing of curing agents, reactivity stabilization and waiting for reprocessing (5 - 110°C);

III.- distribution of the compound and accelerating agents on filling material, carrier or in volume with their prefixing relative to each other and pre-polymerization (15 – 110°C);

IV.- –forming of half-finished material and ultimate structuring of the compound into coatings, work parts, materials or articles (35-120°C).

The above-listed stages and their related transformations of thermosetting oligomers from A-state (resol) through B-state (resitol) into C-state (resite) are shown in *Figure 2* along the axis of ordinates, whereas operations and conventional duration of the reprocessing period are demonstrated along the abscissa axis. Note that 1, 2 and 3 curves depict continuous transformations of the reaction mass despite the fact that in reality the reaction can be stopped using technical means ("process-waiting" - mode) and, after a pause, be continued.

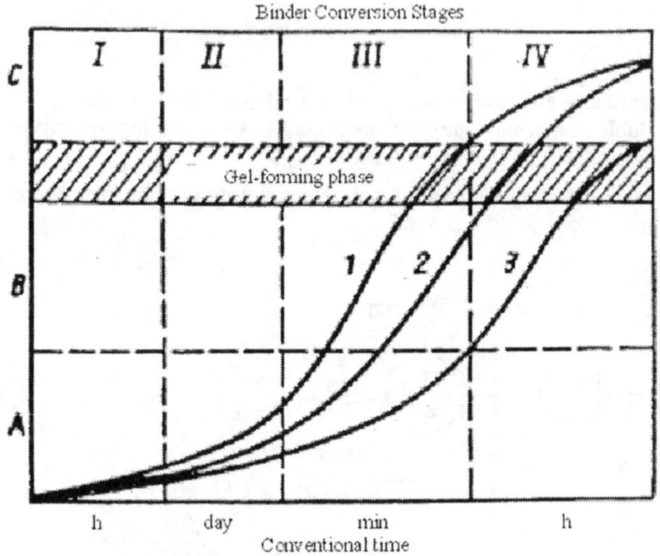

Figure 2. Conventional modifications of the binder fractional conversion (stages A, B and C) in time (h-hours, d-days, m-minutes) when reprocessing

The curves 1 and 3 (see Figure 2) depict inadmissible conversion values. At these values in the case of Curve 1 a high degree of the binder cure is reached during coating and drying operations resulting in losses in fusibility and solubility; i.e. the least possible number of 3D joints is generated (τ_{gel} - gel-forming time) - gelatinization. In the case of Curve 3 a minor fractional conversion results in the binder draining since the gel-forming phase, determining the cessation of flowing capacities, is reached over a longer period of time.

When realizing the process from the initial stage up to the final stage using the above scheme, the variety of potential impacts on the system reduces sharply to a possibility of modifying only wait time and temperature (in most cases towards lowering) after reaching the binder gel-forming phase [5].

Thus, even knowing in advance when the system reaches the "gel-forming phase", technologists are unable either to predict the system behavior at initial process stages or to exert goal-directed actions using available dosed agents.

Many compound-designers agree that such a binder parameter as so-called "adhesiveness time" or, more precisely "microgel-forming time " can be used to describe the reaction mass at initial stages of the curing process, i.e. during oligomer prepolymerization.

For short let us denote the "microgel-forming time" by $\tau_{microgel}$ or τ_{mgel}. This time is determined on polymerization plates as follows: the reaction mass gathers viscosity and, when raising upward a special spatula, resin filaments 20 to 30 mm long are generated; next, these filaments do not break abruptly but drain back to the plate surface.

The microgel (adhesiveness)-forming phenomenon of reactive oligomers is also recorded in optical analyses of the reaction mass for light diffusion asymmetry (in this case the phenomenon is observed long before gel is generated) and it is established as well by dilatometric analyses.

The article authors recorded the "microgel-forming time" in both laboratory and industrial conditions when fabricating phenol resins and different binders; note that τ_{mgel} is determined with less difficulties. The very fact of such a phenomenon is capable for specifying the system, whereas the τ_{mgel} - τ_{gel} relationship gives an estimate of the binder pre-polymerization degree. This circumstance also suggests that the last-mentioned parameter can be changed for controlling the binder flow properties or viability, and, also, for stabilizing the binder parameters from lot to lot. The last statement is very important in the process management.

To our knowledge the "τ_{mgel} - τ_{gel}" relationship was never mentioned in special literature; thus, one suggests designating it as "binder pre-polymerization degree" ("BPD") and finding the possibility of keeping it within 0,25-0,75 with a minor variation of the "gel-forming time".

It is anticipated that at BPD=0,25 the "binder flowability", as a parameter of polymeric compound, is capable for ensuring minimum flowing requirements. As a flowing parameter, the "resin flowability" at BPD=0,75 will ensure filling of deep undercuts providing for full curing in both cases at identical binder composition.

Theoretical views of oligomer curing at initial stages can be summarized as follows: at a test temperature within the binder mass "nucleation" centers add or orient molecules/associated compounds toward them and expand towards each other until the centers are able to "keep" peripheral areas. When the center impact on peripheral areas loses force, one can consider that around "nucleation" centers a microgel has been generated wherein the process of forming a closer-packed structure still continues; this results in extrusion of weakly-connected segments/molecules from the microgel. The extrusion intensifies further due to the growth of microgels towards each other. The extruded/redistributed structures including molecules and their segments possess higher energy due to the extrusion phenomenon. When reaching microgel interface, such structures come into contact with identical structures of other microgel and begin their interactions (onset of the exothermic process). As a result, first, a minimum number of 3D gel cross-linking bonds (typical for the gel-forming process) are generated in volume, next, more active curing starts.

An assumption suggests itself that during initial stages of curing one needs to ensure most regular growth of uniform-sized microgels (of closest packing); in that way faulted areas within conglomerates will be eliminated. From this the relevant requirements for oligomers and curing agents concerning chain-length distribution, binder component compatibility, their structure, reactivity as well as distribution by functionality type arise.

The microgel-forming phenomenon can be explained more accurately as follows: microgels are generated in polymeric substances for any conditions; they are responsible for blend solution stratification, precipitation of individual fractions, et al.

Microgels can be destroyed, but they are generated again with closer packing. This phenomenon is clearly confirmed by investigations of Professor M. Akutin and S. Kasparov [4]. In keeping with their research, after an acoustic impact of a particular frequency and amplitude on phenol oligomers, their viscosity, first, decreases and, next, increases in time up to a value below the initial one; note that the subsequent impacts (up to seven times) produce similar effects.

Apparently, microgels, their growth and mutual interactions contribute to some redistribution of amorphous and crystalline phases and can lead to an increase in viscosity of oligomer solutions without variations in their concentration, functional group consumption or gel-forming time. In such a case gels of higher thickness can be generated due to the creation of permolecular structures (globules, spherulites, et al.), which can be destroyed through mechanical or thermal impacts. These phenomena were studied in detail by M. Magomedova [4].

The process of maturing has its origin in microgel-forming phenomenon that creates necessary conditions for stabilizing the reactivity. In most cases the maturing stage is completed in 18 to 24 hours; thereafter binders can wait for reprocessing over many days. Based on theoretical backgrounds, we carried out an investigation aimed at obtaining initial compounds of a common formula:

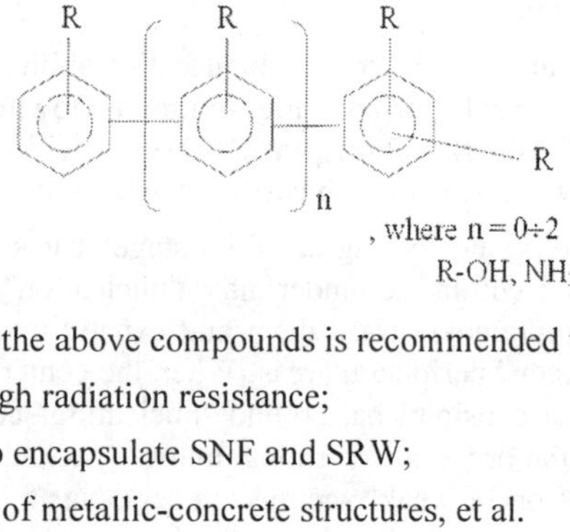

, where $n = 0 \div 2$

R-OH, NH_2

The subsequent use of the above compounds is recommended when designing:

- coatings of high radiation resistance;
- compounds to encapsulate SNF and SRW;
- plasticization of metallic-concrete structures, et al.

Thus, a new technology of obtaining compounds with unique properties clarifies itself.

4. CONCLUSION

Based on specific tasks of ensuring safety during radiation-dangerous operations in the Russian Navy when using polymeric compounds, theoretical backgrounds for creating an oligomeric base were established, and specific procedures and techniques for its reprocessing were determined. Structure simulation confirmed unique characteristics of aromatically conjugated hydroxy-phenylenes able to ensure high radiation, hydrolytic & acid resistance as well as mechanical strength due to compactness of their structure, de-localization of electronic density and polyfunctionality. Modeling of the developed innovative methods of reprocessing the above compounds into gel-forming systems, coatings, compounds, hermetics, adhesives, encapsulated and other systems has demonstrated their ability to ensure preservation of SNF, RW, FA of damaged reactors and their components for a long period of time.

Approaches were determined for designing a portable (transportable) self-sufficient installation for use during both operation and decommissioning of radiation-dangerous units & installations.

5. References

1. Berlin, A.A., Kefeli V.J., Korolev G.L. (1978) *Oligo-ether-acrylates,* Nayka (Science), Moscow (in Russian).

2. Berlin, A.A., Pakhomova L.K. (1990) Polymeric materials for high-strength reinforced composites, *Vysokomoleculiarnye soedinenija (High-molecular compounds),* 7, 1343-1348 (in Russian).

3. Temnikova, T.I. (1963) *Course of organic chemistry theory*, Khimia (Chemistry), Leningrad (in Russian).

4. Mescheriakov, Y.J. (1994) Technological aspects of manufacturing foil-clad glass fiber plastics, *Electrotekhnika (Electrotechnology)* **12**, 6-7 (in Russian).

5. Zagorujko, A.I., Kuntsevich, V.M., Lutov, S.D., et al. (1982) *Method for controlling fabrication of reinforced plastics,* Inventor's certificate USSR 939274, *Moscow* (in Russian).

6. Mescheriakov, Y., Pluzhnikov, V. (2001) Incombustible and gasoline-resistant coatings and materials for installations of potential fire risk, roads and pipeline sanitation, *Nayka i technologija v promyshlennosti (Science and technology in industry)* **3(6),** 70-73 (in Russian).

SUBMARINES WITH LIQUID METAL COOLANT

PROBLEMS OF LONG TERM AND SAFE STORAGE OF UNLOADED AND NON-UNLOADED SPENT NUCLEAR FUEL FOR NUCLEAR SUBMARINES WITH LIQUID-METAL COOLANT

D. PANKRATOV, V. ANDREYANOV, M. BUGREVEV
A. DEDOUL, YE. EFIMOV, B. KOMLEV, L. RYABAYA,
V. SAZONOV, B. SIVAK, G. TOSHINSKY, V. CHITAYKIN
Institute of Physics and Power Engineering
Obninsk, Kaluga rg., Russia

Introduction

During the years 1962 through 1997 several nuclear submarines (NS) with lead-bismuth liquid-metal coolant (LMC) in the primary circuit of the reactor installation were included in Russian Navy.

Reactors of these NSs were developed and designed in Obninsk under IPPE (Institute for Physics and Power Engineering) scientific supervision and are intermediate neutron reactors. The reactor fuel composition includes inter-metallid UBe_{13} with ~ 90% enrichment in U-235 dispersed in the beryllium matrix. There was a two-reactor nuclear power installation (NPI) on board of the first NS (Project 645). After the accident occurred in 1968 during the second lifetime in the left side reactor with partial damage of the core, the NS was left standing at the dock. Then, in 1981 after filling the free chambers of the reactor and the reactor compartment with conservants, the NS was sunk in the Kara Sea near the New Land Island at the depth of ~ 50 m. Each of other LMC cooled NSs (Project 705, Project 705K known as NSs of the "Alpha" class) has one reactor.

Currently, all LMC cooled NSs of Project 705, Project 705K have been withdrawn from operation since the beginning of 90s and on-shore technical maintenance has not been provided. For four NSs the nuclear fuel has been unloaded, for two NSs the nuclear fuel is in the reactors with the "frozen" coolant. The nuclear fuel in the two reactor compartments cut out of NSs (Order 900, Order 105) has also not been unloaded. The coolant in these reactors is "frozen"'. Further it is planned that spent nuclear fuel (SNF) will be unloaded from the other NS reactors, except for reactor (Order 900).

It should be highlighted that Order 900 NS was the first in the series of Project 705, and Project 705K ships. It was put in operation in 1970 [I]. The core of its reactor had operated ~ 10 % of its lifetime. In 1972 the NS was withdrawn from operation. This was caused by failure of the primary circuit's auxiliary pipelines and the need to repair them. The reactor compartment was cut out of the NS, the free chambers of the primary circuit were filled with a conservant based on furfurol and a bitumen layer was placed over the whole surface of the reactor compartment's upper deck including the reactor lid. In this state the reactor compartment of this NS can be stored long term without unloading the nuclear fuel.

A.A. Sarkisov and L.G. LeSage (eds.),
Remaining Issues in the Decommissioning of Nuclear Powered Vessels, 341–348.

Particular features of LMC cooled reactors of NSs are the following: by using special transport-technological equipment, their cores are loaded into the reactors and unloaded out of them as a unified removable unit that includes a core with the inserted control and protection system (CPS) rods, a side beryllium reflector (Project 705, Project 705K) and a upper plug for biological protection. When nuclear fuel is unloaded, the side beryllium reflector is not extracted out of the reactors of the Project 705 NSs. Unloaded spent removable units (SRU) with SNF are in special repositories in the of village Gremikha (the Kola Peninsula). They were placed in steel containers in which there was liquid "pure" (non-radioactive) lead-bismuth eutectic alloy that then solidified ($t < 125°C$).

Lead-bismuth eutectic alloy was used as coolant in the NPI primary circuit. It consists of 44.5% (mass) of lead (mark C00 in compliance with COST 22861-77 [2]) and 55.5% (mass) of bismuth (mark Bi00 in compliance with GOST 10928-75 [3]).

Currently, residual heat release in each core has been less than 2 kW.

There are also two SRUs that were unloaded out of the reactors in 1967. These were the initial cores from Project 645 NSs. The conditions of their storage are similar to those of the SRUs of Project 705 and Project 705K NSs. In the reactors of Project 705 and of Project 705K NSs nuclear fuel was not replaced by the fresh fuel. The operational time of the cores of these reactors was 10 - 100 % of the lifetime.

1. Radiation Potential of LMC Cooled Reactors

Operation of NSs' reactors at power levels was followed by accumulation of radioactivity [41] in their cores, structures and coolant. When handling the SNF, it was necessary to pay careful attention to radioactivity. For each NS the amount of long-lived radioactivity depends on the operational time of its reactors. Below are cited the computational data for evaluation of actinide radioactivity levels, long-lived fission products and long-lived radionuclides resulting from activation of the reactor structures of the Project 705K NS with an entirely expended core. It was presumed that during the lifetime the core operated at nominal power,

TABLE 1. Radioactivity of Actinides, Bq

Radionuclide	**$T_{1/2}$, years**	**The time after shutting down the reactor, years**				
		5	10	20	50	100
U-232	72	3.03E09	2.88E09	2.62E09	1.96E09	1.21E09
U-234	2.48E05	3.29E11	3.29E11	3.29E11	3.29E11	3.29E11
U-235	7.04E08	9.88E09	9.88E09	9.88E09	9.88E09	9.88E09
U-236	2.34E07	2.77E10	2.77E10	2.77E10	2.77E10	2.77E10
U-238	4.47E09	2.06E08	2.06E08	2.06E08	2.06E08	Z06E08
Np-237	2.14E06	2.43E.10	2.43E10	2.43E10	2.43E10	2.43E10
Pu-236	2.85	8.89E09	2.64E09	2.34E08	1.62E05	8.74E-01
Pu-238	87.75	5.37E13	5.12EI3	4/73E13	3.74E13	2.52B13
Pu-239	24380	3.40E12	3.40E12	3.40E12	3.40E12	3.39E12
Pu-240	6537	8,21 Ell	8.20E11	8,19E11	8.17E11	U2En
Pu-241	14.54	1.35E14	1.06E14	6.60E13	1.58E13	1.47E12
Pu.242	3.76E05	2.90E08	2.90E08	2.90E08	2.90E08	2.90E08
Pu-244	8.2E07	8.70E-01	8.70E-01	8.70E-01	8.70E-01	8.70E-01
Am-241	432.8	1.75EI2	2.73E12	4.04E12	5.45E12	5.44E12
Am-242m	152	2.08E10	2.03E10	1.94E10	1.69E10	1.35E10
Am.243	7400	5.02E08	5.02E08	5.02E08	5.02E08	5.02E08
Cm-243	28.5	2.01E09	1.78E09	1.40E09	6.74E08	2.00E08
Cm-244	18.1	7.68E09	6.35E09	4.33E09	1.38E09	2.05E08
Cm-245	8532	5,72.805	5.72E05	5.71E05	5,70E05	5.68E05
Cm-246	4820	9.37E03	9.37E03	9.35E03	9.31E03	9.25E03
Cm-247	1.56E07	2.55E-02	2.55E-02	2.55E-02	2.55E-02	2.55E-02
Cm-248	3.39E05	1.60E-02	1.60E-02	1.60E-02	1.60E-02	1.60E-02
Total, Bq		1.95E14	L65E14	1.22E14	6.33E13	3.67E13

TABLE 2. Radioactivity of Long-lived Fission Products, Bq

Radionuclide	$T_{1/2}$	The time after shutting down the reactor, years				
		5	10	20	50	100
Se-79	6.5E04years	1.47E10	1.47E10	1.47E10	1.47E10	1.47E1Q
Kr-85	10.74 years	2.62E14	1.90E14	9.95E13	1.43E13	5.68EI 1
Rb-87	4.88E10 years	9.13E05	9.13E05	9.13E05	9.13E05	9.13E05
Sr-90	28.5 years	2.53E15	2.24E15	1.76E15	8.46E14	2.5 IE.14
Y-90	61.4h	2.53E15	2.24E15	1.76E15	8.46E14	2.51E14
Zr-93	1.53E06 years	7.11.E10	7.11E10	7.11E10	7.11E10	7.11E10
Nb-93m	13.6 years	3.20E10	3.70E10	4.50E10	6.00E10	6.60E10
Nb-94	2.03E04 years	8.52E05	8.52E05	8.52E05	8,52E05	8.52E05
Tc-99	2.15E05years	4.64E11	4.64E11	4.64E11	4.64E11	4.64E11
Ru-106	368.2 days	1.02E13	3.29EU	3.43E08	3.9E-01	-
Rh-106	29.9 s	1.02E13	3.29E11	3.43E08	3.9E-01	-
Pd-1.07	6.5E04 years	5.39E08	5.39E08	5.39E08	5.39E08	5.39E08
Ag-108m	12 7 years	2.58E05	2.51E05	2.38E05	2.02E05	1.54E05
Ag-108	2.41 min.	2.19E04	2.13E04	2.Q2E04	1.71E04	1.30E04
Cd-113m	13.6 years	1.52E11	1.17E11	7.05E10	1.52E10	1.18E09
Sn-121m	55 years	3.29E09	3.09E09	2.72E09	1.87P;09	9.94E08
Sn-126	1.0E05 years	1.28E10	1.28E10	1.28E10	1.28E10	1.28E10
Sb-126m_1	19 min.	1.28E10	1.28E10	1.28E10	1.28E10	1.28E10
Sb-126	12.4 days	1.79E09	1.79E09	1.79E09	1.79E09	1.79E09
I-129	1.57E07years	8.67E08	8.67E08	8.67E08	8.67E08	8.67E08
Cs-134	2.062 years	1.51E14	2.80E13	9.63E11	3.91E07	1.84EOO
Cs-135	2.3E06 years	4.62E10	4.62E10	4.62E10	4.62E10	4.62E10
Cs-137	30.174yea.rs	2.58E15	2.30E15	1.83E15	9.18E14	2.91E14
Ba-137m	3.53.5 s	2.43E15	2.17E15	1.73E15	8.66E14	2.74E14
Ce-144	284.3 days	3.30E13	3.87E11	5.37E07	1.4E-04	-
Pr-144m	7.2 min.	4.30E11	5.04E09	6.99E05	1.8E-06	-
Pr-144	17.28 min.	3.30E13	3.87E11	5.37E07	1.4E-04	-
Pm-147	2.623 years	5.93E14	1.58E14	1.12E13	4.05E09	7.27E03
Sm-151	87 years	4.39E13	4.22E13	3.9QE13	3-07E13	2.06E13
Eu-152	12.4 years	2.68E12	2.03E12	1.16E12	2.18E11	1.34E10
Eu-154	8.5 years	4.29E13	^E13.	L '•26E13	1.09E12	1.84E10
Eu-155	4 96 years	2.75E13	1.37E13	3.38E12	5.09E30	4.63E07
Total, Bq		1.13E16	9.41 El 5	7.25E.1.5	3.52E15	1.09E15

The CPS absorbing rods of the reactors of Project 705 and Project 705K NSs use a composition that includes boron and europium. That was the cause of rather high accumulated radioactivity of Eu^{152} and Eu^{154} isotopes. It should be highlighted that during the cooling period of 1 - 20 years europium radioactivity in the absorbing elements was a major contribution to residual heat release in the reactor core and exceeded the fission products contribution by a factor of 2.

TABLE 3. Europium Radioactivity in CPS Rods, Bq

Radionuclide	$T_{1/2}$,years	The time after shutting down the reactor, years				
		5	10	20	50	100
Eu-152	13.2	4.74E15	3.65E15	2.16E15	4.48E14	3.24E13
Eu-154	8.5	1.98E15	1.32E15	5.82E14	5.04E13	8.48E11
Total, Bq		6.72E15	4.97E15	2.74E15	4.98E14	3.32E13

Beryllium was used as a moderator in the reactor core and in the reflector. Two sequential reactions of neutron bombardment caused formation of tritium in the reactor.

$$_4Be^9 + n \rightarrow {}_2He^4 + {}_3Li^6$$

$$_3Li^6 + n \rightarrow {}_2He^4 + {}_1H^3$$

Besides, the following reaction caused formation of tritium in the CPS absorbing rods:

$$_5B^{10} + n \rightarrow 2_2He^4 + {}_1H^3,$$

Tritium was also formed by triple uranium fission.

TABLE 4. Tritium Radioactivity, Bq

Elements	**The time after shutting down the reactor, years**				
	5	10	20	50	100
Core	1.85E15	1.40E15	8.00E14	1.45E14	8.50E12
Side Beryllium Reflector	2.80E15	2.10E15	1.20E.15	2.20E14	1.30E13
CPS rods	5.50E13	4.10E13	2.40E13	4.40E12	2.60E11
Total, Bq	4.70E15	3.54E15	2.02E15	3.70E14	2.18E13

TABLE 5. Radioactivity of Steel of Fuel Elements Claddings, Lattices, and a Removable Unit Shell, Bq

Radionuclide	**$T_{1/2}$, years**	**The time after shutting down the reactor, years**				
		5	10	20	50	100
Ni-59	7.5E04	3.82E11	3.82E11	3.82E11	3.82E11	3.82E11
Ni-63	100.1	3.40E13	3.30E13	3.05E13	2.50E.1.3	1.80E13
Co-60	5.27	5.50E14	2.90E14	8.50E13	2.00E12	2.70E09
Fe-55	2.72	1.10E15	3.00E14	2.30E13	1.05E10	3.00E04
Total, Bq		1.68E15	6.23E14	1.39E14	2.74E13	1.84E13

TABLE 6. Radioactivity of Primary Circuit Coolant, Caused by Long-lived Nuclides, Bq/kg

Radionuclide	**$T_{1/2}$**	**The time after shutting down the reactor, years**				
		5	10	20	50	100
Ca-41	1,35E05years	7.00EOO	7.00EOO	7.00EOO	7.00EOO	7.00EOO
Mn-54	312.3 days	1.69E03	2.73E01.	7.12E-03	1..26E-13	-
Co-60	5.27 years	1.41E05	7.34E04	1.98E04	3.88E02	5.54E-01
C.o-58	70.8 days	9.86E-03	1.84E-10	-	-	-
Ni-63	100.1 years	2.30E04	2.22E04	2.07E04	1.68E04	1.19E04
Ag-108m	127 years	2.80E04	2.73E04	2.58E04	2.19E04	1.67E04
Cd-113m	13.6 years	3.18E05	2.47E05	1.48E05	3.20E04	2.50E03
Sn-119m	250 days	6.24E03	8.34E01	1.49E-02	-	-
Tl-204	3.78 years	9.55E04	3.82E04	6.13E03	2.53E01	2.65E-03
Pb-205	1.51E07years	2.56E03	2.56E03	2.56E03	2.56E03	2.56E03
Bi-207	30.2 years	3.16E05	2.82E05	2.24E05	1.12E05	3.54E04
Bi-208	3.65E05 years	8.39E04	8.39E04	8.39E04	8.39E04	8.39E04
Bi-210m	3.6E06 years	1.76E04	1.76E04	1.76E04	1.76E04	1.76E04
Po-210	138.4 days	2.57E06	2.77E02	3.21E-06	-	-
A_α, Bq/kg		2.58E06	1.79E04	1.76E04	1.76E04	1.76E04
A_β, Bq/kg		1.01E06	7.76E.05	5.31E.05	2.70E05	1.53E05

The analysis of Table 1 through Table 6 data has revealed that during the cooling period of 5 - 50 years, after shutting down the reactor the total activity of long-lived radionuclides in nuclear fuel, structures and coolant can reach $2.2 \cdot 10^{16}$ - $3.7 \cdot 10^{15}$ Bq or $6 \cdot 10^5$ - 10^5 Ci. It is also important to highlight that among the actinides accumulated in the core during the lifetime, the mass of Pu^{238} ($T_{1/2}$ = 87.75 years) is ~ 80 g. There is a neutron source operating continuously in the shut down reactor govered by the half-life time of Pu^{238}. Its intensity is ~ $4 \cdot 10^9$ n/s and is caused by (α, n) reaction on beryllium nuclei in the fuel composition. The α particles come from the radioactive decay of Pu^{238}. With due account for multiplication in the subcritical reactor, (e.g. under 10% subcriticality), during the long period after shutting down the reactor, the total intensity of neutron irradiation will be ~ $4 \cdot 10^{10}$ n/s.

2. SNF Storing

Currently, nuclear safety of stored SNF is ensured because the cores of the reactors both in the NS vessels (non-unloaded SNF) and in the on-shore repositories are in a deep subcritical state (K_{eff} < 0.95) due to full insertion of the CPS absorbing elements (AE) into the cores. The CPS drivers on the unloaded cores that are stored in the on-shore repositories have been dismantled. Tight steel caps have been installed on the absorbing elements cladding and welded to them. The absorbing elements cladding have been cut slightly above the reactor lid.

Unloaded and non-unloaded SNF is in "frozen" lead-bismuth alloy. During SNF stored unauthorized insertion of positive reactivity is eliminated by the fact that the lead-bismuth alloy has been "frozen" in the CPS claddings and AEs prevented from moving. The CPS drivers of the NS with non-unloaded SNF have been de-energized, and the alloy in the CPS claddings has been "frozen". Highly improbable unauthorized manual extraction of the four emergency protection (EP) rods installed in the dry channels does not bring the reactor out of the subcritical state (total efficiency of the EP rods is less than 1.5 %). In the NS reactor the probability of forming open pores in solidified eutectic has been eliminated by the technology of coolant "freezing" in the primary circuit. Therefore, in case of tightness failure of the primary circuit in the NS reactor, ingress of condensed or some other moisture into the core is eliminated. Primary circuit tightness has been provided by standard means. For that reason, insertion of additional positive reactivity into the reactor has been eliminated.

In the on-shore repositories the lids of steel containers (i.e. SNF repositories) are tight. Additionally, at the point of joining the SRU and the container there has been applied a bitumen layer preventing ingress of condensate moisture, rain and ground waters into the containers.

Radiation and ecological safety of stored SNF is based on the principles of high-separation shielding for the pathways of probable proliferation of radionuclides into the environment [5], [6]. This is a sequential chain of physical safety barriers. For the fission products these barriers are: a matrix of heat-releasing elements, fuel elements claddings, "frozen" coolant, a tight primary circuit (or a steel container in the on-shore repository), a strong vessel in the NS. In compliance with radiation and technological control data, in both the SNF repositories and NS reactors the fuel element claddings of all the cores of liquid-metal reactors are almost undamaged. In the coolant and in-vessel structures the barriers against radioactive products are as follows: a coolant matrix, the walls of the reactor (container) vessel, a strong vessel in the NS. With proper technology for "freezing" the SRU that eliminates forming of pores and chambers in the eutectic, as it has been performed in the NS reactor, the concept of long-term storage LMC cooled reactors' SNF in "frozen" lead-bismuth alloy is a most drastic measure for ensuring nuclear and radiation safety. A "frozen" SRU is a strong-tight ensemble preventing ingress of noticeable amounts of moisture into the core under any possible scenario. The matrix of solidified coolant possesses good immobilization properties and extremely low speed of leaching radionuclides when contacting with sea or distilled water. Besides, in this case the problem of highly enriched uranium non-proliferation is solved successfully. In conditions of a strong-coupled ensemble of fuel subassemblies (FSA), their unauthorized extraction out of the SRU is almost impossible.

3. Problem of SNF Long Storing

However, when the LMC cooled NPIs were designed, no provisions were made for long-term storage of SNF in the NSs and in on-shore repositories. It was presumed that after several years of cooling in the repositories, the spent cores of liquid-metal reactors would be transported to the nuclear fuel reprocessing plant. The technology used for loading the SRUs into the containers-repositories and further "freezing" did not rule out the possibility of forming pores and caverns in the eutectic. For that reason, in current conditions for storing the SRUs, we cannot rule out condensation or some other probable moisture ingress into the core. In compliance with preliminary computations, ingress of 1 kg of cold water equally distributed in the core pores gives positive reactivity effect of ~ 0.5 β_{eff}. Therefore, at a subcriticality of 10 β_{eff}, ingress of 20 kg of water (~ 10 % of coolant volume in the core) can result in reaching SRU criticality ($k_{eff} = 1$).

4. Proposals on International Cooperation

Currently the RF MINATOM is taking measures for providing reliable, safe long-term storage of unloaded nuclear fuel from NSs with LMC. However, the proposed measures could be carried out within the frameworks of international cooperations. The priorities for this cooperation should be as follows:

- to develop diagram and methodology designs and, using technical control means, organize operative control of changing SRU reactivity in conditions of their on-shore storage;
- to carry out research on identifying effective conservants and develop measures for additional conservation and sealing of repositories that eliminate ingress of moisture into the SRU core;
- to carry out computation and research on the analysis of the radiation consequences of hypothetical accidents that might occur in the SRU repository as a result of extreme external impacts.

5. Conclusions

1. We have evaluated the activity levels of radionuclides in the fuel material, CPS rods, structural materials and lead-bismuth coolant that in the aggregate determine the radiation potential of the reactor compartment. These estimates were performed for the nuclear steam-generating installation of the Project 705K NS with LMC that was withdrawn from operation and for which the core lifetime was expired.
2. For substantiation of nuclear and radiation safety of long-term (100 years and more) storage of the SNF liquid-metal reactors in the NS reactor compartments and particularly in the on-shore container-repositories, it is necessary to carry out in-depth analysis of all possible ways (including highly improbable ones) of affecting subcriticality of storing SRUs. In addition, the possible radiation consequences of extreme external impacts (i.e. fire, flooding, terror actions, etc.) must be evaluated as well. It is also necessary to investigate physical and chemical processes (galvanic and chemical corrosion, phase conversions, etc.) that occur during the storing period in the complex repository system or reactor compartment.
3. The following measures of providing safety for long-term storage of SRUs must be considered obligator)':
   - organization of reactivity control and radiation dose control in the SRU— repository system;
   - periodical inspection of storage conditions and, if necessary, taking measures on ensuring safety.
4. The proposals on international cooperation have been formulated.

6. References

1. Gromov, B.F., Grigoriev, O.G., Dedoul, A.V., Toshinsky, G.I., Stepanov, V.S., and Nikitin, L.V. (1999) Analysis of Experience of Operating Reactor Installation Using Lead-Bismuth Coolant and Accidents Happened, *Heavy-Liquid Metal Coolant in Nuclear Technology,* (Proc. Conf-, Obninsk, Russia), Vol. 1, pp. 63-69.
2. (1978) High pure lead. Technical conditions, COST 22861-77. Publishing House of Standards, Moscow.
3. (1975) '•Bismuth. G08T 10928-75", Publishing House of Standards, Moscow.
4. Usanov, Y.I, Pankratov, D.V., Popov, E.P., Markelov, P.I., Ryabaya, L.D. and Zabrodskaya, S.V (1999) Long-Lived Radionuclides of Sodium, Lead-Bismuth and Lead Coolants in Past Reactors, *Atomnaya En&rgia,* 87, issue 3, pp. 204-210.
5. (1999) Normalization of the Radio-Ecological Situation m the North-West Region of Russia due to Removing Spent Nuclear Fuel of-'Alpha' Class NSs out of the Region, *Conceptual design,* Proposals on international cooperation within the Contract with NNC Limited, SSC RF-IPPE, Obninsk.
6. Sazonov, V.K., Tihomirov, V.I., Tihonenko, V.B., Toshinsky G.I., Shestopalov, A.M., Stepanov, V.S., Vahrushin, M.P., and Alekseev, G.A. (1999) Technology and Experience of Repair Works and Refuelings of the Nuclear Power Installations with Lead-Bismuth Coolant, Heavy-Liquid Metal Coolant in Nuclear Technology, (Proc. Conf., Obninsk, Russia), Vol. 2, pp. 643-649.

CURRENT PROBLEMS OF UTILIZATION OF NUCLEAR SUBMARINES WITH LIQUID-METAL COOLANT

V. SAZONOV, M. BUGREYEV, A. DEDOUL, A. ZABUD'KO,
D. PANKRATOV, G. TOSHINSKY, V. CHITAYKIN
Institute for Physics and Power Engineering
Obninsk, Kaluga rg., Russia

V. STEPANOV, M. VAHRUSHIN
Experimental Design Bureau "GIDROPRESS",
Podolsk, Moscow rg., Russia

S. VERHOVODKO
SPMBM "Malakhit",
St. Petersburg, Russia

1. Introduction

Currently all nuclear submarines (NS) of the Russian Navy with lead-bismuth liquid-metal coolant (LMC) in the primary circuit have been withdrawn from operation.

After unloading the spent cores from the reactor, some of these NSs were utilized.

The last of the spent cores were unloaded in 1992. Since then the specialized on-shore refueling base has not been functioning. A distinguishing feature of the period is the long period the ships stay in the non-standard state and a particular state of the primary circuit coolant — it has been "frozen".

Reactors of NS plant N° 105 with steam-produced installation (SPI) units of plant № 120 and № 125 and reactors of NS plant № 910 are planned to be unloaded. These are NSs of Project 705, Project 705K known as NSs of the "Alpha" class, which are fast, automated, with one reactor, one shaft and with a titanium hull. The chief constructor of the ship is SPMBM "Malakhit". Reactor plants for these NSs were designed by EDO "Gidropress" and OKBM under scientific supervision of Institute for Physics and Power Engineering (IPPE) [4].

Unloading the spent reactor cores is one of the most important stages in the technological scheme of NS utilization. Finding a successful solution to the problems of this stage has a crucial effect on the rates of NS utilization.

With the existence of the developed technology of reprocessing spent nuclear fuel (SNF), the technical means of on-shore base infrastructure, special transport means and the technical state of the NSs at the final stage of their lifetime the NSs should be consistent with the criteria of readiness for the spent reactor cores to be unloaded.

A.A. Sarkisov and L.G. LeSage (eds.),
Remaining Issues in the Decommissioning of Nuclear Powered Vessels, 349–355.

2. NSs with LMC at the Final Stage of Their Lifetime

2.1. THE NS OF PLANT № 105

The NS of plant № 105 includes two power units; a power unit of plant № 120 and a power unit of plant № 125.

2.1.1. The Power Unit of Plant № 120

The power unit of plant № 120 is the SPI of the plant № 105 NS at the first stage of its operation. In 1986 after the accident in which the primary circuit coolant leaked into the compartment, the SPI unit with the reactor compartment was cut out of the NS vessel. The compartment equipped with additional bow and after floating volumes, was launched and nowadays it stays in the village of Saida. The reactor was separated from the primary circuit by the check valves and "frozen". It was a monolithic "lead-bismuth" ingot pierced by the channels of the reactor control and protection system (CPS). All absorbers were completely inserted made immobile by dismantling the electric cables. The compensating rods (KR) and automatic control rods (ACR) were in the "frozen" alloy.

The operational time of the reactor was ~ 50 % of its lifetime. From the moment of cutting out of the reactor compartment, no inspection of the equipment and reactor installation systems operated for heating the reactor before unloading the core had been carried out. Before heating the reactor, pressurization of the heating system's coiled pipelines must be carried out, they must be divided into smaller sections and their scheme must be re-arranged for directed and continuous heating of the "frozen" reactor. Maintaining the operability of the reactor temperature control sensors is not probable. For that reason, it is presumed to use thermo-sensors of the specially designed tensor-meter measuring device. It is planned to install the device in place of on of the emergency protection (EP) rods. To carry out this potentially nuclear- and radiation-hazardous work, corresponding equipment was designed and fabricated. It was not used during operation. It is important to keep this equipment at the Navy warehouse. Thus, the vital task of the preparation period for the spent reactor core to be unloaded is the inspection of the equipment and systems that must function when heating the SPI. If the coiled pipelines of the reactor heating system keep their operational parameters, there will be no special problems preventing reactor heating and, therefore, unloading the core of the reactor unit of plant № 120.

The state of the claddings of the heat releasing elements of the core before their "freezing" was estimated as a normal.

2.1.2. The Power Unit of Plant №125

The power unit of plant № 125 is a part of the plant № 105 NS. In 1997 the reactor was re-drained, cut out of the primary circuit and "frozen". All absorbers were completely inserted and made immobile by dismantling the electric cables. CRs and AC rods were in the "frozen" coolant.

The Operational time of the reactor was 15 % of its lifetime.

Note. The NS was withdrawn from operation for the reasons not associated with a technical state of its equipment and systems.

In 2001 the SPI inspection revealed that the reactor heating system had kept their operational characteristics and the other equipment was in order. However, it has been determined that the standard temperature control system was removed from operation without authorization. However, all temperature sensors of the reactor were in order. In compliance with the control system's state, the opportunity to use the standard system of cooling the equipment was also eliminated. For that reason, at the current stage of preparing the cores to be unloaded, both at the unit of plant № 125 and at the unit of plant № 120 (in which the pump equipment was decommissioned) consideration was given to draining the water out of the containers of lead-water shielding (LWS) and using the high-temperature sensors for controlling the neutron flux. In compliance with the data for apparatus radiation-technological control, the state of the claddings of the heat releasing elements of the core before its "freezing" were estimated to be normal.

2.2. THE NS OF PLANT № 910

The reactor of the plant № 910 NS was "frozen". All absorbers were immersed complete inserted. Coolant for the CR and AC claddings was "frozen" too. All CPS executive mechanisms were de-energized by dismantling their cables.

Operational time of the reactor was ~ 80 % of its lifetime.

A particularity of the state of the NS reactor installation is that the irradiation level of the reactor upper plate is high. It was caused by the fact that, at the level of the reactor upper plate, europium isotopes (Eu^{152}, Eu^{154}) that were included in the absorbing composition of CRs, in the chamber of the executive mechanisms of the three CRs were exhusted. That event happened when the NS was staying at the ship-repair factory. It was caused by ingress of water condensate into the CPS gas system when it was made up. Further violent boiling of water caused ingress of europium isotopes into the chamber of the executive mechanisms. Such phenomenon was not detected for any other NS reactor. The high level of irradiation at the CPS executive mechanisms' region prevented their dismantling and the carrying out other works on the reactor lid that preceded core unloading. It should be highlighted that there is no radioactive contamination in the reactor compartments of plant № 910 NS and other NSs, and as the apparatus control data have revealed, the state of the fuel elements claddings of the reactor core before "freezing" the SPI was normal. At all reactors that are subject to unloading, the residual heat release level is less 2 kW.

3. Technology of Unloading SNF oat of LMC Cooled Reactors

Putting in operation the propulsion reactor installations with lead-bismuth LMC, the cores of which were part of a non-dissembled removable unit of the reactor, required designing and constructing the technical mechanisms for fuel unloading that were different from similar ones used for channel-by-channel fuel replacement in the other type reactors. For that reason, in addition to the refueling equipment itself, the load-lifting mechanisms and a dry dock for stiff mantling the NS had to be included into the complex of refueling technical mechanisms, and a special base for refueling LMC cooled reactors was constructed. It was equipped with a whole complex of the equipment and constructions that provided a technological process for refueling including long term storage of the unloaded fuel (until its transportation to the reprocessing factory) [1]. The refueling equipment used for unloading the spent removable unit (SRU) of LMC cooled reactor of Project 705, Project 705 K NSs has been designed at EDO "Gidropress".

SRU unloading requires a whole complex of preliminary preparation works including installation of basic apparatus for controlling reactor subcriticality and dismantling the CPS executive mechanisms with further reliable fastening of each rod at the lower position.

The basic preparation operations that provide for SRU unloading include the following: heating the alloy, dismantling the standard sealing of the removable unit, sequential installation of the adapter box (AB) and the unloading cask (UC) on the reactor vessel flange and then step-by-step lifting of the SRU into the cask while carrying out physical measurements and radiation dose control.

The UC with the reactor SRU is transported to the preliminary cooling repository, and is installed on its flange. After that the SRU is step-by-step immersed into the storing container that has been filled in advance with the necessary amount of alloy and heated to (150-170) °C by the stationary heating system of the repository. Step-by-step immerse of the SRU into the container is followed by carrying out physical measurements. This operation is completed by sealing the SRU in the storing container and the compulsory air cooling system is switched on.

After the level of residual heat release in the core is reduced to ≤20 kW, the storing container is extracted out of the repository and installed in the box of the long-term cooling repository equipped with an air cooling system with natural circulation. As it was highlighted above, for all SRU that must be unloaded, the current level of residual heat release is less 2 kW.

4. Conditions forStoring Unloaded SNF and Infrastructure Conditions for the on-Shore Technical Base for Refueling the LMC Cooled NS Reactors

Currently all unloaded SRUs of LMC cooled NS reactors have been stored at the long-term cooling repository of the on-shore technical base located in village of Gremikha, Murmansk region. They include:

- the SRU of the reactor of plant № 905 NS that was unloaded in 1989;
- the SRU of the reactor of plant № 915 NS that was unloaded in 1990;
- the SRU of the reactor of plant № 106 NS that was unloaded in 1991;
- the SRU of the reactor of plant № 107 NS that was unloaded in 1992.

In addition, two SRUs of the reactors of the Project 645 NS that were unloaded on completion of their first lifetime in 1968 were placed in that repository several years ago. The concrete well clad with steel has a system for controlling possible ingress of water into the repository. In case of detecting water, it can be removed. The temperature of each SRU installed in the repository is controlled using standard thermo-couples. Today the temperature of different SRUs is within (15 - 50) °C.

There are 8 wells in the long-term cooling repository, two of them being free. For that reason, one of the SRUs that are to be unloaded has to be left in the preliminary cooling repository until SNF is transported to the reprocessing factory. The shielding parameters of the preliminary cooling repository meet the set up requirements.

Nuclear safety for storing SNF is provided by the fact that reactor cores are in a subcritical state (K_{eff} < 0,95) due to complete insertion of the absorbers. For all unloaded SRUs the executive mechanisms of the CPS drivers were dismantled, the tight steel caps were installed on the claddings of the CRs, AC and EP rods and welded to them. The claddings were cut slightly higher than the reactor lid. During SNF storage unauthorized insertion of positive reactivity was eliminated by the fact that the lead-bismuth alloy was "frozen" in the CPS claddings.

Radiation and ecological safety for storing SNF is based on the principles of high-separation shielding for the pathways of probable proliferation of radionuclides into the environment. This is a sequential chain of physical safety barriers which are: a matrix of heat-releasing elements, fuel element cladding, "frozen" coolant, a steel shell of the SRU, and a steel storing container. As was mentioned above, the fuel elements claddings of the cores of the LMC cooled reactors are almost undamaged. However, when the LMC cooled nuclear power installations (NPI) were designed and then a specialized on-shore refueling technical base was designed, no provisions were made for SNF long-term storage at the NSs and in the on-shore repositories. It was presumed that after several years of cooling in the repositories, SNF would be transported to the nuclear fuel-reprocessing factory. For that reason, there is no satisfying substantiation of safety for so long term (~ 35 years) storage of SNF in conditions of on-shore technical base (STB). It is necessary to carry out in-depth analysis of all possible ways (including scarcely probable ones) of affecting subcriticality of stored SRUs including physical and chemical processes that occur during the storage period in the SNF repository. Note. This problem has been considered in detail in the report devoted to the problems of safe storage of unloaded and non-unloaded SNF of NSs with LMC. However, after 10 years being out of service the STB has lost its readiness to immediate operation and a number of problems need solving. They include the following:

1. Failure of the base boiler-room installation that is an outer source of steam for heating the reactor before unloading the SRU.
2. Necessity for repairing the cranes (two traveling cranes and one portal crane).
3. Failure of the base physical control system.
4. Flaws in the construction structures of the long-term cooling repository (cracks in the walls of the ground part of the building). In compliance with a program developed by MINATOM, reconditioning of all STB infrastructure are carried out and the new and old documents are developed and corrected.

5. Solutions to the Problems of Utilization of LMC Cooled NSs

The RF Government charged RF MINATOM with a mission of utilizing the ships with NPIs.

Currently MINATOM has begun realization of the plan of utilizing LMC cooled NSs. The practical task of the current period is preparation of the SRU reactors and

LMC cooled NSs withdrawn from operation to be unloaded at the specialized on-shore refueling technical base located at the village of Gremikha, Murmansk region.

The task of reconditioning the STB infrastructure is the urgent one.

As was mentioned above, one of the STB problems is reconditioning the base source of steam necessary for heating the reactor before unloading the SRU. Our vision of finding a solution to the problem is buying a transportable boiler-room installation operating by using diesel fuel. This boiler-room installation must be located near the dock and use existing pipelines. The requirement for continuous steam transfer to the ship during the period of SRU unloading is solved by mounting the reserve boiler-room.

The electricity-consumers for the two installations receive electricity from two independent power sources.

Reconditioning the standard boiler-room including its construction structures and long heat pipelines for steam transfer to the ship necessary for heating the reactor only (without the primary circuit) and reconditioning the station for preparing the high purity water was not considered to be expedient. The base designer is beginning to develop the accepted decision.

Reconditioning the cranes is in progress. Reducers and electric systems are being replaced. The operational ability of the carrying metallic structures, which were subject to the severe climate of the region, is being reconditioned.

It is necessary to develop almost an new system of physical control for performing the operation of unloading the SRU out of and loading it into the storage container for the preliminary cooling repository. As was mentioned above, due to failure of operational ability of the ship systems for cooling the lead-water shielding (LWS) container, a decision was made to use at the ship the high-temperature units for detecting the neutron flux with a set of pulse channels for measuring physical power of a ZII-02R-type (at entire draining the LWS container). A decision to was made add a system of physical control to the long-term cooling repository (one SRU for the present).

Due to developing precipitate cracks in the over-ground part of the building for the long-term cooling repository, it is necessary to carry out analysis of the state of the whole construction in order to detect ingress of ground water into the reinforced concrete wells with immersed SRUs. At the present time no existence of water has been registered. When preparing the NS, the following tasks should be considered as urgent ones. Design and fabrication of the transportable system of reactor temperature control instead of the standard ship system. The transportable (unit-carried) system must be universal and use standard sensors of the SPI of plant № 105 NS (units of plant № 120, 125) and the SPI of plant № 910 NS. A signal must be transmitted to the refueling control desk.

First of all, inspection of the systems of the cut out reactor compartment of plant №120 SPI m provides for tightness control of the coiled pipelines of the reactor heating system. The equipment has been fabricated. It is necessary to provide conditions for inspecting the unit at the base (where it has been stored).

The task of normalization of the radiation conditions on the lid of reactor of plant №910 NS can be solved stage-by-stage. At the first stage it is necessary to make a re-determining inspection of radiation parameters by using the special compact equipment. A major task is to diminish the dose load on the personnel who disassemble the CPS executive mechanisms, cut away the seams of the reactor removable unit sealing and disassemble the adjusting flange.

Today lack of financing does not allow launching the works for normalization of radiation conditions for their NS. In the future this may cause a hasty decision that its reactor cannot be unloaded (i.e. accident one).

For that cause, inspection of the reactor heating system of the unit of the plant № 120 SPI should not be delayed.

Close attention should be paid to the heating system of this reactor for the following reason. When the unit was produced in factory conditions, the following error occurred: Pipes made of standard stainless steel material that can poorly resist inter-crystallite chloric corrosion were used for assembly. Finally that was a cause of making the decision to replace the power unit of plant .N° 105 NS. The following problems are proposed for consideration within the international cooperation frameworks:

1. Development of the technology for normalizing the radiation conditions on the lid of the reactor of plant № 910 NS with along carrying out preliminary re-determining inspection.
2. Development of the project for decommissioning the reactor of plant .N° 900 NS and provide for its further development with SNF utilization ensuring long and safe ground storage of the reactor with non-unloaded core.
3. Development of the project of disassembling the SRUs to nuclear-safe fuel sub-assemblies (FSA) at the place where they will be stored (village of Gremikha). Then these FSA must be sent to the SNF reprocessing factory. The project is purposed to improve radio-ecological conditions in the Northwest region [5].

6. References

1. Sazonov, V.K., Tihomirov, V.I., Tflionentco, V.B., Toshinsky G.I., Shestopalov, A.M., Stepanov, V.S., Vahrushin, M.P., and Alekseev, G.A. (1999) "Technology and Experience of Repair Works and Refiielings of the Nuclear Power Installations with Lead-Bismuth Coolant", *Heavy-Liquid Metal Coolant in Nuclear Technology,* (Proe. Conf., Obninsk, Russia), Vol. 2, pp. 643-649.
2. Bugreev, M.I., Gromov, B.F., Pankratov, D.V., and Suvorov, G.P. (1999) "Problems of Utilization of Cores and Nuclear Submarine Compartments with Lead-Bismuth Cooled Reactors. Analysis of Risks Associated with Nuclear Submarine Decommissioning, Dismantling and Disposal", in AA. Sarkissov and A. Tournyol du Clos (eds.), Kluwer Academic Publishers, Dordrecht, pp. 261-264.
3. Shmakov, R.A, (1977) "The Small Fast Automated Submarine-Destroyer of Project 705 (705K)", *Military and Technical Literary Miscellany "Typhoon".*
4. Grigoriev, B.V. (1999) "Solutions Determined the Appearance of Project 705 NS", *Military and Technical Literary Miscellany "Typhoon".*
5. (1999) "Normalization of the Radio-Ecological Situation in the North-West Region of Russia due to Removing Spent Nuclear Fuel of "Alpha" Class NSs out of the Region", *Conceptual design,* Proposals on international cooperation within the Contract with NNC Limited, SSC RF-IPPE, Obninsk.

STEP-BY-STEP SOLUTION TO THE PROJECT ON LONG-TERM STORAGE OF SPENT FUEL FROM NUCLEAR SUBMARINES WITH HEAVY LIQUID METAL COOLED REACTORS

MIKHAIL BUGREEV, BORIS GROMOV, EVGUENI EFIMOV, SVIATOSLAV IGNATIEV, DMITRY PANKRATOV, VALERY SAZONOV, ALEXEY ZABUD'KO
Institute of Physics and Power Engineering (IPPE)
Obninsk, Kaluga Region, Russia

1. Introduction

It is Alfa class Nuclear Submarines (NS) with heavy liquid metal lead-bismuth coolant in the primary circuit of the nuclear reactor that have been put into operation in Russia at the height of Cold War. The particular feature of such NSs was that the reactor cores while refueling have been unloaded in the form of a single removable block, including reactor core with inserted control rods, reflector, and the upper closure fitting of the biological shielding. So, in fact, it was the unloading of a sub-critical reactor. The unloaded spent removable blocks (SRB) are currently stored at one of the North naval bases on the Kola Peninsula. Our expertise shows that a) a rather high-level of radioactivity is accumulated in the single SRB; b) the stored SRB have a relatively low sub-criticality that is sensitive to the exposure of adverse external conditions; c) the technology of SRB storage preparation does not exclude the possibility of water penetration into the SRB volume. Thus, the stored SRB are the source of potential nuclear and radiation danger. The International Project of the step-by-step solution of this problem is disscussed and the technological and other challenges that the Project faces in its realization are analyzed.

2. Status of Problem

During 1960 – 1990 a number of NSs with liquid metal lead-bismuth coolant in the reactor primary circuit were in operation in the former Soviet Union Navy. Besides, a couple of the heavy liquid metal test facilities (HLMTF) were created at IPPE in Obninsk and in Sosnovy Bor and operated as a prototypes of such NSs.

All of the operated reactors in question are of the intermediate neutron reactor type. The reactor fuel composition included the inter-metallic compound UBe^{13} with U^{235} enriched up to ~90% and dispersed into the beryllium matrix. Metallic beryllium served as reflector in these reactors.

The particular feature of such NSs was that the reactor cores when refueling were unloaded in the form of single removable blocks, including reactor core with dropped control rods, reflector, and the upper closure fitting of the biological shielding.

At present the unloaded spent removable blocks (SRB) are currently stored at one of the North naval bases on the Kola Peninsula.

Figure 1 represents the storage conditions of one single SRB. It was put into the steel montejus (0), which contains a pure (non-used and non-radioactive) lead-

A.A. Sarkisov and L.G. LeSage (eds.),
Remaining Issues in the Decommissioning of Nuclear Powered Vessels, 357–364.

bismuth eutectic alloy (LBEA) heated up to a temperature between ~150 to ~160°C. The lower vessel (3) is a supporting metallic structure composed of concentric ferrule, as well as upper and lower flanges and bottr-pan. The metallic structure has the sockets (4, 5) to connect the airways (9, 11) that make an air natural circulation circuit. The lower vessel also has the radiation shielding (6) to prevent penetration of thermal neutrons emitted by concrete shielding (17) under its irradiation. In order to provide the air intake and align the air speed through the montejus perimeter six leading ribs (7) and the throttling ring (8) are welded inside the vessel. Besides, a special tube ending at the connecting face pipe is welded to the vessel supporting ferrule to secure the connection with basic drainage system. The montejus airways are composed of inlet airway (9) of box-like cross-section with baffle (10) and outlet airway (11), the cross-section of which becomes round-like in the baffle (12) junction. Thus, airways (9, 10) along with an annular gap between the tank and the lower vessel ferrule supported by ribs (7) and ring (8) form the air natural circulation to cool the montejus during storage. The natural circulation is due to the difference between the of cold air density in the inlet airway (9) and that heated in the lifting section of the system, as well as due to the high height of the lifting section. The airways are fastened to the upper vessel (1)

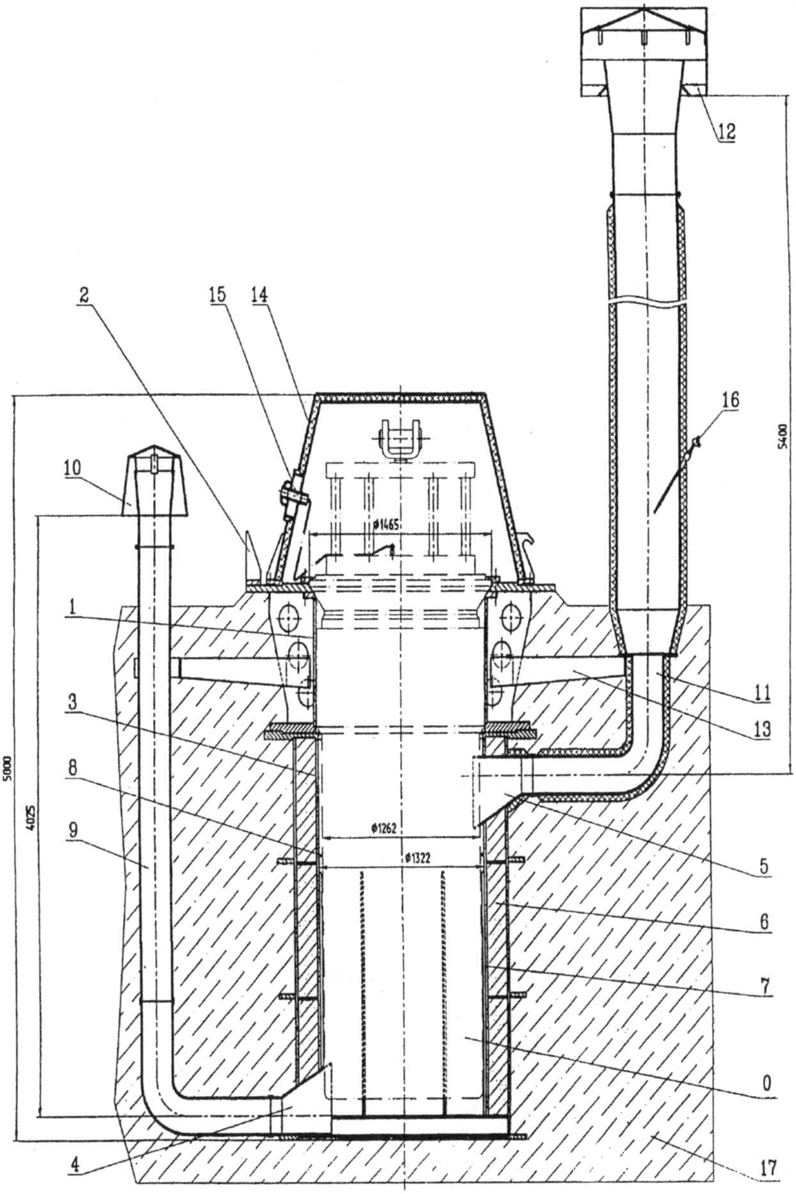

Figure 1. Storage conditions of a separate SRB

using the corbels. The cowl (14) is a protective metallic structure with heat insulation that prevents the upper section of the SRB from mechanical and atmosphere effects. It is the transfer box with built-in two-pin plugs for connection of thermoelectric converter (16) with the instrumentation dashboard that is placed in the cowl. All of the system was placed into the concrete well.

As soon as the SRB was loaded into the container and the LBEA was cooled the alloy congealed. The estimated residual heat release in all SRB (both unloaded and not unloaded) is as much as 3 kW. It was originally assumed that the spent SRB would be stored at a costal storage site before being reprocessed. Long-term storage was not assumed. The technology of SRB storage preparation did not excluded the possibility of water penetration into the SRB volume. Meanwhile, accumulation of water and it penetration for unknown reasons into the SRB introduces a positive reactivity into the system.

Under a certain (even if improbable) conditions the latter can increase the reactivity of the system and essentially deteriorate the regional radio-ecological situation. That is why the study, validation and choice of reasonable ways for treatment of unloaded and non-unloaded SRBs are an important and relevant problem.

The status and possible solutions of this problem, as well as different aspects on radiation safety analysis and emergency condition scenarios have been repeatedly considered and proposed earlier [1-6]. At the same time the approaches suggested did not include the description of concrete procedures on its practical implementation under the existing economical and geopolitical situation. This paper proposes for the first time a realistic and reasonable approach on how to start the step-by-step solution of this problem and analyzes the technological and other challenges that this solution faces in its realization. The authors expect to get understanding and support of the science community from the countries that most of all are interested in early solution of the problem.

The goal of this paper is to develop the Pilot Project for the transport-technology step-by-step solution of the problem in question, including dismantling of the SRB to separate cassettes and fuel element assemblies; loading them into special transportation containers; delivery to fuel reprocessing; reprocessing and extraction of high-enriched uranium; and securing the waste storage.

The Project is based of the infrastructure that exists at the IPPE heavy liquid metal test facility in Obninsk, as well as experience gained while dismantling the first run SRB that had been implemented at that test facility.

The proof of the Project could be realized through the dismantling of spent active zone of the KM-1 heavy liquid metal test facility in Sosnovy Bor that completed 100% of it run and was unloaded in 1987. The SRB storage conditions in Sosnovy Bor are the same as that at Kola Peninsula. So, the proof the Project will provide an to estimate of the real state of the SRB at the Kola Peninsula and provide a basis for deciding either on removal of SRB or keeping them in the current storage conditions for some additional time.

3. Radiation Safety

As was mentioned above, the unloading of a sub-critical reactor has taken place while refueling of the NS. Thus, storage of the SRB made special demands on security of nuclear material and radiation safety that had no analogues in the world reactor building.

Despite the fact that long-term storage had not been assumed originally, the real "temporary" storage of SRB has been prolonged for an indefinite time. For example, the unloaded SRB of the NS of Project 645 has been stored ~35 years while other SRBs ~10 years and more.

It was originally presumed that nuclear safety during SRB storage need not be provided because reactor cores of both unloaded and non-unloaded reactors are in the deep sub-critical state, due to significant burn out of the fuel and full insertion of the neutron-absorbing control rods into the core. It was also presumed that unauthorized input of positive reactivity during the SRB storage was excluded by the "frozen" LBEA, which prevents any repositioning of control rods inside the core. However, one cannot exclude completely that some open interstice and cavities could appear in the mass of the solidified coolant during preparation of the SRB for temporary coastal storage. The filling of interstice and cavities by water, that can take place in case of depressurization of the protection cowl and compaction of the montejus seat bottom, could lead to a system reactivity increase.

During long-term storage one cannot also exclude some other possible ways to affect system sub-criticality, such as, for example, the consequences of passing of physical and chemical processes inside the montejus "frozen" coolant (galvanic and chemical corrosion, phase conversion), as well as possible diversion and terrorist acts, and on-site unfavorable natural phenomena and cataclysms (earthquakes, fires).

In these circumstances the requirements for the securing of nuclear safety of the long-term storage of the sub-critical reactors, i.e. the unloaded SRB, must be toughened. In the first place it is necessary to organize periodical monitoring of the SRB radioactivity level. But the best solution would be the early dismantling of the SRB to single fuel rod arrays that are not of nuclear danger.

4. Main Stages and Tasks of Project Implementation

The Project implementation is divided into three main stages:

Stage #1. Development of the pilot technological scheme for SRB dismantling into single cassettes and fuel element groups (this could be done on the basis of the 27/BT HLMTF in Obninsk, Russia); loading the SRB into containers and transportation to the reprocessing plant for high-enriched uranium extraction;

Stage #2. The proof of technological scheme by dismantling spent active zone that is currently stored at the KM-1 HLMTF in Sosnovy Bor, Russia;

Stage #3. Project appraisal and development of recommendations on treatment of the SRBs that are stored at the Kola Peninsula.

The main tasks of each stage are presented in Tables 1,2, and 3.

TABLE 1. Main Tasks of the Stage #1

Number	Task Description
1.1	Feasibility study and ecology validation of the Project of pilot transport-technology scheme.
1.2	Getting permissions, approvals and licenses from federal and local regulation authorities, including Minatom, Gosatomnadzor, and signing of agreements with Ministry of communications and Mayak Production Association for transportation and reprocessing of the dismantled SRB.
1.3	Revision of infrastructure of dismantlement station that exists at the 27/BT HLMTF and specification of additional equipment necessary for the Project implementation.
1.4	Modernization of the dismantlement station at the 27/BT HLMTF and rent or purchase of TUK-19 transportation containers.
1.5	Dismantling of the SRB.
1.6	Packing of the dismantled fuel elements and rod arrays and loading into the TUK-19 transportation container.
1.7	Delivery of cargo in railroad cars to Mayak for reprocessing, uranium extraction, and final waste disposal.
1.8	Issue the detailed report on the Project implementation.

TABLE 2. Main Tasks of the Stage #2

Number	Task Description
2.1	Feasibility study and ecology validation of the transport-technology scheme of delivery of the SRB from Sosnovy Bor to Obninsk.
2.2	Getting permissions, approvals and licenses from federal and local regulation authorities to implement the delivery as described in the paragraph 2.1.
2.3	Signing of agreements with Ministry of communications and Mayak Production Association for transportation and reprocessing of the dismantled SRB.
2.4	Development of a special container for transportation of the SRB from Sosnovy Bor to Obninsk.
2.5	Placement of order, fabrication of container, its testing and certification, and commissioning.
2.6	Dismantling of the SRB.
2.7	Detailed analysis of the dismantled SRB and its component, including metallographic analysis, evaluation of the results.
2.7	Packing of the dismantled fuel elements and rod arrays and loading into TUK-19 transportation containers.
2.8	Delivery of cargo in railroad cars to Mayak for reprocessing, uranium extraction, and final waste disposal.
2.9	Issue of the detailed report on the Stage #2 implementation.
2.10	Holding of the International workshop for the purpose of international evaluation of the results of Stage #2 implementation and final decision-making on implementation of the Stage #3.

TABLE 3. Main Tasks of the Stage #3

Number	Task Description
3.1	External investigation and radiation monitoring of the residuary 9 SRB stored at Kola Peninsula.
3.2	Feasibility study of the Project for a dismantling station building at Kola Peninsula as a prototype for the Project of pilot transport-technology scheme on SRB dismantling that is implemented in the course of Stage #1.
3.3	Feasibility study of the Project for transportation of the residuary 9 SRBs (without dismantling) from the Kola Peninsula to Mayak for on-site dismantling and subsequent reprocessing, uranium extraction, and final waste disposal.
3.4	Feasibility study of the Project unloading of the SRBs, that are currently not unloaded and stored in the cut reactor compartment at Kola Peninsula, and their following disposal.
3.5	Evaluation of a capability to prolong the long-term storage of both unloaded and not unloaded SRBs under existing storage conditions or validation and acceptance of other specific solutions on their utilization.
3.6	Issue the detailed report on the Stage #3 implementation.
3.7	Holding of the International workshop for the purpose of international evaluation of the results of Stage #3 implementation.

5. Conclusion

The decision-making concerning the disposal and utilization of the unloaded and not unloaded SRB that are currently stored at the Kola Peninsula has been postponed for an very uncertain time due to financial and economical difficulties.

Our expertise shows that a) a rather high-level of radioactivity is accumulated in a single SRB; b) the stored SRBs have a relatively low sub-criticality that is sensitive to the exposure to adverse external conditions; c) the technology of SRB storage preparation does not exclude the possibility of water penetration into the SRB volume. The latter can increase the reactivity of the system and (under a certain conditions) essentially deteriorate the regional radio-ecological situation. Hence, the SRB stored are source of potential nuclear and radiation danger. Thus, the SRB must be unloaded from the reactors and coastal temporary repositories and finally reprocessed. That is why the cardinal solution of this problem is of important. Clearly, the Russian Federation is unable to solve this problem independently. The assistance of the world community could not only eliminate this center of potential radioecology danger as a Cold War legacy, but also reduce the social and psychological tension in the population of the nearest European country.

6. References

1. Bugreev, M.I., Gromov, B.F., Pankratov, D.V., Sazonov, V.K., Suvorov, G.P. (1999) Problems of Utilization of Cores and Nuclear Submarine Compartments with Lead-Bismuth Cooled Reactors, in A.A.Sarkisov and A.Tournyol du Clos (eds.), *Analysis of Risks Associated with Nuclear Submarine Decommissioning, Dismantling and Disposal*, Kluwer Academic Publishers, pp. 261-264.
2. Efimov, E.I., Ignatiev, S.V., Pankratov, D.V., Bugreev, M.I. (1999) Analysis of Potential Danger of Spent Reactor Cores of Nuclear Submarines with Heavy Liquid Metal Coolant, in Per Strand and Torun Jolle (eds.), *Proc. of the 4th International Conference on Environmental Radioactivity in the Arctic, Edinburgh, Scotland, 20-23 September 1999,* Norwegian Radiation Protection Authority, Norway, September, p. 301.
3. Eefimov, E.I., Ignatiev, S.V., Pankratov, D.V. Box Model of Radionuclide Concentration Calculation in the SeaWater, ibid, pp. 302-304.

4. Gromov, B.F., Toshinsky, G.I., Chekunov, V.V., et al. (1999) Designing the Reactor Installation with Lead-Bismuth Coolant for Nuclear Submarines. The Brief History. Summarized Operation Results, *Proc. of International Conference "Heavy Liquid Metal Coolants in Nuclear Technology"*, Obninsk, SSC RF – IPPE, vol. 1, p. 14.
5. Suvorov, G.P., Bugreev, M.I., Kuzko, O.V. Experience of 27/VT Facility Construction and Operation, ibid, p. 67.
6. Sazonov, V.K., Tikhomirov, V.I., Tihonenko, V.B., et al. Technology and Experience of Repair Works and Refueling for the Nuclear Power Installations with Lead-Bismuth Coolant, ibid, vol. 2, p. 599.

SPECIAL TOPICS IN NUCLEAR SUBMARINE
DECOMMISSIONING AND UTILIZATION

SUBSTANTIATION OF THE NECESSITY TO CONSTRUCT CENTERS FOR LONG-TERM REACTOR COMPARTMENT STORAGE AND PRINCIPAL TECHNICAL SOLUTIONS ON ENVIRONMENTALLY SAFE STORAGE OF REACTOR UNITS OF DECOMMISSIONED NUCLEAR SUBMARINES

V. MAZOKIN, M. NETECHA
FSUE Russian Design Institute of Power Engineering
Moscow, Russia

Multi-purpose utilization of Russian Nuclear Submarines (NSs) is a system of organizational, scientific, technological and industrial activities ensuring the whole cycle of NS utilization-related work from their decommissioning up to disposal of radioactive and other wastes (that cannot be processed). These activities also include the environmental rehabilitation of radiation-dangerous Coastal Servicing Enterprises (CSEs) and Floating Servicing Vessels (FSVs) of the Russian Navy. The basic problem and principal solutions in the technical policy, as applied to the actual economic and political situation in Russia, are stated in "The State Concept of Multi-Purpose Utilization of Nuclear Powered Vessels".

The Concept was developed on the basis of: -research including studying the accumulated activity within the equipment and structures of reactor compartments during NS operation; -comparative technical and economic assessments of possible options regarding handling of both Spent Nuclear Fuel (SNF) and radiation-dangerous equipment from reactor facilities; -analysis of conditions ensuring nuclear, radiation and environmental safety and consideration of foreign experience in the area under consideration.

The implementation of "The State Concept..." resulted from the transfer of the managerial functions related to these activities from the Russian Navy to the RF Minatom. But first of all this was due to the necessity of: -identifying optimal technical solutions the realization of which would be able to eliminate (reduce) wasteful expenses of both budgetary and off-budgetary funds as well as of finances obtained within the framework of international agreements; -ensuring nuclear & radiation safety during the utilization process; -determining most expedient directions of investments when constructing facilities and creating technologies for safe management of SNF, radiation-dangerous equipment in reactor facilities and Radioactive Wastes (RWs).

The "Concept..." was developed in the Dollezhal Russian Design Institute of Power Engineering (RDIPE) with the participation of other Russian research institutes & design offices. The Concept was agreed upon with the Ministry of Defense of Russia, the Ministry of Economic Development, the Ministry of Industry & Science, the Ministry of the Environment and Natural Resources, Ministry of Health, Gosatomnadzor (Federal Nuclear & Radiation Safety Authority), Rossudostroenie (Shipbuilding) and approved by the Science & Technical Panel of the Minatom of Russia. Thereafter the Concept was ratified by Minatom leaders and put into action by the RF Government Order № IK-P7-02738 of February 17, 2001.

A.A. Sarkisov and L.G. LeSage (eds.),
Remaining Issues in the Decommissioning of Nuclear Powered Vessels, 367–371.

The State Concept provides for a “deferred” utilization of the radiation-dangerous equipment from reactor facilities. This means that their utilization is to be performed only following long-term storage within reactor compartments.

Such a type of storage is to be carried out at specially constructed Long-Term Storage Centers (LTSCs). The location and individual equipping of these centers depend on the totality of technical and economic factors, environmental safety requirements of every LTSC-location region with consideration for potential accidents of both natural and technogenic origin.

Unfortunately, no coastal center to store Reactor Compartments (RCs) has been built in Russia yet; therefore, NS utilization is being performed via a temporary flowsheet providing for forming of floating reactor units (in most cases these are three-compartment units: a reactor compartment plus two adjacent compartments) after SNF unloading from NS reactors. Reactor Units (RUs) are temporarily stored afloat in Saida Bay (the Northwest Region) and in Razboinik Bay (the Far East Region).

The integral mass of an RU reaches 3000 t, the length up to 36 m, the width up to 18 m and the height up to 13 m. The lifetime of such an RU is 25 years under conditions of its docking examination/repair every 10 years.

Presented in Table 1 are general data on RUs under waterborne storage, as of the beginning of the year 2002: there are 69 RUs in total including 48 RUs in the Northwest region and 21 RUs in the Far East region. Among one-compartment units (OCUs) stored in Saida-Bay one OCU still contains SNF in the reactor. This OCU has been transferred to an environmentally safe status in plant conditions using a special technology.

TABLE 1. General Data on RUs under Waterborne Storage, as of the Beginning of the Year 2002

RU type	Total amount of RUs, total / Northwest/Far East	RUs under waterborne storage over 10 years, units	RUs after dock repair and examination, unit
One-compartment	8 8/0	2 2/0	1 1/0
Three-compartment	49 28/21	5 0/5	0 0/0
Multi-compartment	12 12/0	6 6/0	0 0/0
Total	69 48/21	13 8/5	1 1/0

For lack of coastal RC storage centers a new problem arises: instead of storing NSs, one is forced to store afloat RUs comprising an important residual radioactivity even after SNF removal (about 100 000 Ci in every RU). This is obviously a grave potential radiation hazard, considering the hypothetical impacts of such RUs.

Furthermore, one has additional (permanently growing) expenses for both RU maintenance and the development of an appropriate infrastructure to support RU temporary storage sites. In the near future the accumulation of “afloat” stored RUs can become the determining factor in NS utilization paces. Thus, at present the construction of coastal facilities to store RCs of the former NSs represents a high-priority task.

By now the needed investigations as well as technical & economic assessments to choose the site, type and technical appearance of the regional facilities to store RCs have been performed. They also considered an option of constructing a coastal site for RC long-term storage (about 70 years) at SMBE "Zvezdochka". Such an option provided for resolving the task in question at a local scale for the Severodvinsk region (the site design capacity being limited by local conditions, whereas schedule times & costs would be comparable to those for the construction a regional-scale facility, i.e. about 3 years).

Based on the results of technical & economic assessment, the creation of regional centers for RC long-term storage in the form of open-air coastal sites above sea level in Saida Bay and in Ustrichnyi (Oyster) Cape between Chazhma Bay and Razboinik Bay was accepted as most favorable.

Actually the first stage of the project regarding the construction of the above facilities is completed: the feasibility studies have been prepared, undergone environmental expert examination and are to be approved in the immediate future.

The facility in Saida-Bay represents an open-air coastal site 270x150m in area comprising 120 slip places to arrange one-compartment defueled RCs.

The projected LTSC site on Saida-Bay shore is distant from localities & military bases and also from industrial & agricultural lands and water supply sources.

The possibility of RC site flooding is ruled out for it is located above the upper level of syzygial tide (4,06 m), the storm-area (1,2 m) and the wind-induced surge area (0,75 m): the planning level mark of the site is 6,15 m. RC slip places are located at parallel reinforced concrete sites equipped with vessel-conveying tracks (altogether 15 routes). LTSC is equipped with a wharf 34,5 m in length and 8 m in width to moor the transport-intermediate dock, arrange a self-propelled vehicle to ensure RC transfer from the wharf to storage area, and also with technological & engineering systems including the collection and decontamination of rain (melted snow) waters.

A similar approach was used to choose the LTSC area in the Far East Region. The coastal site (6 ha) in Ustrichnyi Cap is located at a distance of 16 km from Fokino-town, at the 1,3-m level mark that is appreciably above the syzygial tide level (0,5 m). To perform the sea–dry land transport operations, the use of a floating dock of 3500-t tonnage is suggested.

The environmental safety of the operation of RC regional storage facilities during the designed lifetime (60 to 70 years) is ensured by both engineering solutions and technical-organizational procedures. The principle of "deferred" RC utilization is based on the radioactivity decay phenomena leading to a decrease of the accumulated activity over time. Thus, during RC long-term storage "natural" decay of the dominant radionuclides will result in a gradual & permanent decrease of the accumulated radioactivity in both the equipment and structures. 70 years following the reactor shutdown, when gamma dose rate levels within the RC compartments decrease by about 10 000 times (Co^{60} is the main dose-forming radionuclide), one will have an opportunity to perform RC dismantling operations observing maximum permissible dose levels for the involved personnel. Moreover, neither the RC equipment nor its materials & constructions (save for the reactor vessel) will be considered as Solid Radioactive Wastes (SRWs) and, thus, their reuse as "normal" scrap metal will be possible.

A RC prepared for long-term storage represents a sealed metallic construction that, in keeping with the Concept, is to be considered as "a radiation source in the form of an article containing the contaminated equipment and materials ".

Within RC the radioactivity is distributed non-uniformly. The largest fraction is the SNF (up to 10^6 Ci 0,5 years after the reactor shutdown). SNF unloading results in a decrease of the integral accumulated activity by a factor of 10. However the remaining radioactivity of the RC equipment & materials makes up a rather high value of about 10^5 Ci.

About 99% of the radioactivity is concentrated within the reactor vessel and reactor internals. Biological protection materials within the reactor baffle comprise about 1,0%, and about 0,1% falls to corrosion products on inner surfaces of both the equipment & pipelines.

To eliminate radionuclide release out of the RC, special protective barriers are constructed, such as: the reactor vessel, walls of both the equipment & primary circuit pipelines, sealed baffles and biological protection constructions and, finally, the RC strong vessel including the end partitions. Under design (normal) RC storage conditions all the above barriers (save for RC end partitions) are to be operable throughout the storage period, i.e. 70 years. In keeping with "Prometej" CRI CM data, corrosive wear of the RC outer cover during 100-150 years will result neither in its depressurization nor in strength losses.

To ensure the same wear resistance properties of RC end surfaces, additional tail partitions are to be mounted. It should be stressed that, when preparing for storage, the RC outer surface undergoes special examinations; if necessary, local protection is installed to ensure a non-excess of the gamma dose rate of 100 μSv/h at 1-m distance in any direction is not exceeded.

Thus, under normal operating conditions the LTSC will not constitute a source of inadmissible radiation load on the region environment.

Regional LTSCs are to be constructed in a few stages, the cost of every milestone being estimated at about 30% of the whole design cost. It is planned that the LTSC starting unit in Saida Bay will be comprised of 40 slip places (40 OCUs), its cost being estimated at about 40 million USD and the construction time at 36 month.

The first construction stage of the LTSC in Ustrichnyi Cap will be comprised of 28 OCUs (the estimated cost about 40 million USD and the construction time is about 30 months).

The building of regional-scale LTSCs necessitates rather important capital outlays from the State Budget (e.g. only the construction of starting regional LTSC units requires 15 to 16 million USD/year during about 3 years). Therefore without international assistance the construction process will be very long. In such a case the period of waterborne RU storage will also increase, and conditions for the occurrence of radiation incidents will be created because of a higher probability of RU-sinking due to degradation processes in their buoyancy systems. There is no doubt that the consequences of such incidents would appear outside the territorial waters of Russia.

International cooperation is expedient regarding the following components of the integral task of regional LTSC construction:

- Construction of a floating vessel (floating dock 3500-t tonnage or another vessel) to perform RU transportation–conveying to shore;
- Construction of an offshore terminal to accept a RU from the floating dock;
- Construction of industrial and civil buildings;
- Building of power and heat supply systems;
- Building of a waste disposal plant;
- Construction of access roads and inner roads;
- Creation of the radioecological monitoring system;

- Supplying of building & assembly machinery;
- Development and construction of a system ensuring physical protection of the concerned installations.

The issues of potential cooperation can be worked out in more detail in the course of negotiations between the RF Minatom and potential investors.

SPECIFIC FEATURES OF THE NS "KURSK" UTILIZATION

V. NIKITIN
FSUE "Onega"
12 Mashinostroitelei pr., Severodvinsk 164509, Russia

Utilization of nuclear-powered submarines (NPS) is carried out according to the utilization technology developed by the research and design enterprise "Onega".

Technological concepts of the complex utilization of NPSs assume the following stages:

- NPS preparation for utilization;
- defuelling of the nuclear core;
- vessel arrangement on the slip;
- cutting out of the three-compartment block;
- preparation of the three-compartment block for interim storage;
- cutting out and dismantlement of the missile compartment;
- dismantlement of the equipment, devices, pipelines, and power cables;
- hull cutting into large sections, cutting sections, equipment, and cables;
- launching of the three-compartment block for transportation to the interim storage site.

Utilization of the NPS "Kursk" will be carried out according to the same technological scheme with certain specific features.

Figure 1 displays the utilization scheme for the NPS "Kursk".

The specific technological features of the utilization scheme for the NPS "Kursk" are caused by the following reasons:

- The "Kursk" NPS is a damaged vessel and the scope of destruction is unpredictable.
- Long-term storage of the NPS "Kursk" in floating slip #50 would encumber systematic docking of ships and vessels of the Arctic Fleet.

Because of these reasons, the enterprises and organization of the Russian Navy and industry considered various variants of transport and technological scheme for NPS "Kursk" handling after its arrangement at the floating slip #50 in order to choose the optimal variant with respect to safety, duration, and cost. Based on these considerations, the variant which included utilization at SRE "Nerpa" was chosen; the plant is the closest one in the region that is capable of carrying out the utilization of the NPS in question.

From an economic standpoint, the most appropriate utilization scheme for the NPS "Kursk" contains the following stages:

A.A. Sarkisov and L.G. LeSage (eds.),
Remaining Issues in the Decommissioning of Nuclear Powered Vessels, 373–378.

1. partial welding of holes in the light hull of the NPS in order to improve its floatability and restore impermeability of driving ballast tanks (DBT);
2. transition of the NPS from slip #50 to slip #42;
3. hauling of the NPS in floating slip #42 to the SRE "Nerpa";
4. cutting out the of the three-compartment block;
5. dismantlement of the nose and stern part for scrap;
6. transition of the three-compartment block to the slipway, defuelling, and formation of the three-compartment module of the reactor compartment (RC);
7. transportation of the three-compartment module to the interim storage site (ISS).

Ship-repairing plant "Nerpa" is a state enterprise of the Russian Agency for Ship-Building. It is located close to the bases of the Arctic Fleet at the entrance to the Kola gulf and has a nonfreezing bay.

Starting in 1995, the plant has carried out NPS utilization. As of the beginning of 2002, "Nerpa" has completely utilized 12 NPS; four NPS are at various stages of utilization.

According to the Decision of the Government of Russia #518, the plant provides maintenance for the NPSs decommissioned from the Navy and supports the ground site for the ISS at Saida bay for the RC blocks formed in the course of the NPS utilization.

The SRE "Nerpa" has specialized floating and coastal equipment for managing spent nuclear fuel (SNF) and radioactive wastes (RW):

- floating technical bases for defuelling of NPS reactors;
- auxiliaries and fixtures for defuelling;
- floating tanks PEK-50 for collecting liquid radioactive wastes (LRW);
- containers for storing solid radioactive wastes (SRW);
- site of interim storage of SRW;
- sector for deactivation of radioactive equipment.

In collaboration with the NPS designer, CCO "Rubin", and the designer of the power engine, OKBM, the research and design office "Onega" developed the utilization project for the NPS "Kursk". The project provides industrial, ecological, nuclear, and radiation safety; the specific feature of the project is the unique operation for vessel lifting, placing in the skip, missile unloading, transition from one slip to another one, and a new transport and technological scheme of the reactor defuelling.

Currently, the SRE "Nerpa" uses two schemes for the SNF management.

Scheme 1:

A NPS is located in the water area of the plant at the landing stage of the specialized embankment. The floating maintenance base (FMB) PM-12 performs defuelling of the nuclear core. The spent fuel assemblies are placed in storage at PM-12. Later, according to the preliminary agreed schedule, the FMB "Lotta" of Murmansk steam navigation arrives at the SRE "Nerpa", takes the SNF from PM-12 storage and transports it to the water area of the "Atomflot" enterprise, where SNF is loaded into the transportation and protection container TK-18 by the special loading section at the FMB "Lotta". The filled containers are then loaded onto special train and shipped to IA "Mayak" for further reprocessing.

Scheme 2:

Defuelling is carried out similarly to scheme 1. Then PM-12 is transported to the water area of the "Atomflot" enterprise, where FMB "Lotta" reloads the SNF from the PM-12 storage to TK-18 containers.

Owing to the damage in the vessel hull, it was decided to reject the afloat defuelling and to carry out this procedure in the slipway after cutting out of the three-compartment RC block.

Four possible variants of the transportation and technological schemes for defuelling of the NPS "Kursk" located in the slipway are considered in detail.

Variant 1:

SNF is unloaded in the slipway by the lifting crane of 100 t carrying capacity using a standard reloading container (assembly 02SP). The container, which carries one spent fuel assembly (SFA), is placed on the specially manufactured transportation cart. The FMB PM-12 stays in dock PD-42 near the slipway. The filled container on the cart is moved from the slipway to the dock, where PM-12 places the container in the SFA storage using its own crane. The SFA is unloaded from the container to the storage case (type ChT11). The empty container is turned back to the slipway to carry the next SFA.

Variant 2:

The SFAs are unloaded from the nuclear core using the reloading container 02SPC and placed into the case located at the loading station. One case carries 3–5 SFA. The case is placed in the base container KB-650BA. The container is secured on the transportation cart and moved from the slipway to the floating dock with FMB PM-12. The case from the container is placed in storage at PM-12 using its own lifting crane. The unloaded container is placed back on the cart and transported to the slipway to carry the next case.

Variant 3:

The SFA are unloaded by the reloading container 02SP similarly to variants 1, and 2 and placed in the cases of the transportation container TK-18, which can carry up to 7 cases with SFA. The filled container TK-18 is sealed with the standard cover, secured on the transportation cart, moved to the embankment, and lifted by the 100 t carrying capacity floating crane on board the FMB "Lotta" of the Murmansk steamline staying at the embankment. The FMB "Lotta" can carry up to 7 filled TK-18 containers.

Variant 4:

The FMB "Imandra" of the Murmansk steamline is placed at the slip parallel to the 3-compartment block of the NPS "Kursk". It has on board all required equipment for SNF unloading from the NPS reactor and is certified by the supervisory authorities to carry out these operations. The SNF (one by one) are reloaded from the reactors to the cases of the "Imandra" storage using the reloading container 02SP and the 100 t carrying capacity overhead crane.

The research and design office "Onega" considers the third variant to be the safest because:

- it provides the minimum potentially hazardous operations of SNF reloading;

- experience with loading SFA into TK-18 by the reloading container 02SP was gained the in course of the liquidation of the Navy training center in Paldiski (Estonia);
- the considered scheme corresponds to the handling scheme developed by FSUE "Zvezdochka" for the SNF from a utilized NPS; the developed scheme passed through all required expert reviews and approvals;
- if required, the scheme can be corrected according to the implementation results obtained by FSUE "Zvezdochka".

However, this variant is rather expensive and time-consuming. Therefore, it was decided to carry out defuelling following the fourth variant schematically displayed in Fig. 2.

The conditionally satisfactory state of the reactor compartment and the primary circuit of the steam-generating facility of the NPS "Kursk" indirectly confirm the feasibility of the suggested scheme.

The research and design office "Onega" develops the projects for NPS utilization. At the moment, the utilization projects are developed for three NPS constructions and four enterprises. The utilization project for the NPS "Kursk" is under development.

According to the state standard, all technological documentation concerning the NPS utilization should pass through state ecological expert review.

Prior to this, the developed documentation should be approved by the state supervisory authorities.

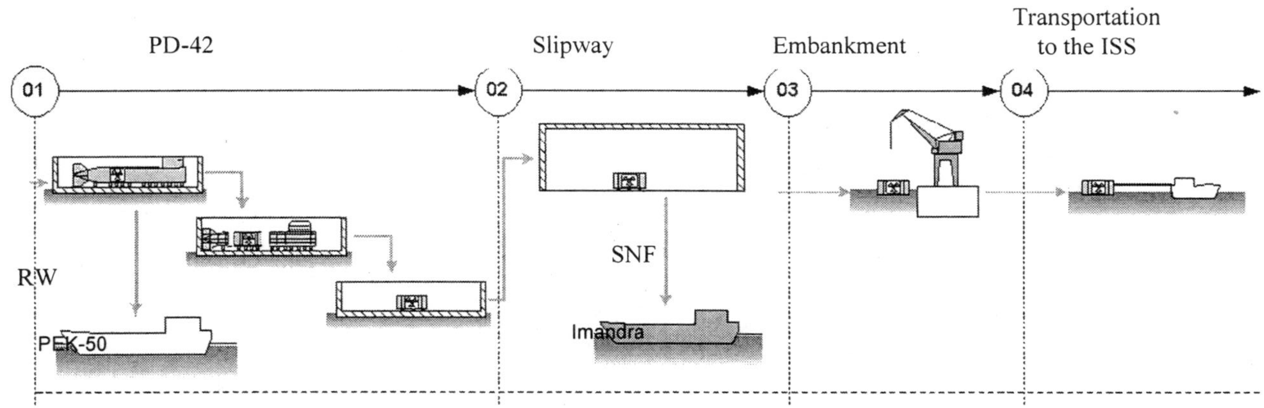

1. NPS preparing for utilization
2. RW collecting
3. Cutting into blocks
4. Utilization of nose and stern blocks
5. Formation of 3-compartment block
6. Launching of 3-compartment block (when required, formation of high slip)

1. Preparing for defuelling
2. Defuelling
3. Formation of 3-compartment block

1. Finishing of 3-compartment block
2. Commissioning of 3-compartment block

1. Block transportation to the ISS (Saida bay)

Activities of the SRE "Nerpa"

1. Section cutting into marten pieces
2. Cable reprocesing
3. Equipment disassembling
4. Extraction of precious metals
5. RW processing
6. SRW processing
7. SNF management

Figure 1. Utilization scheme of the NPS "Kursk" at the SRE "Nerpa"

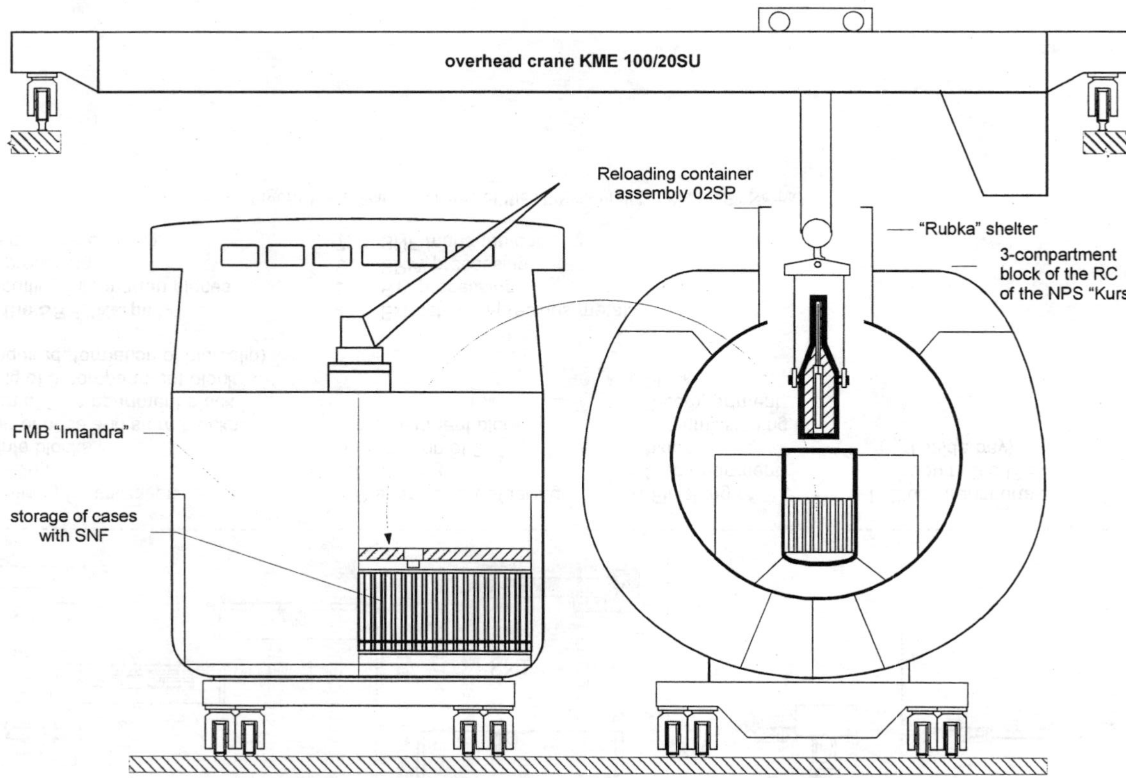

Figure 2. Reloading of the SNF from the reactor of the NPS "Kursk" to the FMB "Imandra"

DEVELOPMENT OF COMPUTER MODELS AND INFORMATIONAL SUPPORT SYSTEMS FOR UTILIZATION OF DECOMMISSIONED NUCLEAR SUBMARINES

S. BOGATOV, V. VYSOTSKY, V. DANILYAN , R. KALININ, V. KISSELEV, D. TOKARCHUK, A. SARKISSOV, AND P. SHVEDOV
IBRAE RAS
12 Bolshaya Tulskaya, 115191 Moscow, Russia

One of the most important problems related to decommissioning and utilization of nuclear-powered submarines (NPS) or surface vessels (NPSV) and management of their reactor compartments is providing nuclear and radiation safety during long-term forced afloat storage of the compartments. IBRAE RAS develops specialized computer informational and modeling system aimed at the solution of problems of on-line control over safety at radiation-hazardous objects involved in the processes of utilization of nuclear-powered submarines and surface vessels. The system is based on state-of-art geoinformational technologies.

The development of the informational and modeling system includes the following stages:

- collecting and analysis of the available information, determining the content and structure of the database of the informational system;
- analysis of the existing models and determining the content of the model bank of the informational and modeling system;
- choice of the software toolkit and development of the application software for the informational system;
- preparing the required electronic maps;
- assessment of the radiation potential of the decommissined objects;
- database filling by the attributive information;
- computer realization of the chosen models.

The chosen background DBMS was MS ACCESS, which offers simple production of convenient forms for the information presentation. The chosen background geoinformational system is the well known system MAPINFO, which completely corresponds to modern requirements.

Currently, the developed informational and modeling system includes the following subsystems:

- The subsystem on the decommissioned NPS includes the list of the NPS storage sites, estimates of their radiation potential, electronic maps of the neighboring territories, description of the modern technical state of each NPS (based on the utilization logbooks). The subsystem is supplemented

A.A. Sarkisov and L.G. LeSage (eds.),
Remaining Issues in the Decommissioning of Nuclear Powered Vessels, 379–391.

by descriptions of the main technical characteristics of all designs of decommissioned NPS.

- The database system on the infrastructure objects includes descriptions of the coastal maintenance bases and plants involved in the utilization process; information about the location of solid and liquid radioactive wastes in Arctic and Far-East regions; lists and descriptions of the floating maintenance bases and auxiliary vessels.
- Modeling block to describe radionuclide migration in the atmosphere.

Within this work, we analyzed the available information sources, prepared the required electronic maps, started collecting and preparing the required descriptive information, in particular, ecological information related to the problems of the NPSV and NPS utilization.

For each decommissioned NPS (reactor compartment), the following information on its history and modern state is available:

- Serial and tactical numbers.
- NPS design description (design number and its description).
- Date of commissioning.
- Date of decommissioning.
- Date of defuelling.
- Date of construction of the reactor block.
- Present NPS location.
- Description of the modern technical state of the NPS (based on the passport of the utilization logbook, see Fig. 1).
- Calculated estimate of the residual activity of the fuel and constructions of nuclear power engines of the NPS.

The informational system works in the environment of the operational systems Windows 98, Windows NT, or Windows 2000. The software has a friendly interface “user—PC” and is meant for the unskilled PC user.

The loaded system starts with the welcome screen (Fig. 2) followed by the main menu of the informational system (Fig. 3). Pushing a button, the user calls the corresponding subsystem.

Паспорт №

3-х отсечного реакторного блока утилизированной АПЛ

Тактический номер	Проект	Заводский номер	Спуск на воду	Вывод из эксплуатации,	Дата формирования реакторного блока	Место базирования
К-245	**667АУ**	**451**	**1971**	**1992**	**1994**	**ПВХ губа Сайда**

Корпус, корпусные конструкции

Прочный корпус	Легкий корпус	Запас плавучести	Проектные				Фактические			
			Тн	Тк	θ	ψ	Тн	Тк	θ	ψ
Герметичен	Герметичен	13%	8,6	8,6	0	0	8,6	8,3	0	0
Дополнительные сведения	Отсутствует формуляр объекта									

Реакторная установка

Активные зоны реакторов	Дата вывода реакторов из действия	Наработка реакторов, час	Nср %	Фильтры активности	Состояние I контура
Выгружены в 1993г.	ПБ – 1991г. ЛБ – 1991г.	ПБ = 39066 ЛБ = 42697	ПБ=23,4 ЛБ=22,7	Загружены	Герметичен, осушен.
Дополнительны е сведения	Паспорт спецпомещения отсутствует. Сведений о дополнительно загруженных ТРО и выгруженном оборудовании РО нет. Данных о радиационной обстановке внутри и снаружи блока нет.				

ТАБЛИЦА

расчета суммарной активности в реакторном отсеке АПЛ
за весь период эксплуатации, Бк.

№ п/п	Радио-нуклиды	Наведенная активность		Активность отложений на поверхностях I контура		Сумма активности	
		2002	2012	2002	2012	2002	2012
1	**Co-60**	1,95E+13	5,23E+12	1,40E+11	3,75E+10	1,96E+13	5,26E+12
2	**Fe-55**	6,88E+13	5,28E+12	2,57E+10	1,97E+09	6,88E+13	5,28E+12
3	**Ni-63**	2,14E+13	1,99E+13	7,57E+10	7,05E+10	2,15E+13	2,00E+13
4	**Ni-59**	2,07E+11	2,07E+11	5,50E+08	5,50E+08	2,08E+11	2,08E+11
5	**Mo-93**	1,29E+08	1,28E+08			1,29E+08	1,28E+08
6	**Mn-54**	2,69E+09	8,08E+05	3,04E+08	9,15E+04	2,99E+09	9,00E+05
7	**Sr-90**			4,92E+10	3,88E+10	4,92E+10	3,88E+10
8	**Cs-137**			1,66E+11	1,32E+11	1,66E+11	1,32E+11
	Всего	1,10E+14	3,07E+13	4,57E+11	2,81E+11	**1,10E+14**	**3,09E+13**

Figure1. Example of the passport for 3-compartment reactor block from a decommissioned NPS.

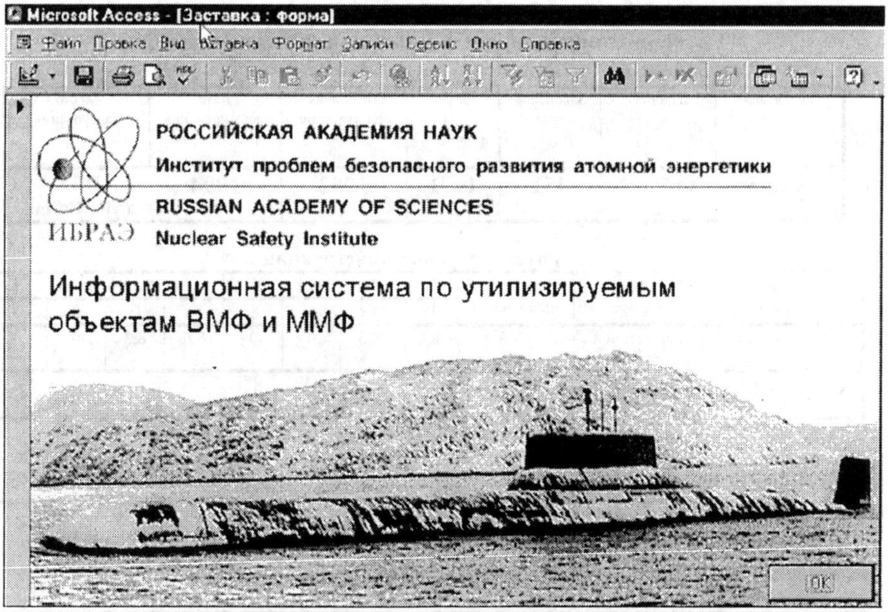

Figure 2. Welcome screen of the informational system.

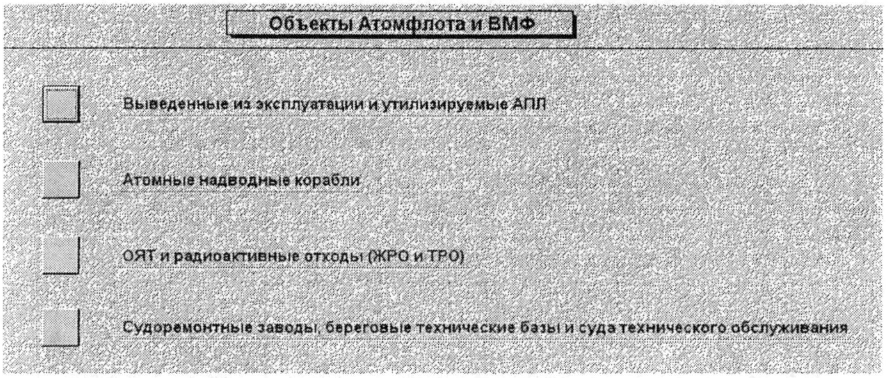

Figure 3. Main menu of the informational system.

The most important susbsystem of the informational system is the one for the decommissioned NPS. It contains general information for the Arctic and Far-East regions, including information about dynamics of the MPS decommissioning, defuelling, and formation of the reactor blocks (RB), and the review information about the disposition of the NPS and RB in the Arctic and Far-East regions. Also available are the list of the storage sites, electronic maps of the neighboring territories with the specified number of the decommissioned NPS, the number of NPSs at each storage site, and information about the modern state of each NPS (Figs. 4-8).

The subsystem for the decommissioned NPS is supplemented by the descriptions of the main technical characteristics of the design projects for the decommissioned NPS (Figs. 9-10).

The subsystem for NPSV at the moment includes brief descriptions of the existing nuclear-powered surface vessels and their technical characteristics (Fig. 11).

The subsytem for the infrastructure objects includes a description of the coastal and floating maintenance bases involved in the utilization activities and a list and characteristics of the floating maintenance bases and auxiliary vessels (Figs. 12-13). The subsystem on SNF, LRW, and SRW includes review information about the location of the SNF and radioactive wastes in the Arctic and Far-East regions.

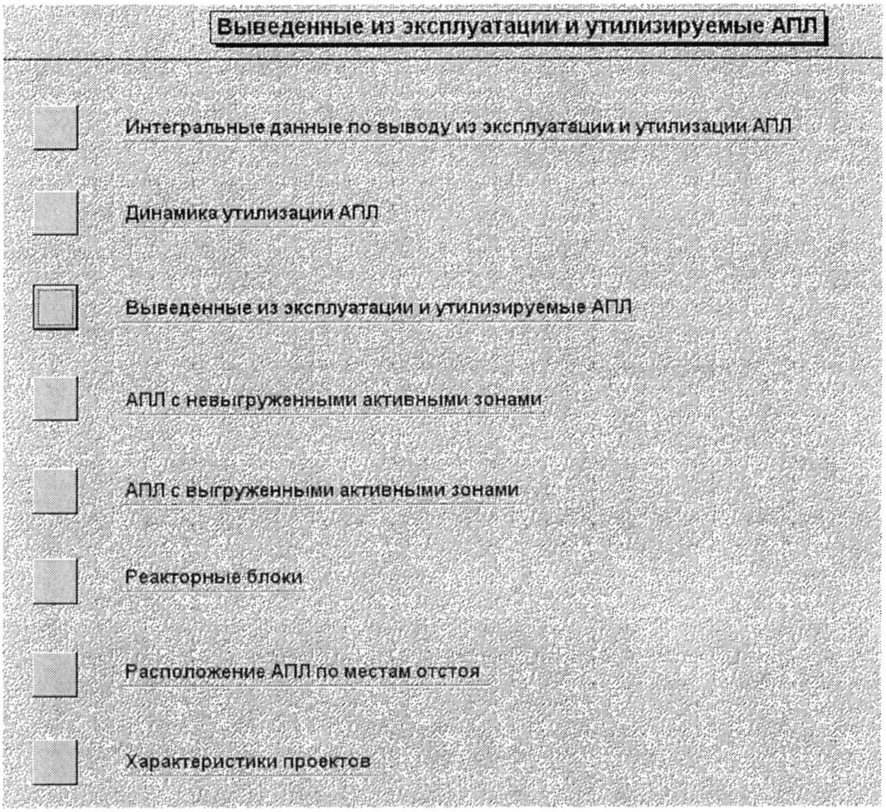

Figure 4. Menu of the subsystem for the decommissioned NPS.

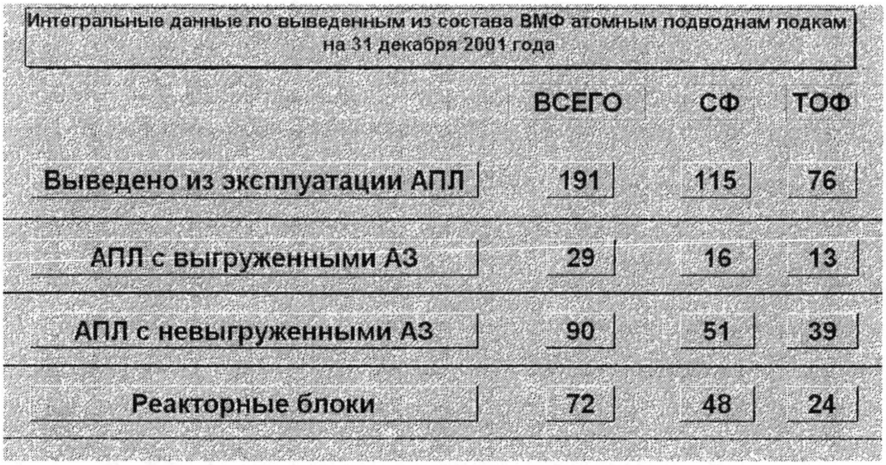

Интегральные данные по выведенным из состава ВМФ атомным подводнам лодкам на 31 декабря 2001 года

	ВСЕГО	СФ	ТОФ
Выведено из эксплуатации АПЛ	191	115	76
АПЛ с выгруженными АЗ	29	16	13
АПЛ с невыгруженными АЗ	90	51	39
Реакторные блоки	72	48	24

Figure5. Screen form for the integrated information on the deommissioned NPS.

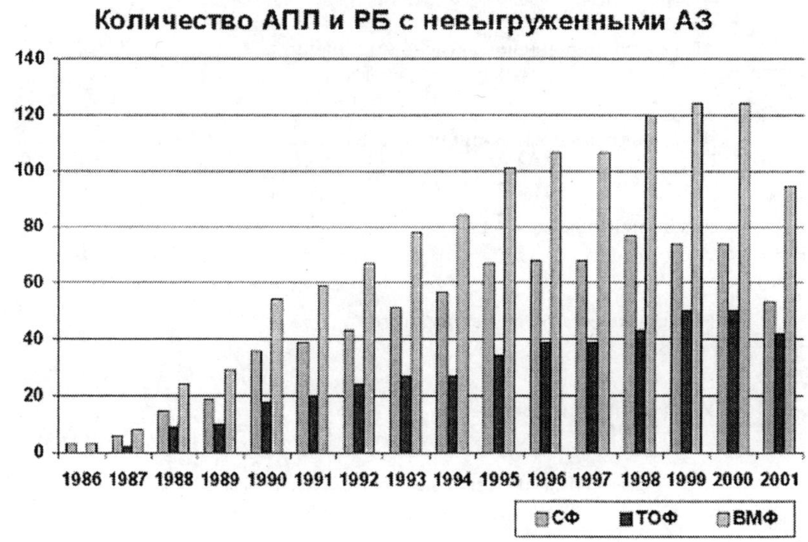

Figure 6. Time dynamics of the non-defuelled NPS and reactor blocks.

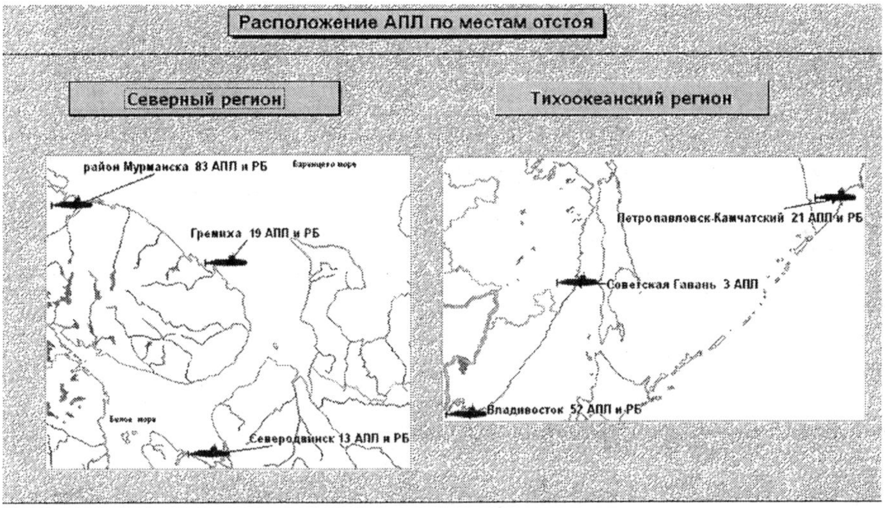

Figure7. Screen form for NPS distribution in the storage sites.

Figure 8. Map of the suburbs of Severodvinsk city with a list of the anchored NPSs.

№	Характеристика	Пр 627	Пр 627А	Пр 659	Пр 675	Пр 658
1	Год спуска на воду	1957	1958	1960	1963	1959
2	Общее количество	1	11	5	29	8
3	На СФ	1	6	-	11	6
4	Выведено на СФ	1	6	-	11	6
5	На ТОФ	-	5	5	18	2
6	Выведено на ТОФ	-	5	5	18	2
7	Водоизм. норм., м3	3065	3101	3731	4500	4030
8	Водоизм. полное, м3	4750		4920	5760	5300
9	Длина, м	107,4	107,4	111,2	115,4	114
10	Ширина, м	7,9	7,9	9,2	9,3	9,2
11	Скорость надв., узлы	15	15	21		18
12	Скорость подв., узлы	30	30	29	29	26
13	Число отсеков	9	9	9	10	10
14	Вооружение	8 533 ТА	8 533 ТА	6 КР, 4 н. 533ТА , 4	8 КР, 4-533, 2- 400 Т/	3 БР, 4 н. 533ТА, 2 к
15	Экипаж, чел.	104	110	120	137	104
16	Пред. глубина погр., м	300	300	300	300	300
17	Бюро проектант ПЛ	СКБ 143	СКБ 143	СКБ 18 "Рубин"	СКБ 18 "Рубин"	СКБ 18 "Рубин"
18	Гл. конструктор	Перегудов В.Н.	Перегудов В.Н.	Пустынцев П.П.(Кли	Пустынцев П.П.	С.Н.Ковалев

Figure 9. Example of the information representation on the design projects for the decommissioned NPS.

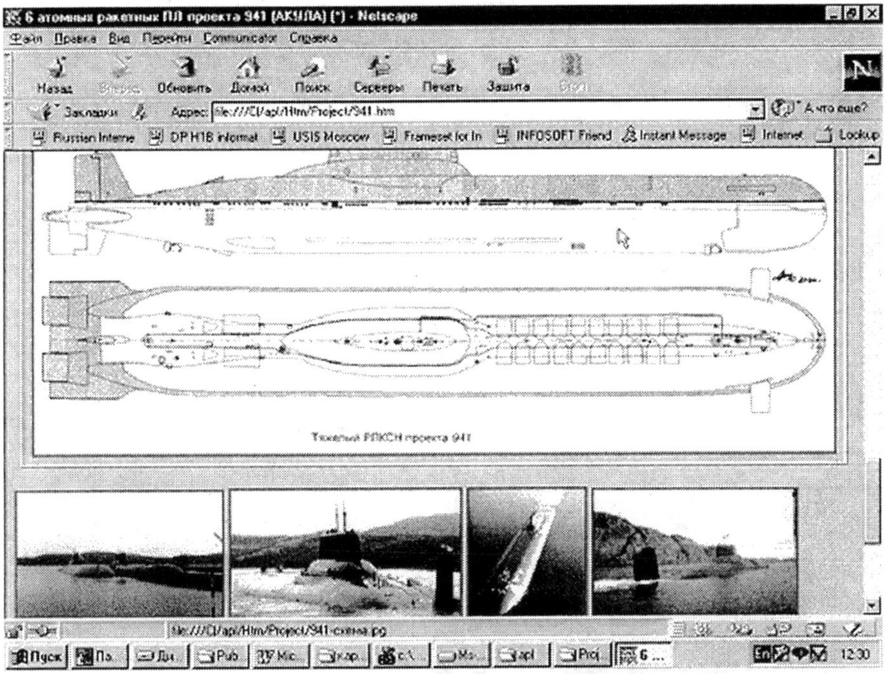

Figure 10. General information on a design project for the decommissioned NPS.

Атомное транспортное судно "Севморпуть"

Основное предназначение первого российского ледокольно-транспортного судна с атомной энергетической установкой — лихтеровоза-контейнеровоза "Севм
лихтерах и контейнерах в северные районы страны. Судно не случайно построено в комбинированном варианте, этого требуют условия транспортировки
оборудованных портов.

Атомный лихтеровоз "Севморпуть" был построен в Керчи на судостроительном заводе "Залив" имени Б. Е. Бутомы и сдан в эксплуатацию в 1988 году
требованиям Международной конвенции о безопасности торговых судов на ядерном топливе.

Основные характеристики.

Длина наибольшая — 260,3 м

Ширина наибольшая — 32,2 м

Осадка— 10,68 м

Высота борта — 18,3 м

Полное водоизмещение — 61000 т

Тип главной установки — атомная турбинная

Мощность — 40000 л. с.

Скорость хода на чистой воде — 20 уз,

Лихтеровместимость и контейнеровместимость — 74 лихтера (типа ЛЭШ)

или 1336 20-футовых контейнеров

Figure11. Brief information about nuclear-powered lighter-aboard vessek "Sevmorput'".

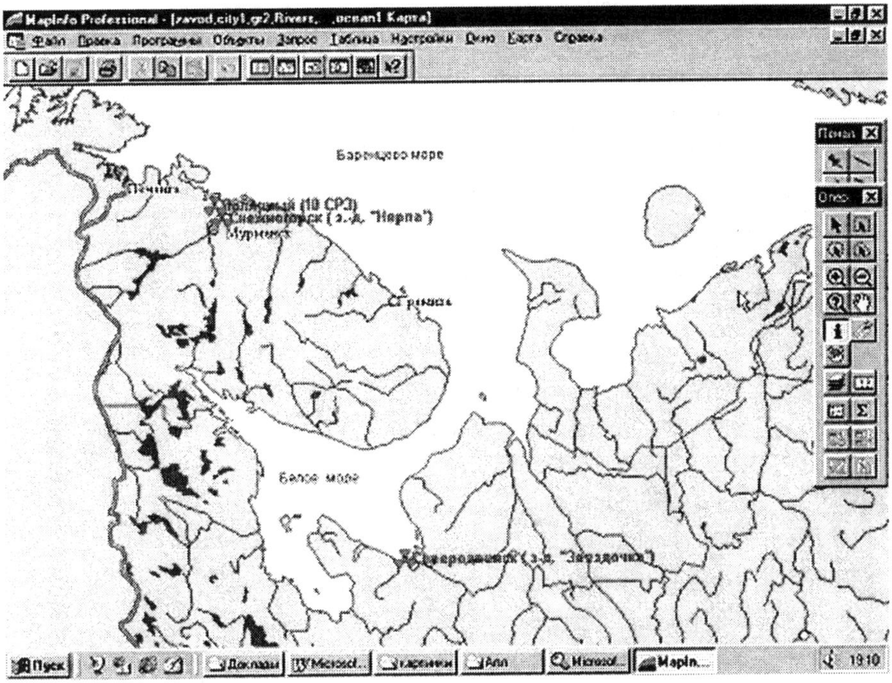

Figure 12. Map of the NPS utilization infrastructure for the Arctic Fleet.

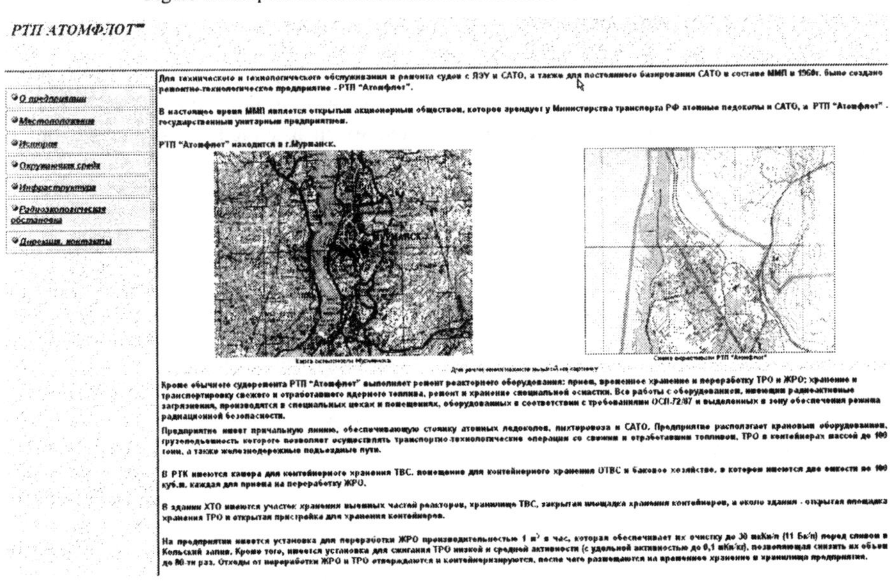

Figure 13. Fragment of information on the repairing and maintenance enterprize “Atomflot”.

At the moment, the system can generate a number of reports that provide a user with the required information in the forms of MS Word file and a hardcopy.

The following reports are available:

- Integrated data.
- Dynamics of NPS utilization.
- NPS classification.
- General vessel data.
- Dislocation at the Arctic Fleet.
- Dislocation at the Pacific Fleet.
- Distribution over the storage sites.
- NPS designs of the 1st generation.
- NPS designs equipped with a rector with liquid-metal coolant.
- NPS designs with cruise missiles.
- Nuclear-powered surface vessels.
- Coastal maintenance bases.
- Ship-repairing plants.
- NPS under utilization (information for individual NPS).
- NPS under utilization (cumulative table).
- LRW storage.
- SNF storage.
- SRW storage.

The modeling block is based on the software complexes “TRACE” and “NOSTRADAMUS” earlier designed in IBRAE and aimed at modeling of radionuclide migration in the atmosphere as a result of possible radiation accidents. Figure 14 displays an example of the information representation in the “NOSTRADAMUS” computer system. The primary event is related to an assumed missile strike on a building at the coastal maintenance base in Sysoeva bay.

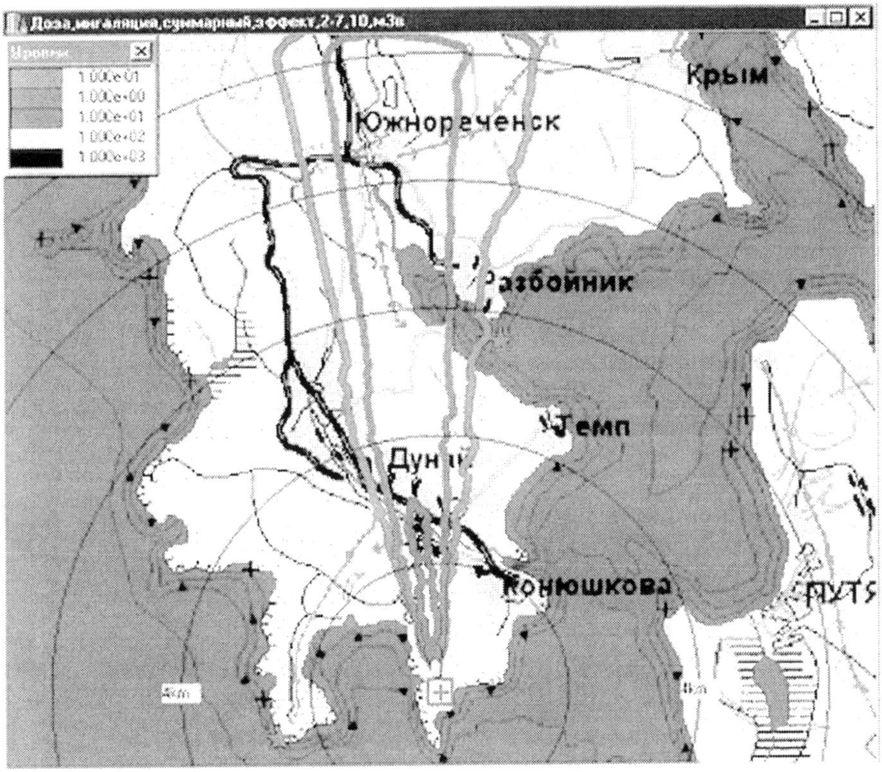

Figure 14. Example of the event modeling by the computer system "NOSTRADAMUS".

TOWARD A SYSTEM DYNAMICAL MODEL OF RUSSIAN NUCLEAR SUBMARINE DISMANTLEMENT BARRIERS

CHARLES D. FERGUSON
Scientist-in-Residence
Center for Nonproliferation Studies
Monterey Institute of International Studies
11 Dupont Circle, NW, 9th Floor
Washington, DC 20036, USA

1. Abstract

System dynamics, a management modeling tool with roots in the disciplines of systems engineering, control theory, and nonlinear dynamics, has helped to resolve problems in diverse fields, such as engineering, time management, business development, manufacturing, and public health. For instance, managers have used system dynamics modeling to determine how to complete projects on time and under budget. Policymakers have employed such modeling to reason through the impact of their policies and to learn how to develop more effective policies to solve problems. However, such modeling can only complement, not replace, existing planning tools. For these primary reasons, system dynamics has relevance for understanding barriers to Russian nuclear submarine dismantlement. This paper will try to motivate development of a model that will aim toward resolving some important problems in decommissioned Russian nuclear submarine dismantlement. Because system dynamics modeling is an iterative process, the research presented here will point toward a prototypical model that will require further input and development.

2. Why Use System Dynamics for Submarine Dismantlement?

Safely dismantling dozens of decommissioned Russian nuclear submarines is a complex and dynamic process. Recognizing this situation, the Cooperative Threat Reduction (CTR) program's management team adopted a systems approach for its successful Russian ballistic missile submarine (SSBN) dismantlement project. [1] Under this approach, CTR developed a system that integrated all components of SSBN dismantlement. The components included infrastructure equipment, shipyard preparation, de-fueling, launch tubes removal, reactor disposition, submarine decontamination, and low-level radioactive waste storage. The SSBN dismantlement system is the process that interlinked all the components into a functioning whole. Moreover, the management team ensured that good management practices were integrated into the SSBN dismantlement system.

A.A. Sarkisov and L.G. LeSage (eds.),
Remaining Issues in the Decommissioning of Nuclear Powered Vessels, 393–402.

But the system was not fully running and functioning effectively during the start up period when CTR mainly provided equipment to Russian shipyards. Dismantlement rates were too low, and Russia complained that CTR was providing little direct benefit to the shipyards. The shipyards were short of money to pay for the operation of the CTR equipment. Unable to come up with enough money from Moscow, shipyard workers' wage arrears piled up. On the U.S. side, the Congress did not permit CTR to pay these workers' salaries. The system improved significantly after an intensive project review in 1996. CTR then decided to provide direct contracts to the shipyards, which would receive funds upon verification of work completion. Dismantlement rates increased, and CTR funds to the shipyards paid the workers' wage arrears. Also, CTR procurement rules changed from "buy American" to "buy the best and the cheapest." Thus, Russian companies could compete on a fair basis with American companies. [2] With an integrated, systemic approach in place, the CTR management team could more readily identify any barriers that could prevent efficient SSBN dismantlement.

Such a systems approach would be useful for the safe dismantlement of the decommissioned Russian general purpose nuclear submarine fleet, which includes both attack (SSN) and guided cruise missile (SSGN) submarines. However, SSN/SSGN dismantlement is even more vexing than SSBN dismantlement because it faces more barriers than CTR's SSBN dismantlement project. In particular, decommissioned SSN/SSGNs waiting to be dismantled outnumber the number of decommissioned SSBNs that have been dismantled under CTR by about three-to-one. Moreover, SSN/SSGNs lack a comprehensive dismantlement system. Contributing problems include inadequate spent nuclear fuel storage facilities and insufficient long-term storage for reactor compartments, which take up valuable pier space at shipyards. [2] Further, unlike the CTR legal agreement, which provides strong legal protections for the United States, the Multilateral Nuclear Environmental Program in the Russian Federation (MNEPR) agreement, as presently formulated, would not provide as rigorous a level of legal protections. The MNEPR umbrella agreement, among nine European countries, the European Commission, the United States, and Russia, is intended to facilitate nuclear safety work, such as submarine dismantlement activities, within Russia. [3]

From the U.S. perspective, SSBN dismantlement was and remains a compelling problem. The U.S. wanted to remove the latent strategic threat posed by submarine launched ballistic missiles (SLBMs) on the SSBNs; therefore, CTR provided sufficient support to ensure that Russia met its SSBN dismantlement commitment under the Strategic Arms Reduction Treaty (START I). In contrast, from the U.S. perspective, SSN/SSGN dismantlement is not nearly as compelling. These general-purpose submarines cannot launch SLBMs. Thus, the direct strategic threat from these submarines to the U.S. homeland is nonexistent. Nonetheless, the SSN/SSGNs present a multitude of potential threats, including environmental contamination, proliferation of fissile materials, radioactive materials for "dirty" bombs, and technology transfers to other countries seeking nuclear-powered submarine technology.

The essential point here is that none of these potential threats alone has stood out as a convincing justification for a comprehensive dismantlement program, either funded by the United States or a coalition of donors. (However, donor interest is growing, and donors have funded some modest projects dealing with Russian general-purpose submarine dismantlement.) Currently, Russia apparently has insufficient resources to dismantle these submarines on its own, or at least Russia has not reallocated enough resources from other activities in its military-industrial complex. Confronted with these barriers, both Russia and donor countries have been unable to develop a comprehensive plan for SSN/SSGN dismantlement. Nonetheless, such a plan could be developed. Perhaps a new set of planning and policy tools could help.

2.1. INTRODUCTION TO SYSTEM DYNAMICS

System dynamics could provide those planning tools and complement, but not replace, other analytic tools. Over the past four decades, since Jay W. Forrester of the Massachusetts Institute of Technology founded the field in 1961, [4] system dynamics has improved the quality of thinking about problem solving by offering fresh perspectives on reality. Ideally suited for complex (multiple variable) and dynamic (changing over time) problems, it focuses on exposing longer term patterns, not examining particular events. Moreover, it seeks to elucidate underlying relationships among the variables within a system. In particular, it exposes the causal links between the variables.

Rather than viewing external forces as the main determinants of a system's behavior, system dynamics incorporates most forces into the internal structure of a system. This viewpoint places responsibility on people who manage the problem. When used effectively, system dynamics allows management to more clearly see what they can control. [5][6]

System dynamics factors in both "hard" and "soft" variables. "Hard" variables are things, such as submarines, spent nuclear fuel, storage space, and funding. In principle, these things can be precisely measured. Engineers and scientists are comfortable working with hard variables but have less familiarity with "soft" variables. These variables embody the human element in problems. Including soft variables in a model of a system is essential because otherwise the model would neglect the role of the decision maker who sets policy and allocates resources. Although "soft" variables are difficult to measure precisely, they can be quantified.

The quantification of soft variables may be arbitrary, but it is not ambiguous. [5] For instance, assigning a precise number to donor motivation to devote funding toward submarine dismantlement may be impossible, but quantifying this motivation on a scale from 0 to 100 (lowest to highest) can be very useful information in a system dynamics model of submarine dismantlement. The real power of system dynamics comes into play when the "soft" variables interact through feedback with other relevant "hard" and "soft" variables. For example, if the donor governments' domestic publics express strong interest in preventing environmental contamination from decommissioned Russian submarines, the value of the motivation variable would increase. To the contrary, if the Russian government does not provide access to facilities and adequate legal protections for nuclear safety work, donor motivation would decrease.

To develop a manageable model of the whole system, system dynamics cannot focus on every detail but instead examines "the view from 10,000 meters." [5] First and foremost, system dynamics seeks to lay bare the relationships among the major variables affecting a problem. These relationships extend throughout a vast space. In the Russian submarine context, the web of relationships spans the globe from Japan and the Russian Far East through Northwest Russian and Europe to the United States. Also, the relationships cover a long period of time from the Cold War period through the post-Cold War transition period to a time when Russia will likely be even more aligned with democracies. Attempting to capture all the details would be an onerous task. Fortunately, it is not necessary when constructing a useful system dynamics model. It is not necessary when constructing a useful system dynamics model. As Albert Einstein said, "Things should be made as simple as possible, but no simpler."

3. System Dynamics Modeling

System dynamics modeling begins with specifying the problem or issue, which leads to constructing a hypothesis or a model. The next step is to test the hypothesis or model. Testing will not determine whether the model is an exact description of the real system, but whether it is useful for providing insight into solving the problem. Insights should result in implementing positive policy changes. These changes feed back to the problem specification stage to further refine the description of the problem. The process keeps iterating until sufficient insight is achieved about how to effectively solve the problem.

3.1 CAUSAL LOOP DIAGRAMS

At the start of the modeling process, the endogenous variables (those internal to the system) are specified. Causal loop diagrams show the causal links among these variables. These diagrams illustrate the system's interdependencies and feedback processes. Moreover, these diagrams demonstrate one of the key principles of system dynamics. That is, causes and effects do not proceed solely from cause to effect in a chain of events, but rather they form a cycle or loop in which, because of feedback, "effects" can directly or indirectly affect "causes." [7]

Figure 1 illustrates three interconnected causal loop diagrams showing some of the major influential factors on the rate of de-fueling Russian submarines. Each causal link is represented with an arrow going from the cause to the effect. If an increase (decrease) in the value of a cause would tend to, keeping all other factors constant, increase (decrease) the value of an effect, the causal link is labeled with a plus sign or positive polarity. But if an increase (decrease) in the value of a cause would tend to, keeping all other factors constant, decrease (increase) the value of an effect, the causal link is labeled with a minus sign or negative polarity.

The loop labeled B1 includes the factors: rate of de-fueling submarines, de-fueled submarines, spent nuclear fuel (SNF) at shipyards, and SNF transportation capacity. If the rate of de-fueling goes up, the number of de-fueled submarines would go up. Then the amount of SNF at the shipyards would tend to also increase. This would tend to fill up or decrease the transportation capacity for SNF. As transportation capacity decreases, the rate of de-fueling would tend to decrease.

B1 is a balancing loop because if any factor, such as rate of de-fueling, would initially tend to increase (decrease), the feedback from the other factors would tend to decrease (increase) the rate of de-fueling. (When making a full cycle around a loop, add up the number of negative polarities. If this number is odd (zero or even), the loop is balancing (reinforcing).) In the case of B1, if there were no other factors influencing the rate of de-fueling, the rate would be driven to a steady-state value, which could be close to zero – an undesirable state – depending on the values of the other factors. However, there are other influential factors on loop B1. These other factors could drive up the rate of de-fueling.

The loop labeled R1 includes the factors: rate of de-fueling submarines, de-fueled submarines, spent nuclear fuel (SNF) at shipyards, and pressure to build transportation capacity, and SNF transportation capacity. (To see this fact, just pick a starting point, the tail of an arrow, and follow the arrows from tail to head until arriving back at the initial tail.) Compared to B1, R1 contains the additional factor of pressure to build transportation capacity. This factor is influenced, among other factors that are not specified in Figure 1, by the amount of SNF at the shipyards. As the SNF piles up, pressure to remove it should tend to increase, putting aside the influence of other factors. Pressure to build up transportation would tend to increase the transportation capacity. However, there could be a substantial delay before capacity starts to increase due to the time required to reallocate resources. This delay shows that the system has inertia, which is common to many systems, e.g., an everyday experience is the delay in reaching a desired temperature from a shower.

R1 is a reinforcing loop because there are zero negative polarities. In other words, the rate of de-fueling would tend to go up if loop R1 were the only causal loop acting in the system. This result arose from the pressure to build transportation capacity. From a decision maker's perspective, effective management would be to quickly identify the potential for SNF build up at the shipyards and anticipate the need for increased transportation capacity. When analyzing systems, decision makers should be on the alert for points of leverage, such as transportation capacity. Judicious use of leverage can move the system to a desired state.

The caveat to this preliminary policy recommendation is that the model as it now stands is incomplete. Other factors, such as availability of workers and shipyard capacity, should be included in the model.

The next section will introduce the concepts of stocks and flows. These concepts will provide the additional framework that will allow quantitative analysis. Such analysis is essential for determining how much additional capacity would be needed.

The loop labeled B2 includes the factors: pressure to cleanup the environment, cleanup effort, and environmental quality. If the pressure to cleanup the environment increases, the cleanup effort would tend to increase. As the cleanup effort increases, environmental quality would tend to increase. However, increasing environmental quality would tend to decrease the pressure to cleanup the environment. Therefore, B2 is a balancing loop which would eventually lead to a steady state value of the pressure to cleanup the environment. B2 positively affects B1 because as the pressure to cleanup the environment increases (decreases), the rate of de-fueling would tend to increase (decrease), isolating the influence of other factors.

Due to geographic proximity to the decommissioned Russian submarines in the Northern Fleet, Northern European pressure to prevent environmental contamination from these submarines is large. Therefore, from this perspective, loop B2 plays a major role in exerting pressure to increase the rate of Russian submarine de-fueling. However, the preliminary model neglects another major factor, i.e., the availability of sufficient funding. Northern European countries have smaller amounts of funding available relative to the United States. Nonetheless, a coalition of European donors, possibly working through the MNEPR agreement, could pool together sufficient funds. But from the U.S. perspective, loop B2 plays a minor role because pressure for environmental cleanup of decommissioned Russian submarine activities is much lower than concerns about potential proliferation and other national security threats to the United States.

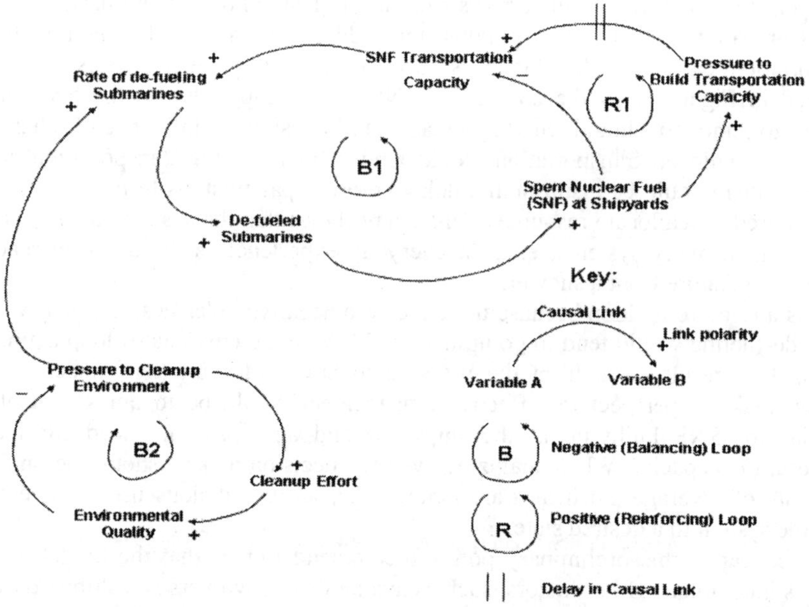

Figure 1. Causal loop diagrams illustrating some major factors influencing the rate of de-fueling Russian general-purpose nuclear submarines.

3.2 STOCKS AND FLOWS

Causal loop diagrams can form the framework of a mental model. Quite often, this mental model is sufficient to provide insights to formulate effective policy. Nonetheless, these diagrams can only go so far. To perform quantitative analysis, a system dynamics modeler would have to track the stock and flow of the relevant material, such as de-fueled submarines or spent nuclear fuel.

Stocks represent the accumulated material. To quantify the state of the system, a modeler would have to know how much stock is in each relevant location and the rate of change or flow of this stock from one location to another. Mathematically, stocks are expressed as integral equations, and flows are derivatives.

Figure 2 depicts a simple stock and flow diagram. In this situation, the stock is the amount of spent nuclear fuel at some location, such as a shipyard. SNF flows from some source or sources into this stock. The source is some other stock, such as SNF residing on decommissioned submarines or SNF assemblies piled up at Andreeva Bay or some other location. The SNF can also flow out of the stock in Figure 2 to a sink or some other stock. The sink could be interim SNF storage at Mayak.

From the policy perspective, the most interesting feature of Figure 2 is the valves. The valves regulate the amount of material flowing into and out of the stock. Policy decisions, such as whether to build additional transportation capacity, can affect the flow rate. For instance, if transportation capacity were to increase (assuming all other factors remain constant), the valve controlling the outflow would tend to open up more, allowing SNF to leave the stock and head toward interim storage.

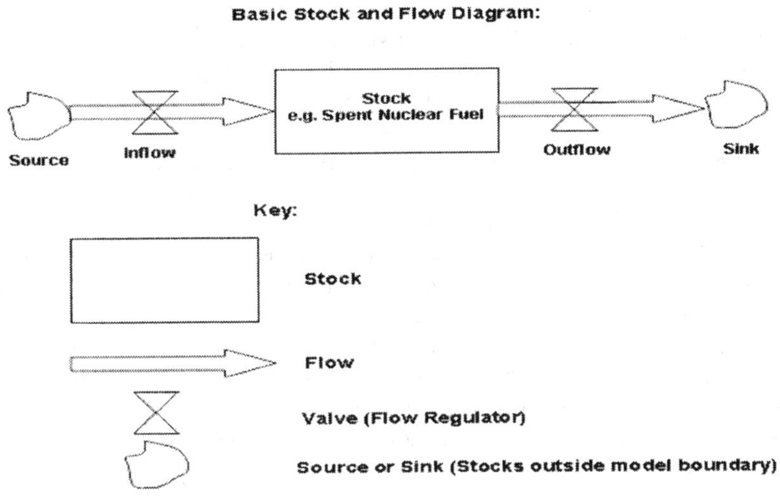

Figure 2. Illustration of the components of a stock and flow diagram.

4. What are the Major Problems Posed by Decommissioned Russian Nuclear Submarines?

"Models can be mesmerizing." [8] Creating a model for its own sake is useless. Models should be directed toward solving problems. As mentioned above, system dynamics is geared toward building models to solve problems. Before launching any further into system dynamics modeling for decommissioned Russian nuclear submarines, we need to know what the outstanding problems are. The next subsections will briefly outline these problem areas.

4.1 ENVIRONMENTAL SECURITY AND NUCLEAR SAFETY

In October 2001, the United States Department of State, in conjunction with the Departments of Energy and Defense, and the U.S. Environmental Protection Agency, submitted a report to Congress on environmental security threats posed by decommissioned Russian nuclear submarines, naval and civilian marine spent nuclear fuel, and associated radioactive waste. [3] The report identified "[s]afely disposing of spent nuclear fuel" as the "key environmental challenge." It also recognized that "Accidents involving decommissioned Russian marine nuclear reactors and associated spent nuclear fuel, radioactive waste, and contamination could threaten U.S. interests abroad by contamination of the air, ground, and water in both Russia and neighboring regions. Such contamination could harm population safety and health as well as cause regional economic and political instability."

Based on this report, a major problem for system dynamics modeling to consider would clearly be environmental security threats to Northern Europe from decommissioned submarines in Northwest Russia. Japan and Russia also have concerns over environmental security threats from submarines in the Russian Far East. Many of the submarines in both the Northwest and Far East are in danger of sinking and leaking radioactive contamination into the environment.

4.2 NONPROLIFERATION AND RADIOLOGICAL TERRORISM

If the potential proliferation risk is narrowly defined as diversion of weapons grade uranium or plutonium, decommissioned Russian nuclear submarines and associated spent nuclear fuel would probably not stand out as major problems. Only a small fraction of these submarines is believed to have used initial fuel enrichments close to weapons grade. Although the other submarines did not use weapons grade enrichments, they still used highly enriched uranium. Moreover, the spent nuclear fuel would all contain varying amounts of plutonium. Even if these materials were not readily useable for nuclear weapons, they might be useful for radiation dispersion devices (RDDs), so-called “dirty bombs” that terrorists may construct and use. Though more information is required to quantify these potential threats, enough might already be known to justify a donor program to provide physical security for the spent nuclear fuel.

4.3 IS THERE A COMMON DENOMINATOR TO THE PROBLEMS?

The environmental security, nonproliferation, and radiological terrorism threat areas have the common denominator of concern over the spent nuclear fuel. “While spent nuclear fuel is only about 5% of the radioactive waste volume from Russian submarines, it comprises more than 99% of the radioactivity from these vessels.” [3] Concentrating on safely securing spent nuclear fuel could solve the bulk of the problems posed by these submarines.

Perhaps each problem area alone is not sufficient to justify donor support for constructing a comprehensive dismantlement system for general-purpose Russian submarines. Instead, each donor country or a coalition of donors could commit to working on segments of each problem. [9] By doing this, an integrated approach could form. The donors and Russia would have to work diligently to ensure that the separate elements within the dismantlement system do not interfere with each other.

Nevertheless, the formation of this coalition will continue to be stymied unless both donors and Russia work to resolve differences. Russia should be more forthcoming with needed information. Such information would allow potential donors to determine the extent of each problem area. Information is the lifeblood of policy. Without it, bureaucratic sclerosis can occur. Russia also must ensure that adequate legal protections are in place.

The donors also share responsibility for clearly communicating to Russia their motivation and commitment to solving these problems. Setting aside funds is one of the most tangible ways to express this commitment. Of course, donors also have to work to surmount internal barriers within their own bureaucracies.

Figure 3 illustrates the interplay of donor motivation and commitment, Russian commitment, and Russian bureaucratic barriers. These reinforcing loops could either create a vicious cycle, in which motivation and commitment spiral downward, or a virtuous cycle, in which both sides work together for win-win goals.

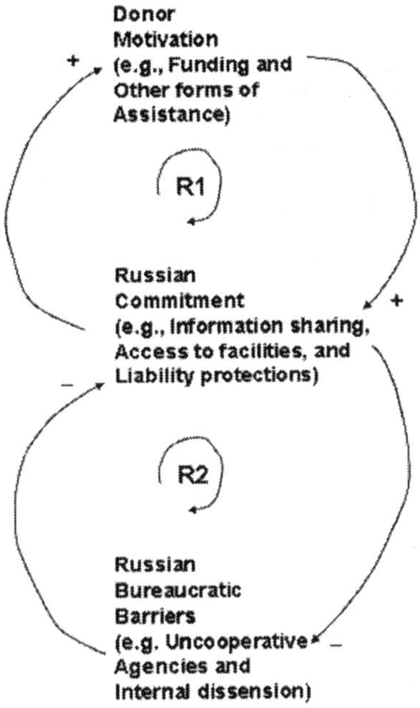

Figure 3. Causal loop diagram of the interactions among donor motivation, Russian commitment, and Russian bureaucratic barriers.

5. Conclusion

This paper has introduced and hopefully motivated the continued usefulness of system dynamics modeling and analysis for the problems posed by decommissioned Russian general-purpose nuclear submarine dismantlement. The next research step would be to develop a complete prototypical model focused on solving these pressing problems. Then the model would be ready for coding into a computer. The quantitative muscle of system dynamics flexes its full strength when stock and flow diagrams are coupled with causal loop diagrams in computer simulations. These simulations would allow policymakers to easily and safely explore policy options. However, they can gain useful insights just with systems analysis by employing mental modeling.

6′. References

1. Kuenning, Thomas E. (2002) Remarks at Russian Nuclear Regionalism conference, April 5, Washington, DC.
2. Moltz, James Clay (2000) Russian nuclear submarine dismantlement and the naval fuel cycle, *The Nonproliferation Review*, Spring.
3. U.S. Department of State (2001) Environmental security threats from decommissioned Russian marine nuclear reactors and associated spent nuclear fuel, radioactive waste and contamination, Report to Congress, October. The report is available at http://www.state.gov/p/eur/rls/rpt/2001/.

4. Forrester, Jay W. (1961) *Industrial Dynamics*, Pegasus Communications, Inc., Waltham, Massachusetts.
5. Richmond, Barry (2000) *The "Thinking" in Systems Thinking: Seven Essential Skills*, Pegasus Communications, Waltham, Massachusetts.
6. Senge, Peter M. (1994) *The Fifth Discipline: The Art & Practice of the Learning Organization*, Doubleday & Company.
7. Sterman, John D. (2000) *Business Dynamics: Systems Thinking and Modeling for a Complex World*, Irwin McGraw-Hill, Boston, Massachusetts.
8. Miller, Marvin (2002) Personal communication.
9. Moltz, James Clay (2000) Ideas presented at a Russian submarine conference, Washington, DC, December 5; Moltz, James Clay (2001, 2002) Conversations with the author.

CREATING ADEQUATE PUBLIC OPINION IN RUSSIAN REGIONS AS A TOP-PRIORITY TASK OF NUCLEAR SUBMARINE UTILIZATION & RADIOACTIVE WASTE MANAGEMENT

R. KALININ
Nuclear Safety Institute (IBRAE) of the Russian Academy of Sciences
Moscow, Russia

B. PAPKOVSKIY
Russian Research Center "Kurchatov Institute"
Moscow, Russia

Public opinion is a very powerful factor, which has a profound effect on leaders of the whole country, regional & departmental authorities and, even, on chefs of individual enterprises when making practical decisions.

This statement is especially true with respect to the nuclear industry including issues of nuclear vessel utilization and Radioactive Waste (RW) management.

Regional authorities, being influenced by inadequate public opinion, are often forced to make decisions or take measures that do not complywith interests of technological progress and real improvement of the environmental safety; as a result both money & time are spent to resolve issues of minor importance.

For example, the construction of a temporary site to store casks with Spent Nuclear Fuel (SNF) at SRE "Atomflot" (Murmansk region) is able to not only accelerate the rate of SNF removal from the region but also improve the safety of its temporary storage. This project met with powerful resistance from the Murmansk-town administration. Not long ago the town authorities forwarded some letters to the Presiding Commissioner of the Russian State Environmental Commission of Experts demanding the disapproval of the site construction; instead, many claims of secondary-importance were put in.

Actually, in Russia a negative attitude of the population to the nuclear industry, as a whole, and RW management (including Nuclear Submarine (NS) utilization techniques & products), in particular, is prevailing. However this negative perception does not correspond to the real threats and risks issuing from the above technogenic activities. Here is a typical example: some Russian regional public health & environment protection bodies carried out an inquiry wherein different technogenic risks were assessed in conventional scores. These data were compared with the relevant results of investigations performed by specialists of the Nuclear Safety Institute (IBRAE) of the Russian Academy of Sciences [1]. A concurrence in opinion of different groups of respondents was obtained with regard to all subjects of the questionnaire save for the nuclear industry: in the latter case a discrepancy was reached several times.

Such results can be explained as follows:

1. When managing radiation-dangerous materials, a potential hazard, surely, exists: if no potential danger existed, no problem would arise at all!

A.A. Sarkisov and L.G. LeSage (eds.),
Remaining Issues in the Decommissioning of Nuclear Powered Vessels, 403–406.

However the danger is much exaggerated. (It is appropriate to mention here that any moving transport facility also presents a high potential hazard especially when comparing its specific characteristics with those of RWs. However no proposal is made to reduce or eliminate vehicular traffic).

2. Various "green" movements are actually rather active. At the same time one should be aware that their activities are quite often guided by subjective interests of different sponsors (customers). Unfortunately, the "green" actions (publications, statements, etc.) are not always answered adequately in mass media by competent institutions and/or noted scientists.
3. Local authorities have no economic inducement to arrange for either the facilities or the infrastructure of RW management & NS utilization in their territory.
4. The educational and the information level of both the population and managerial personnel regarding environment protection, nuclear engineering & industry, radiation safety and dosimetry is very low.

In the authors opinion both popular literature and the periodic press contribute significantly (possibly the most) to creating a negative perception of the nuclear industry and their related activities by the population. To prove this thesis, two spectacular examples are worthy of consideration:

Example 1.

In an issue of the Isvestija daily paper (March, 2002) an interview by B. Reznik, a deputy of the Russian State Duma, was published under the name of "Military Men Keep back Nuclear Waste Leakage". This interview contained at least 3 false statements, such as: -"a NS with operating reactors sank near the Kamchatka shore ", -"no radiation measurement was taken" and "the NS reactor was comparable in power with those of the Chernobyl Nuclear Power Plant."

In reality the following sequence of events took place: two decommissioned NSs of the Pacific Navy (both de-fueled and without crews) lost their buoyancy and sank but were salvaged immediately afterwards. During salvaging operations all necessary measurements were taken. The power of the NS reactor was below that of a Chernobyl NPP unit by several orders of magnitude.

Example 2.

The title of a brochure "Floating NPPs of Russia: Threats to the Arctic & the World Ocean and also to the Nonproliferation Control" by A. Jablokov, et al. [2] is characterized by being completely negative to the vary idea of considering floating NPPs as a potential energy resource. This idea is suggested to everyone who has only read the brochure title. Moreover, to the question of how many NPPs are enough to create such a threat, the author indicates only a few tens. After reading the brochure the reader cannot even guess that during the last 50 years several hundred floating NPPs (under the name of nuclear submarines) were in operation, and no environmental threat was created!

Another brochure by A. Jablokov goes under the name of "The Myth on the Necessity of Constructing Nuclear Power Plants"[3]. With consideration for the aforesaid, this title needs no comment.

To create an adequate public opinion, both the trustworthiness and every nuance of official statements are of prime importance. To illustrate this thesis, one more example is opportune. After NS "Kursk" collapse a high-ranking Navy official in his interview with TV reporters declared "a decisive role of the NS crew in preventing nuclear atastrophe." Though we piously revere the memory of the fallen NS crew, we also assert that the "Kursk" crew had nothing to do with preventing a catastrophe. No "nuclear catastrophe" could have occurred because both automatic devices and engineering solutions used in the design of "Kursk" steam producing facility ensured reliable reactor protection even in conditions of an explosion with enormous destructive effect. Thus, the Kursk NS catastrophe confirmed the high survivability and safety of its power reactor facility, for which the credit, first, should be given to its creators.

Let us now take a quick look at the so-called "Green" movements.

There is no doubt that "Green" movements have a right to exist; moreover, these activities can be of great benefit to the society. But, unfortunately, when focusing attention of the public on topical environmental problems, they often exaggerate real facts and sometimes go too far. In such a case rapid and trustworthy answer provided by competent research institutes and/or individual scientists is advisable. However here one faces another problem: in the first case the work is paid for by its customer, in the second case it is not paid for at all.

Under conditions of a market economy the 'struggle' for the public opinion requires important funds. Therefore, overhead charges for any investment in the nuclear industry (including NS utilization and SNF & RW handling operations) should include expenses for explanatory work with the population aimed at elucidating goals and potential consequences of actions to be undertaken.

There is also another important circumstance. The time of secrecy as well as the lack of publicity are one with; however the methods used for work with the population have not progressed satisfactorily. Because of inertia and/or by virtue of the past mentality the authorities sometimes try to conceal from the public certain facts declassified long ago. Such unjustified secrecy provokes distrust among the population to all activities related to management of radiation-dangerous materials.

Today the following data should be accessible to the population & the general public as an obligatory minimum:

- radioactivity amount contained within every decommissioned NS stored afloat and in Spent Fuel Assemblies (SFAs) unloaded from NS reactors;
- radioactivity remaining within reactor compartments (which are not to be cut out) after de-fueling and the pattern of its decrease (natural decay) in time;
- all kinds of protective barriers (units & structures of the biological protection system, decontamination centers, separated protection zones, etc.) to prevent potential radionuclide spreading during NS utilization process;
- standardized and real dose loads during NS defueling, SNF handling (crane operations, reloading and spent fuel placement into casks) and also when preparing reactor compartments for long-term storage;
- amounts of produced Liquid & Solid RWs produced during NS utilization, and details on their storage and processing;
- potential consequences of both natural & technogenic accidents/incidents in the course of NS utilization operations (earthquakes, tornados, NS fire, aircraft crash, NS sinking, etc.) and provided countermeasures;
- technical and organizational measures undertaken to ensure physical protection of nuclear fuel in storage centers.

Recommendations and elucidations concerning most of the above subjects are available in special reports dealing with NS utilization by specialists of: Research Center "Kurchatov Institute", Russian Design Institute of Power Engineering

(RDIPE), Experimental Design Office of Machine-Building (EDOMB), Research and Design Technical Office "Onega" and IBRAE.

To inform workers, the population and the general public, various resources should be used:

- reports and radioecological data demonstrating the impact of NS utilization-related work on the environment in the region under consideration (e.g., in 2000-2001 special reports containing much relevant data were issued by RC “Kurchatov Institute” as applied to specific conditions of shipyards "Zvezda" and "Zvezdochka");
- special information centers attached to shipyards wherein demonstrations can be organized to show: -prototypes of the equipment used (NS defueling systems, casks & railcars to transport SFAs, etc.); -flowsheets for managing the contaminated equipment and nuclear fuel, - radiation background data and information on additional radiation doses due to NS utilization operations;
- thematic lectures and visits to work areas & information centers for personnel of the concerned shipyards, workers at nearby enterprises, students and citizens, etc.

Special information programs should be developed, implemented and controlled at all enterprises/centers for NS storage/utilization focusing their attention on radioecological effects of NS utilization-related operations.

Thus, well-organized information work should become a necessary element of the whole NS utilization process. One believes that, in addition to its main purpose, this work will also be able to eliminate frequent wrong/prejudicial views in the general public (including some mass media) of the problem of NS utilization & RW management.

References

1. *Proceedings of Symposium: "15 Years after Chernobyl: Lessons, Estimates and Prospects"* (2002) Komtekhprint, Moscow, (in Russian).
2. Kuznetsov, V., Jablokov, A., Desiatov, V., et al. (2000) *Floating Nuclear Power Plants of Russia: Threats to the Arctic, the World Ocean and the Nonproliferation Control*, Service Publishers, Riazan' (in Russian).
3. Jablokov, A. (2002) *A Myth on the Necessity of Constructing New Power Plants.* Russian Center of Environmental Policy, Moscow (in Russian).